An Introduction to Prograr
and Numerical Methods in MATLAB

S.R. Otto and J.P. Denier

An Introduction to Programming and Numerical Methods in MATLAB

With 111 Figures

S.R. Otto, BSc, PhD
The R & A
St Andrews
Fife
KY16 9JD
Scotland

J.P. Denier, BSc (Hons), PhD
School of Mathematical Sciences
The University of Adelaide
South Australia 5005
Australia

British Library Cataloguing in Publication Data
Otto, S. R. (Stephen Robert)
An introduction to programming and numerical methods in MATLAB
1. MATLAB (Computer file) 2. Numerical analysis — Data processing
I. Title II. Denier, J. P.
518′.02855
ISBN 1852339195

Library of Congress Control Number: 2005923332

ISBN-10: 1-85233-919-5
ISBN-13: 978-185233-919-7
Springer Science+Business Media
springeronline.com

Typesetting: Camera-ready by authors
Printed in the United States of America
12/3830-543210 Printed on acid-free paper SPIN 11317333

For
Julie
and
Jill and Megan.

Preface

This text provides an introduction to the numerical methods that are typically encountered (and used) in science and engineering undergraduate courses. The material is developed in tandem with MATLAB which allows rapid prototyping and testing of the methods. The package MATLAB (matrix laboratory) provides an environment in which students can learn to programme and explore the structure of the numerical methods. The methods included here are of a basic nature and only rely on material which should have been explored prior to the first year undergraduate stage.

The methods presented are supplemented with a set of tasks at the end of each chapter (full solutions of these are given in Appendix C). The tasks are introduced in such a way as to allow students to explore the topics as they evolve. Some are of a mathematical nature, but in the main they involve manipulating codes which are given in the text of the chapter (or section). Those tasks which we regard as being harder are marked with an asterisk. Throughout the text MATLAB commands appear using `this font` in the text. In the main the names of MATLAB commands reflect the commands rôle. We have taken particular care to highlight many of the problems that occur with interpreting the syntax of MATLAB commands. In Appendix B we provide a glossary of all MATLAB commands within the text and short examples of how these commands can be used. Reference is made to the comprehensive help facility within MATLAB: however examples are given which are similar to those given in other places within the text.

Throughout the text we derive the numerical techniques we use, but also emphasise that MATLAB's rich vocabulary provides commands for performing most of the fundamental tasks encountered in numerical methods. This approach serves to introduce students to the methods and also provides an understanding of their inner workings.

Why Do We Need Numerical Methods?

Mathematics is an elegant and precise subject: however when numerical answers are required one sometimes needs to rely on approximate methods to obtain useable answers. There are many problems which simply do not have analytical solutions, or those whose exact solution is beyond our current state of knowledge. There are also many problems which are too long (or tedious) to solve by hand. When such problems arise we can exploit numerical analysis to reduce the problem to one involving a finite number of unknowns and use a computer to solve the resulting equations.

The text starts with a description of how we could perform some very basic calculations (that is, simply using the computer as a calculator). It then moves on to solving problems which cannot, in practice, be solved by hand. Sometimes the solution of these problems can become as intricate and involved as the original problems and requires almost as much finesse and care to obtain a solution. There are several options available to us, both in terms of language and also overall approach. In this book we elect to express our ideas in terms of the syntax of the computer package MATLAB. Once you have mastered the syntax of MATLAB it will be easier for you to learn other languages, if you should decide you need to.

The Structure of This Text

This text is designed to be used as a source of reference for MATLAB commands (mainly through the glossary in Appendix B) and the examples given in the chapters. This is an approach we have found works well with our students. The text gives an introduction to numerical methods and the manipulation of the quantities used therein (for instance matrices). At each stage, short codes are given to allow the reader to try their own examples.

Examples of code which can be typed at the MATLAB prompt will appear within the general text in `this font` (this `font` is also used for the names of MATLAB codes), while longer examples will be written

```
>> commands to be entered

   Results
```

Those codes which are designed to be saved to a file will appear in boxes

```
a = 1;
```

(sometimes wider boxes will be used for codes with longer lines).

At the end of each section, or chapter, there are a variety of tasks which are designed to help the reader understand the topics they have just read. Solutions of these are given in Appendix C.

Chapter 2 concentrates on some aspects of programming. We also introduce another of MATLAB's powerful tools, namely its ability to produce high quality plots of data. Students are shown how to create codes and functions, which serve to augment MATLAB's built-in vocabulary. This chapter concludes with a discussion of the important topic of errors, both from the perspective of classical numerical analysis and also from human interaction which arise in the action of coding. We have found that students benefit enormously from debugging programmes and it is hoped that this will help them to hone these skills, which we consider to be essential.

The third chapter contains a discussion of looping and logical structures within MATLAB. Again the mathematics is developed in tandem. Finally concluding the first part of the book we give some examples of how MATLAB can be used to solve problems (using just algebra and exploiting MATLAB as an advanced calculator).

In Chapters 4 and 5 we meet some classical numerical methods, in the form of root finding and interpolation (and extrapolation). Although MATLAB has intrinsic functions which will perform most of these operations (`fzero` and `polyfit`) we have included a detailed description of both topics. This is aimed at helping the students to understand how these methods work, and where they can potentially fail. In Chapter 5 we discuss the general form of Newton forward differences, which are used in this chapter and subsequent ones (for instance to derive the formula for integration). In the next two chapters we explore the numerical analysis associated with integration and differentiation.

In Chapter 6 we return to the topic of matrices. We start by discussing the mathematical concepts of rank and linear independence. We also discuss eigenvalues and eigenvectors (and their interpretation for 2-by-2 matrices). The topic of numerical integration is taken up in Chapter 7 and in Chapter 8 methods for the numerical solution of ordinary differential equations are explored. Finally in Chapter 9 we use MATLAB to calculate some basic statistical quantities and also to explore some maps, some of which may be exhibit chaotic behaviour.

The text finishes with appendices containing an introduction to the basics of matrix algebra, a glossary of useful terms and solutions to all the tasks contained within the text. These appendices have deliberately been made quite wordy, since we believe that the material they contain is one of the important aspects of the book.

At the outset we expect readers to be mathematically literate to the level of being able to: and solve algebraic equations (for instance determine the roots of a quadratic and solve simultaneous equations); integrate and differentiate simple functions; solve separable differential equations (although examples are given within the text). Although exposure to complex numbers and matrices would be useful at the start of the text, it is not necessary as these concepts are introduced in some detail in the introductory chapter.

We would like to thank students of the University of Birmingham whose patience and comments have been invaluable in the construction of this text. In particular we would like to identify Sukhjinder Chana and Rob Ackeroyd for their careful proof reading and comments.

St Andrews, UK S. R. Otto
Adelaide, Australia J. P. Denier

Contents

1
Simple Calculations with MATLAB

1.1 Introduction and a Word of Warning

MATLAB is an incredibly powerful tool, but in order to use it safely you need to be able to understand how it works and to be very precise when you enter commands. Changing the way you enter a command, even subtly can completely change its meaning.

The main aim of this text is to teach you to converse with MATLAB and understand its responses. It is possible to interact with MATLAB using a "phrase book" approach, which is fine if the answer is what you expect. However it is far better to learn the language so that you can understand the response. As well as learning the language it is essential that you learn the grammar or syntax; this is perhaps even more important with computer languages than conventional languages! MATLAB uses an interpreter to try to understand what you type and this can come back with suggestions as to where you might have gone wrong: sometimes what you have written makes sense to MATLAB but does not mean what you expect! So you need to be careful. It is crucial that you formulate ideas clearly in your head (or on paper) before trying to translate them into MATLAB (or any other language).

We begin by discussing mathematical operations performed on scalars[1]. It is crucial that the material in this chapter is understood before proceeding, as it forms the basis of all that is to follow[2].

---

[1] That is numbers.

[2] MATLAB has a wealth of introductory material available to the user that can

We shall start by introducing MATLAB commands which can be typed at the MATLAB prompt; these will ultimately form part of our vocabulary of MATLAB commands. MATLAB already has an extensive vocabulary: however we will learn that we can expand this set. As the name MATLAB (**MAT**rix **LAB**oratory) suggests, most of the commands work with matrices and these will be discussed in due course. We shall start with scalar operations, for which MATLAB acts like a very powerful calculator.

1.2 Scalar Quantities and Variables

We will begin with the basic ideas of equations and variables. Try entering the commands as they are given. Consider the following two commands:

```
>> a = 3

a =

     3

>> b = 4;
```

[3] These two commands are entered on separate lines; the MATLAB prompt is denoted by `>>` (which does not need to be typed), as distinguished from the standard *greater than* sign `>`. The command on the first line sets the variable `a` to be equal to three (3) and that on the second line sets the variable `b` to be equal to four (4). The two commands also differ because the second one ends with a semicolon. This instructs MATLAB to execute the command but suppress any output; whereas above we can see that the value of `a` has been set to 3. These commands can be read as

```
set a equal to 3
set b equal to 4 (and suppress output)
```

Reading the commands in this way it should be clear that it is not possible to have a command of the form `7 = x (set 7 equal to x)`, whereas we could have `x = 7 (set x equal to 7)`. These variables can now be used again, for instance

be accessed using the commands `demo` or `tour`. There is also a good help facility which, unsurprisingly, can be accessed by typing `help` followed by the command in question. There is also a facility to use a web browser (`helpdesk` or `helpbrowser`).

[3] Here, you would type `a = 3`, and then press RETURN, and then type `b = 4;` and press RETURN again. The spaces are included purely for clarity.

```
>> a = 3;
>> b = a+1;
>> x = a+b;
```

The first line sets the variable a to be equal to 3, the semicolon instructing MATLAB to execute the command but to suppress the output. The second line sets b to be equal to a plus one, namely 4: again the semicolon suppresses output. The third line sets x to be a+b which is 7 (again output is suppressed).

MATLAB can be used as a very powerful calculator and its operations fall into two basic groups: *unary* and *binary*, the former operating on one quantity and the latter on two. We shall begin by considering simple arithmetic operations, which are *binary*. For instance typing 3*4 generates

```
>> 3*4

ans =

    12
```

Notice here that we have multiplied the two integers 3 and 4, and the answer has been returned correctly as 12. MATLAB uses the variable ans to store the result of our calculation, in this case the value 12, so that it can be used in the subsequent commands. For instance the command ans*3 will generate the result 36 (and now the variable ans will have the value 36). We could also have used the commands a = 3; b = 4; x = a*b which can be typed on one line and read as

```
set a equal to 3 (don't output anything),
set b equal to 4 (don't output anything)
and set x equal to a times b
```

Division works in exactly the same way as in the multiplication example above. If we try the command 3/4, MATLAB returns the value 0.75.

It is a good idea to use meaningful variable names and we shall shortly discuss valid forms for these.

Example 1.1 *Try entering the following commands into MATLAB, but before you do so try to work out what output you would expect.*

```
>> 3*5*6
>> z1 = 34;
>> z2 = 17;
>> z3 = -8;
>> z1/z2
```

```
>> z1-z3
>> z2+z3-z1
```

Hopefully you should get the answers, 90, 2, 42 *and* -25.

Example 1.2 *Here we give an example of the simple use of brackets:*

```
>> format rat
>> a = 2; b = 3; c = 4;
>> a*(b+c)
>> a*b+c
>> a/b+c
>> a/(b+c)
>> format
```

In this example you should get the answers, 14, 10, 14/3 and 2/7. Hopefully this gives you some idea that brackets make MATLAB perform those calculations first. (The command `format rat` *has been used to force the results to be shown as rationals, the final command* `format` *reverts to the default, which happens to be* `format short`*.)*

1.2.1 Rules for Naming of Variables

In the examples we have seen so far we have simply used variable names which seemed to suit the task at hand with no mention of restrictions on allowable variable names in MATLAB. The rules for naming variables in MATLAB can be summarised as follows:

1. Variable names in MATLAB must start with a letter and can be up to 31 characters long. The trailing characters can be numbers, letters or underscores (some other characters are also available but in this text we shall stick to these). There are many choices which are forbidden as variable names, some for very obvious reasons (such as `a*b` which signifies a multiplication of the variables `a` and `b`) and others for more subtle reasons (a good example is[4] `a.b`).

   The rules for naming variables also hold for naming MATLAB files. However, in this case a single dot is allowed within the name of the file; everything after the dot is used to tell MATLAB what type of file it is dealing

[4] The reason this is not a valid variable name lays in the fact that MATLAB supports object orientated programming. Because of this `a.b` refers to the value of the "`b`" component of the object `a`.

with (whether it be a file containing MATLAB code, or data etc). We will see more on this later in the section on *script* files.

2. Variable names in MATLAB **are** case sensitive, so that `a` and `A` are two different objects.

3. It is good programming practise to employ meaningful variable names. In our initial examples we have only used very simple (but appropriate) names: however as the examples become more complex our variable names will be more informative.

4. Variables names should not coincide with a predefined MATLAB command or with any user-defined subroutines. To see whether a variable name is already in use we can use the command `type variable_name`, but it may be better to use the command `which variable_name` (this will tell you whether the name `variable_name` corresponds to an existing code or intrinsic function.

1.2.2 Precedence: The Order in Which Calculations Are Performed

This represents one of the most common sources of errors and it is often the most difficult to detect. Before proceeding we briefly comment on the question of **precedence**, or the order in which commands are executed. Consider the mathematical expression $a(b+c)$ which you might read as "a times b plus c" which would appear to translate to the MATLAB command `a*b+c`. Hopefully you can see that this actually is equal to $ab+c$. The correct MATLAB command for $a(b+c)$ is `a*(b+c)`. The brackets have been used to force MATLAB to first evaluate the expression `(b+c)` and then to multiply the result by `a`. We should avoid falling into the trap of assuming that commands are performed from left-to-right, for instance `c+a*b` is equal to $c+ab$ (not $(c+a)b$ as if the addition was performed first).

At this point we should pause briefly and make sure the ideas of brackets are firmly in place. Brackets should always appear in pairs and the mathematics contained within brackets (or equivalently MATLAB) will be evaluated first. Hopefully this concept is familiar to you: however it is worth reiterating, since one of the most common problems in using MATLAB occurs due to either unbalanced or incorrectly placed brackets. For example the commands `(3+4/5)` and `(3+4)/5` are obviously different, the former being $3\frac{4}{5}$ and the latter being $\frac{3+4}{5}$.

The most critical use of brackets, which circumvents another popular source of error, is in terms of division. We should note that in the syntax of MATLAB `a/b*c` is **not** equal to $\frac{a}{bc}$ but $\frac{a}{b}c$. In order to ensure that the denominator of the fraction is calculated first we would need to use `a/(b*c)`, which is equal to $\frac{a}{bc}$. Similarly for examples like `a/b+c` versus `a/(b+c)`.

Example 1.3 *Determine the value of the expression* $a(b + c(c + d))a$, *where* $a = 2$, $b = 3$, $c = -4$ *and* $d = -3$.

Although this is a relatively simple example it is worth constructing the MATLAB statement to evaluate the expression:

```
>> a = 2; b = 3; c = -4; d = -3;
>> a*(b+c*(c+d))*a
```

This gives the answer `124`. *It is worth pausing here to consider the syntax of these commands. In the first line of this code we initialize the four variables* `a`, `b`, `c` *and* `d` *to have the values* 2, 3, −4 *and* −3 *respectively. The commands each end with semicolons; we have chosen to place all four commands on one line: however they could just as easily be placed on separate lines. With the variables assigned values we can now use them to perform calculations, such as in the second line where we form the mathematical expression* $a(b + c(c + d))a$. *Note all multiplications must be denoted by an asterisk and brackets have been used to force precedence of the operation; of course the brackets must balance (for each left bracket there is a corresponding right bracket) for the expression to make sense.*

Example 1.4 *Evaluate the MATLAB expressions*

```
1+2/3*4-5
1/2/3/4
1/2+3/4*5
5-2*3*(2+7)
(1+3)*(2-3)/3*4
(2-3*(4-3))*4/5
```

by hand and then check answers with MATLAB.

Recall that the operations of division and multiplication take precedence over addition and subtraction (type `help precedence` *at the MATLAB prompt for more details).*

The expressions are given by

$$
\begin{aligned}
\texttt{1+2/3*4-5} &= 1 + \frac{2}{3}4 - 5 = -\frac{4}{3},\\
\texttt{1/2/3/4} &= (((1/2)/3)/4) = \frac{1}{24},\\
\texttt{1/2+3/4*5} &= \frac{1}{2} + \frac{3}{4}5 = \frac{17}{4},\\
\texttt{5-2*3*(2+7)} &= 5 - 6(9) = -49,\\
\texttt{(1+3)*(2-3)/3*4} &= \frac{4 \times (-1)}{3}4 = -\frac{16}{3},\\
\texttt{(2-3*(4-3))*4/5} &= (2 - 3 \times 1)\frac{4}{5} = -\frac{4}{5};
\end{aligned}
$$

which can be verified in MATLAB; we can use the command `format rat` to force MATLAB to output the results as rational numbers (that is, fractions).

We mention here MATLAB has a number of intrinsic constants which the programmer can use, for instance `pi` and `eps`. The former is merely $\pi = 3.14159265\cdots$ and the latter is the distance from unity to the next real number in MATLAB[5]. It is also possible to enter numbers using the exponent-mantissa form. This uses the fact that numbers can be written as "mantissa $\times$ 10^{exponent}", for example

Number	mantissa - exponent	MATLAB form
789.34	7.8934×10^2	7.8934e2
0.0001	1×10^{-4}	1e-4
4	4×10^0	4
400000000000	4×10^{11}	4e11

Example 1.5 *Write 3432.6 in exponent-mantissa form and write* 100×10^{10} *in normal form.*

We have

$$3432.6 \equiv 3.4326 \times 10^3$$

and

$$100 \times 10^{10} \equiv 1,000,000,000,000.$$

[5] The smallest positive number that MATLAB can store which is different from zero is `realmin` which is approximately 10^{-308}, whilst the largest number is `realmax` which approximately 10^{308}. These intrinsic constants may be dependent upon your version of MATLAB and/or your computer's operating system.

Example 1.6 *Use MATLAB to calculate the expression*

$$b - \frac{a}{b + \frac{b+a}{ca}}$$

where $a = 3$, $b = 5$ and $c = -3$.

The code for this purpose is:

```
a = 3;
b = 5;
c = -3;
x = b-a/(b+(b+a)/(c*a));
```

with the solution being contained in the variable **x**.

Example 1.7 *Enter the numbers $x = 45 \times 10^9$ and $y = 0.0000003123$ using the exponent-mantissa syntax described above. Calculate the quantity xy using MATLAB and by hand.*

This is accomplished using the code

```
x = 45e9;
y = 3.123e-7;
xy = x*y;
```

Notice that here we have used a variable name **xy** *which should not be confused with the mathematical expression xy (that is $x \times y$).*

We can now set the values of variables and perform basic arithmetic operations. We now proceed to discuss other mathematical operations.

1.2.3 Mathematical Functions

Before we proceed let us try some more of the "calculator" functions (that is, those which are familiar from any scientific calculator).

Arithmetic functions `+`, `-`, `/` and `*`.

Trigonometric functions `sin` (sine), `cos` (cosine) and `tan` (tangent) (with their inverses being obtained by appending an `a` as in `asin`, `acos` or `atan`). These functions take an argument in radians, and the result of the inverse functions is returned in radians. It should be noted these are functions and as such should operate on an input; the syntax of the commands is `sin(x)` rather than `sin x`.

Exponential functions `exp`, `log`, `log10` and `^`. These are largely self explanatory, but notice the default in MATLAB for a logarithm is the natural logarithm $\ln x$. The final command takes two arguments (and hence is a binary operation) so that `a^b` gives a^b.

Other functions There are a variety of other functions available in MATLAB that are not so commonly used, but which will definitely be useful:

`round(x)`	Rounds a number to the nearest integer
`ceil(x)`	Rounds a number up to the nearest integer
`floor(x)`	Rounds a number down to the nearest integer
`fix(x)`	Rounds a number to the nearest integer towards zero
`rem(x,y)`	The remainder left after division
`mod(x,y)`	The signed remainder left after division
`abs(x)`	The absolute value of `x`
`sign(x)`	The sign of `x`
`factor(x)`	The prime factors of `x`

There are many others which we will meet throughout this book. We note that the final command `factor` gives multiple outputs.

We now construct some more involved examples to illustrate how these functions work.

Example 1.8 *Calculate the expressions:* $\sin 60°$ *(and the same quantity squared),* $\exp(\ln(4))$, $\cos 45° - \sin 45°$, $\ln \exp(2 + \cos \pi)$ *and* $\tan 30°/(\tan \pi/4 + \tan \pi/3)$.

We shall give the MATLAB code used for the calculation together with the results:

```
>> x = sin(60/180*pi)

x =

    0.8660

>> y = x^2

y =

    0.7500

>> exp(log(4))
```

```
ans =

     4

>> z = 45/180*pi; cos(z)-sin(z)

ans =

   1.1102e-16

>> log(exp(2+cos(pi)))

ans =

     1

>> tan(30/180*pi)/(tan(pi/4)+tan(pi/3))

ans =

    0.2113
```

The values of these expressions should be $\sqrt{3}/2$, $3/4$, 4, 0, 1 and $1/(3+\sqrt{3})$. Notice that zero has been approximated by `1.1102e-16` *which is smaller than the MATLAB variable* `eps`*, which reflects the accuracy to which this calculation is performed.*

It is worth going through the previous example in order to practise the command syntax. Getting this right is crucial since it is only through mastering the correct syntax (that is, the MATLAB language) that you will be able to communicate with MATLAB. When you first start programming it is common to get the command syntax confused. To emphasise this let's consider some of the commands above in a little more detail. Let us start with $f(x) = x \sin x$: the MATLAB command to return a value of this expression is `x*sin(x)` and not `x*sinx` or `xsin(x)`. The command `x*sinx` would try to multiply the variable `x` by the variable `sinx`; unless the variable `sinx` is defined (it isn't) MATLAB would return an error message

```
??? Undefined function or variable 'sinx'.
```

Similarly the command `xsin(x)` tries to evaluate the MATLAB function `xsin`, which isn't defined, at the point `x`. Again MATLAB would return an error message, in this case

```
??? Undefined function or variable 'xsin'.
```

In cases such as these MATLAB provides useful information as to where we have gone wrong; information we can use to remedy the syntax error in our piece of code. This simple example emphasises the need to read your code very carefully to ensure such syntax errors are avoided.

IMPORTANT POINT

> It is essential that arguments for functions are contained within round brackets, for instance `cos(x)` and that where functions are multiplied together an asterisk is used, for instance $f(x) = (x+2)\cos x$ should be written `(x+2)*cos(x)`.

Example 1.9 *The functions we used in the previous example all took a single argument as input, for example* `sin(x)`*. Mathematically we can define functions of two or more variables. MATLAB has a number of intrinsic functions of this type (such as the remainder function* `rem`*). To see how these are employed in MATLAB we consider two examples of such functions, one of which takes multiple inputs and returns a single output and the other which takes a single input and returns multiple outputs.*

Our first example is the MATLAB function `rem`*. The command* `rem(x,y)` *calculates the remainder when* `x` *is divided by* `y`*. For example* $12345 = 9 \times 1371 + 6$*, so the remainder when 12345 is divided by 9 is equal to* 6*. We can determine this with MATLAB by simply using* `rem(12345,9)`*.*

An example of a command which takes a single input and returns multiple outputs is `factor` *which provides the prime decomposition of an integer. For example*

```
>> factor(24)

ans =

     [2 2 2 3]
```

Here the solution is returned as an array of numbers as the answer is not a scalar quantity. We could just as easily used the command `x = factor(24)` *to set* `x` *equal to the array* `[2 2 2 3]`*. We can now check MATLAB has correctly* *determined the prime decomposition of the number 24 by multiplying the elements of the array x together; this is most readily achieved by using another intrinsic function* **prod(x)**.

1.3 Format: The Way in Which Numbers Appear

Before we proceed we stop to discuss this important topic. This can be simply illustrated by the following example:

Example 1.10 *Consider the following code*

```
s = [1/2 1/3 pi sqrt(2)];
format short; s
format long; s
format rat; s
format ; s
```

this generates the output

```
>> format short; s

s =

    0.5000    0.3333    3.1416    1.4142

>> format long; s

s =

   0.50000000000000   0.33333333333333   3.14159265358979   1.41421356237310

>> format rat; s

s =

      1/2           1/3          355/113       1393/985

>> format ; s

s =

    0.5000    0.3333    3.1416    1.4142
```

There are other options for `format` which you can see by typing `help format`. The default option is `format short` (which can be reverted back to by simply typing `format`). The above options are

`short` – 5 digits

`long` – 15 digits

`rat` – try to represent the answer as a rational.

You should note that whilst `format rat` is very useful, it can lead to misleading answers (in the above example clearly π is not equal to 355/113). At the start of a calculation it is a good idea to ensure that the data is being displayed in the appropriate format. In this example we have performed an operation on four numbers at once using the vector construction in MATLAB. We now proceed to discuss this further.

1.4 Vectors in MATLAB

One of the most powerful aspects of MATLAB is its use of vectors (and ultimately matrices) as objects. In this section we shall introduce the idea of initiating vectors and how they can be manipulated as "MATLAB objects".

1.4.1 Initialising Vector Objects

We shall start with simple objects and construct these using the colon symbol:

```
r = 1:5;
```

This sets the variable `r` to be equal to the vector `[1 2 3 4 5]` (and the semi-colon suppresses output, as normal). This is a row vector, which we can see by typing `size(r)` (which returns `[1 5]`, indicating that `r` has one row and five columns). This simple way of constructing a vector `r = a:b` creates a vector `r` which runs from `a` to `b` in steps of one. We can change the step by using the slightly more involved syntax `r = a:h:b`, which creates the vector `r` running from `a` to `b` in steps of `h`, for instance

```
r = 1:2:5;
s = 1:0.5:3.5;
```

gives `r = [1 3 5]` and `s = [1 1.5 2 2.5 3 3.5]`. We note that if the interval `b-a` is not exactly divisible by `h`, then the loop will run up until it exceeds `b`, for instance `t = 1:2:6` gives `t = [1 3 5]`. We can also initiate vectors by typing the individual entries; this is especially useful if the data is irregular, for instance `t = [14 20 27 10];`. There are many other ways of setting up vectors and for the moment we shall only mention one more. This is the command `linspace`: this has two syntaxes

```
s = linspace(0,1);
t = linspace(0,1,10);
```

Here `s` is set up as a row vector which runs from zero to one and has one hundred elements and `t` again runs from zero to one but now has ten elements. Note here that to set up a vector which runs from zero to one in steps of $1/N$, we can use `w = 0:1/N:1` or `W = linspace(0,1,N+1)`. (For example trying typing `s=0:0.1:1.0; length(s)`. You will find that `s` has eleven elements!). The command `linspace` is especially useful when setting up mathematical functions as we shall discover in the next section.

1.4.2 Manipulating Vectors and Dot Arithmetic

We shall now talk about the idea of calculations involving vectors and for this purpose we shall discuss dot arithmetic. This allows us to manipulate vectors in an element-wise fashion rather than treating them as mathematical objects (in fact for addition and subtraction this is the same thing).

To see how dot arithmetic works let's consider a simple example:

```
>> a = [1 2 3];
>> 2*a;

ans =

     2     4     6
```

Suppose now we try to multiply a vector by a vector, as in

```
>> a = [1 2 3];
>> b = [4 5 6];
>> a*b
??? Error using ==> *
Inner matrix dimensions must agree.
```

An error message appears because both a and b are row vectors and therefore cannot be multiplied together. Suppose however that what we really want to achieve is to multiply the elements of vector a by the elements of vector b in an element by element sense. We can achieve this in MATLAB by using **dot arithmetic** as follows

```
>> a = [1 2 3];
>> b = [4 5 6];
>> a.*b

ans =

     4    10    18
```

A glance at the answer shows that MATLAB has returned a vector containing the elements

$$[a_1 b_1, a_2 b_2, a_3 b_3].$$

The . indicates to MATLAB to perform the operation term by term and the * indicates we require a multiplication. We can also do a term by term division with

```
>> a = [1 2 3];
>> b = [4 5 6];
>> a./b

ans =

    0.2500    0.4000    0.5000
```

The result is, as we would expect,

$$\left[\frac{a_1}{b_1}, \frac{a_2}{b_2}, \frac{a_3}{b_3}\right].$$

Example 1.11 *We shall create two vectors running from one to six and from six to one and then demonstrate the use of the dot arithmetical operations:*

```
s = 1:6;
t = 6:-1:1;
s+t
s-t
s.*t
s./t
s.^2
1./s
s/2
s+1
```

This produces the output

```
>> s+t

ans =

     7     7     7     7     7     7

>> s-t

ans =

    -5    -3    -1     1     3     5

>> s.*t

ans =

     6    10    12    12    10     6

>> s./t

ans =

    0.1667    0.4000    0.7500    1.3333    2.5000    6.0000

>> s.^2

ans =

     1     4     9    16    25    36
```

```
>> 1./s

ans =

    1.0000    0.5000    0.3333    0.2500    0.2000    0.1667

>> s/2

ans =

    0.5000    1.0000    1.5000    2.0000    2.5000    3.0000

>> s+1

ans =

     2     3     4     5     6     7
```

These represent most of the simple operations which we may want to use.

We note that in order for these operations to be viable the vectors need to be of the same size (unless one of them is a scalar – as in the last three examples).

1.5 Setting Up Mathematical Functions

Following on from the previous section we discuss how one might evaluate a function. It is crucial that you understand this section before you proceed.

We revisit the topics introduced in the previous section and discuss the ways in which you can set up the input to the function

Example 1.12 *Set up a vector x which contains the values from zero to one in steps of one tenth.*

This can be done in a variety of ways:

```
% Firstly just list all the values:
x = [0 0.1 0.2 0.3 0.4 0.5 0.6 0.7 0.8 0.9 1.0];

% Use the colon construction
x = 0:0.1:1.0;

% Or use the command linspace
x = linspace(0,1,11);
```

As noted previously we note that there are eleven values between zero and one (inclusive) for a step length of one tenth. You may want to try `linspace(0,1,10)` and see what values you get.

Each of these methods are equally valid (and more importantly will produce the same answer) but the latter two are probably preferable, since they are easily extended to more elements.

We now wish to set up a simple mathematical function, say for instance $y = x^2$. Initially you may want to type `x^2` but this will generate the error message

```
??? Error using ==> ^
Matrix must be square.
```

This is because this operation is trying to perform the mathematical operation $\mathbf{x} \times \mathbf{x}$ and this operation is not possible. Instead we need to use `y=x.^2` which gives

```
>> y = x.^2

y =

  Columns 1 through 7

         0    0.0100    0.0400    0.0900    0.1600    0.2500    0.3600

  Columns 8 through 11

    0.4900    0.6400    0.8100    1.0000
```

Here we see that each element of `x` has been squared and stored in the array `y`. Equivalently we could use `y = x.*x;`.

Example 1.13 *Construct the polynomial $y = (x+2)^2(x^3+1)$ for values of x from minus one to one in steps of 0.1.*

Here it would be laborious to type out all the elements of the vector so instead we use the colon construction. We shall also define $f = (x+2)$ *and* $g = x^3+1$, *so that we have the code:*

```
x = -1:0.1:1;
f = x+2;
g = x.^3+1;
y = (f.^2).*(g);
```

In the construction of `g` *we have used the dot arithmetic to cube each element and then add one to it. When constructing* `y` *we firstly square each element of* `f` *(with* `f.^2`*) and then multiply each element of this by the corresponding element of* `g`.

You should make sure that you are able to understand this example.

Example 1.14 *Construct the function* $y = \dfrac{x^2}{x^3+1}$ *for values of* x *from one to two in steps of* 0.01.

Here we merely give the solution:

```
x = 1:0.01:2;
f = x.^2;
g = x.^3+1;
y = f./g;
```

(We could have combined the last three lines into the single expression `y = x.^2./(x.^3+1);`*).*

For the moment it may be a good idea to use intermediate functions when constructing complicated functions.

Example 1.15 *Construct the function*

$$y(x) = \sin\left(\frac{x\cos x}{x^2+3x+1}\right),$$

for values of x *from one to three in steps of 0.02.*

Here, again, we use the idea of intermediate functions

```
x = 1:0.02:3;
f = x.*cos(x);
g = x.^2+3*x+1;
y = sin(f./g);
```

NB MATLAB will actually calculate `f/g` and in this case it will return a scalar value of -0.1081. Unfortunately this will not generate an error but it will mean that the answer is not a vector as we should be expecting.

1.6 Some MATLAB Specific Commands

We shall now introduce a couple of commands which can be used to make calculations where the input can take a variety of forms. The first command is `polyval`. This command takes two inputs, namely the coefficients of a polynomial and the values at which you want to evaluate it. In the following example we shall use a cubic but hopefully you will be able to see how this generalises to polynomials of other orders.

Example 1.16 *Evaluate the cubic* $y = x^3 + 3x^2 - x - 1$ *at the points* $x = (1, 2, 3, 4, 5, 6)$. *We provide the solution to this example as a commented code:*

```
% Firstly set up the points at which the polynomial
% is to be evaluated
x = 1:6;

% Enter the coefficients of the cubic (note that
% these are entered starting with the
% coefficient of the highest power first
c = [1 3 -1 -1];

% Now perform the evaluation using polyval

y = polyval(c,x)
```

Note that in this short piece of code everything after the % is treated by MATLAB as a comment and so is ignored. It is good practice to provide brief, but meaningful, comments at important points within your code.

IMPORTANT POINT

> It is important that you remember to enter the coefficients of the polynomial starting with the one associated with the highest power and that zeros are included in the sequence.

We might want to plot the results of this calculation and this can be simply accomplished using the `plot` command. Consider the following example:

Example 1.17 *Plot the polynomial* $y = x^4 + x^2 - 1$ *between* $x = -2$ *and* $x = 2$ *(using fifty points).*

```
x = linspace(-2,2,50);
c = [1 0 1 0 -1];
y = polyval(c,x);
plot(x,y)
```

This produces the output

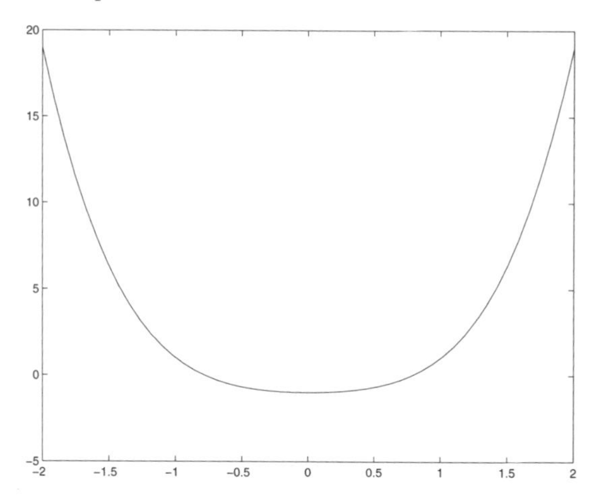

In the next chapter we shall discuss plotting in more detail and show how plots can be customised.

There are many other commands which allow us to manipulate polynomials: perhaps one of the most useful ones is the `roots`. The polynomial is defined in the same way as in the previous examples. The input to the routine is simply these coefficients and the output is the roots of the polynomial.

Example 1.18 *Find the roots of the polynomial* $y = x^3 - 3x^2 + 2x$ *using the* *command* **`roots`**.

```
c = [1 -3 2 0];
r = roots(c)
```

This returns the answers as zero, two and one.

In fact the converse command also exists, which is `poly`. This takes the roots and generates the coefficients of the polynomial having those roots (which is monic, that is the coefficient of the highest term is unity).

1.6.1 Looking at Variables and Their Sizes

Before we proceed we mention a couple of useful commands for seeing which variables are defined. To list the variables which are currently defined we can use the command `whos`. This will give a list of the variables which are currently defined (a shorter output can be obtained by using the command `who`). This command can be used to list certain variables only, for instance `whos re*` lists only the variables whose names start with `re`.

Example 1.19 *The following code*

```
clear all
a = linspace(0,1,20);
b = 0:0.3:5;
c = 1.;
whos
```

gives the output

```
  Name        Size             Bytes  Class

  a           1x20               160  double array
  b           1x17               136  double array
  c           1x1                  8  double array

Grand total is 38 elements using 304 bytes
```

Here we have used the **`clear all`** *command to remove all previously defined variables. To look at the size of one variable we can use the command* **`length`**, *for instance with the previous example* **`length(a)`** *will give the answer 20. We note that the command* **`size(a)`** *will give two dimensions of the array, that is* *in this case [1 20]; this will be particularly useful when we consider matrices in due course.*

1.7 Accessing Elements of Arrays

This is one of the most important ideas in MATLAB and other programming languages which is often misunderstood. Let us start by considering a simple array `x = 0:0.1:1.;`. The elements of this array can be recalled by using the format `x(1)` through to `x(11)`. The number in the bracket is the index and refers to which value of `x` we require. A convenient mathematical notation for this would be x_j where $j = 1, \cdots, 11$. This programming notation should not be confused with $x(j)$; that is x is a function of j. Let us consider the following illustrative example:

Example 1.20 *Construct the function* $f(x) = x^2+2$ *on the set of points* $x = 0$ *to* 2 *in steps of* 0.1 *and give the value of* $f(x)$ *at* $x = 0$, $x = 1$ *and* $x = 2$. *The code to construct the function is:*

```
x = 0:0.1:2;
f = x.^2+2;

% Function at x=0
f(1)
% Function at x=1
f(11)
% Function at x=2
f(21)
```

Note that the three points are **not** *f(0), f(1) and f(2)!*

In this example we have noted that $x_j = (j-1)/10$ and hence $x_1 = 0$, $x_{11} = 1$ and $x_{21} = 2$. These three indices are the ones we have used to find the value of the function.

IMPORTANT POINT

In MATLAB `f(j)` the value of `j` refers to the index within the array rather than the function `f(.)` evaluated at the value `j`!

The expression `end` is very useful at this point, since it can be used to refer to the final element within an array. In the previous example `f(end)` gives the value of `f(21)` since the length of `f` is 21.

Example 1.21 *We now show how to extract various parts of the array* x*.*

```
x = linspace(0,1,10);
y = x(1:end);          % Whole of x
y = x(1:end/2);        % First half
y = x(2:2:end);        % Even indices only
y = x(2:end-1);        % All but the last one
```

1.8 Tasks

In this introductory chapter we shall give quite a few details (at least initially) concerning these suggested tasks. However, as the reader's grasp of the MATLAB syntax develops the tasks will be presented more like standard questions (the solutions are given at the back of the book in Appendix C).

Task 1.1 *Calculate the values of the following expressions (to find the MATLAB commands for each function you can use the Glossary, see for instance the entry for* `tan` *on page 386 or the* `help` *command,* `help tan`*).*

$$\begin{aligned} p(x) &= x^2 + 3x + 1 \;\; \textit{at } x = 1.3, \\ y(x) &= \sin(x) \;\; \textit{at } x = 30^\circ, \\ f(x) &= \tan^{-1}(x) \;\; \textit{at } x = 1, \\ g(x) &= \sin\left(\cos^{-1}(x)\right) \;\; \textit{at } x = \frac{\sqrt{3}}{2}. \end{aligned}$$

Task 1.2 *Calculate the value of the function* $y(x) = |x| \sin x^2$ *for values of* $x = \pi/3$ *and* $\pi/6$ *(use the MATLAB command* `abs(x)` *to calculate* $|x|$*).*

Task 1.3 *Calculate the quantities* $\sin(\pi/2)$*,* $\cos(\pi/3)$*,* $\tan 60^\circ$ *and* $\ln(x + \sqrt{x^2+1})$ *where* $x = 1/2$ *and* $x = 1$*. Calculate the expression* $x/((x^2+1)\sin x)$ *where* $x = \pi/4$ *and* $x = \pi/2$*. (If you are getting strange answers in the form* *of rationals you may well have left the format as* `rat`*, so go back to the default by typing* `format`*).*

Task 1.4 *Explore the use of the functions* `round`*,* `ceil`*,* `floor` *and* `fix` *for the values* $x = 0.3$*,* $x = 1/3$*,* $x = 0.5$*,* $x = 1/2$*,* $x = 1.65$ *and* $x = -1.34$*.*

Task 1.5 *Compare the MATLAB functions* `rem(x,y)` *and* `mod(x,y)` *for a variety of values of* x *and* y *(try* $x = 3, 4, 5$ *and* $y = 3, 4, -4, 6$*). (Details of the commands can be found using the* `help` *feature).*

Task 1.6 *Evaluate the functions*

1. $y = x^3 + 3x^2 + 1$
2. $y = \sin x^2$
3. $y = (\sin x)^2$
4. $y = \sin 2x + x \cos 4x$
5. $y = x/(x^2 + 1)$
6. $y = \frac{\cos x}{1+\sin x}$
7. $y = 1/x + x^3/(x^4 + 5x \sin x)$

for x *from 1 to 2 in steps of 0.1*

Task 1.7 *Evaluate the function*

$$y = \frac{x}{x + \frac{1}{x^2}},$$

for $x = 3$ *to* $x = 5$ *in steps of* 0.01*.*

Task 1.8 *Evaluate the function*

$$y = \frac{1}{x^3} + \frac{1}{x^2} + \frac{3}{x},$$

for $x = -2$ *to* $x = -1$ *in steps of* 0.1*.*

Task 1.9 (D) *The following code is supposed to evaluate the function*

$$f(x) = \frac{x^2 \cos \pi x}{(x^3 + 1)(x + 2)},$$

for $x \in [0, 1]$ (using 200 steps). Correct the code and check this by evaluating the function at $x = 1$ using `f(200)` *which should be $-1/6$.*

```
x = linspace(0,1);
clear all
g = x^3+1;
H = x+2;
z = x.^2;
y = cos xpi;
f = y*z/g*h
```

Task 1.10 (D) *Debug the code which is supposed to plot the polynomial $x^4 - 1$ between $x = -2$ and $x = 2$ using 20 points.*

```
x = -2:0.1:2;
c = [1 0 0 -1];
y = polyval(c,x);
plot(y,x)
```

Task 1.11 (D) *Debug the code which is supposed to set up the function $f(x) = x^3 \cos(x + 1)$ on the grid $x = 0$ to 3 in steps of 0.1 and give the value of the function at $x = 2$ and $x = 3$.*

```
x = linspace(0,3);
f = x^3.*cos x+1;
% x = 2
f(2)
% x = 3
f(End)
```

2
Writing Scripts and Functions

2.1 Creating Scripts and Functions

With the preliminaries out of the way we now turn our attention to actually using MATLAB by writing a short piece of code. Most of the commands in this section have purposely been written so they can be typed at the prompt, `>>`. However, as we develop longer codes or ones which we will want to run many times it becomes necessary to construct scripts. A *script* is simply a file containing the sequence of MATLAB commands which we wish to execute to solve the task at hand; in other words a script is a computer program written in the language of MATLAB.

To invoke the MATLAB editor[1] we type `edit` at the prompt. This editor has the advantage of understanding MATLAB syntax and producing automatic formatting (for instance indenting pieces of code as necessary). It is also useful for colour coding the MATLAB commands and variables. Both of these attributes are extremely useful when it comes to debugging code. The MATLAB editor also has the feature that once a piece of code has been run the values of variables can be displayed by placing the mouse close to the variable's location within the editor. This is extremely useful for seeing what is going on and provides the potential to identify where we might have made a mistake (for instance, if we had set a variable to be the wrong size).

[1] You can of course make use of any other editor you have available on your computer. We have chosen to use the built-in MATLAB editor. Its implementation may differ slightly from platform to platform. If you are unsure of its use try typing `help edit` at the MATLAB prompt.

Example 2.1 *We begin by entering and running the code:*

```
a = input('First number ');
b = input('Second number ');
disp([' Their sum is ' num2str(a+b)])
disp([' Their product is ' num2str(a*b)])
```

This simple code can be entered at the prompt, but that would defeat our purpose of writing script files. We shall therefore create our first script and save it in a file named `twonums.m`*. To do this, first we type* `edit` *at the MATLAB prompt to bring the editor window to the foreground (if it exists) or invoke a new one if it doesn't. Along the base of the typing area are a set of tabs. These allow you to switch between multiple codes you may be simultaneously working on. Since this is our first use of the editor, MATLAB will have given this code the default name* `Untitled.m`*.*

To proceed we type the above code into the editor and then use the File Menu (sub item Save As) to change the name of the code and Save it as `twonums.m`*. We will need to erase the current default name (*`Untitled.m`*) and type the new filename. After a code has been named we can use the save icon (a little picture of a disc) to save it, without the need to use the File Menu. If we now return to the MATLAB window and enter the command* `twonums` *at the prompt, our code will be executed; we will be asked to enter two numbers and MATLAB will calculate, and return, their product and sum. The contents of the file can be displayed by typing* `type twonums`*.*

This simple example has introduced two new commands, `input` and `num2str`. The `input` command prompts the user with the flag contained within the quotes `' '` and takes the user's response from the standard input, in this case the MATLAB window. In the first example it then stores our response in the variable `a`. The second command `num2str` stands for *number-to-string* and instructs MATLAB to convert the argument from a number, such as the result of `a+b`, to a character string. This is then displayed using the `disp`.

IMPORTANT POINT

> It is very important you give your files a meaningful name and that the files end with `.m`. You should avoid using filenames which are the same as the variables you are using and which coincide with MATLAB commands. Make sure you do not use a dot in the body of the filename and that it does not start with a special character or a number.

For instance `myfile.1.m` and `2power.m` are not viable filenames (good alternatives would be `myfile_1.m` and `twopower.m` respectively). Filenames also have the same restrictions which we met earlier for variable names (see page 4). When processing a command, MATLAB searches to see if there is a user-defined function of that name by looking at all the files in its search path with a `.m` extension. If you are in any doubt whether something is a MATLAB command use either the command `help` or the command `which` in combination with the filename, for instance `help load` or `which load` for the MATLAB command `load`.

You should consider creating a directory for your MATLAB work. For instance on a Unix machine the sequence of commands (at the Unix prompt)

```
mkdir Matlab_Files
cd Matlab_Files
matlab
```

will create a directory (which obviously only needs to be done once), the second command changes your working directory to `Matlab_Files` and the third command invokes MATLAB (to check that you are in the correct directory use the command `pwd` to 'print working directory'). In a Windows environment you can use Explorer to create a new folder and then invoke MATLAB by clicking on the MATLAB icon. On a Macintosh you can simply create a new directory as you would a new folder[2]. Depending upon your computing platform you will probably need to change to your new folder; the `cd` within MATLAB allows you to change directories. In MATLAB6 you can use the `...` symbol at the top of the control environment to change the working directory and this is displayed to the left of this symbol. It is also possible to access files from other directories by augmenting MATLAB's search path. Again, this is platform-specific; the following works on our Unix platforms

```
path(path,'/home/sro/MyMatlabFiles')
```

On a Macintosh you can set the path by choosing `Set Path` from the MATLAB `File` menu. With the search paths set it is possible to create and manage a central resource of user written library functions which can be accessed from any directory on your computer.

One of the most common sources of problems for the novice programmer occurs when the wrong script is being run or when the computer cannot find the program you have just entered. This is usually because the file isn't in the correct directory, or is misnamed. You can check that a file is the correct

[2] Under MacOS X you can use the standard Unix commands given above.

one by using the command `type`. The syntax for this command is simply `type file1`; this will produce a listing of the MATLAB file `file1.m`. If the text is not what you are expecting you may well be using a filename which clashes with an existing MATLAB command. In order to see which file you are looking at, type the command `which file1`. This will tell you the full pathname of the listed files which can be compared with the current path by typing `pwd`. You can also list the files in the current directory by typing `dir` or alternatively all the available MATLAB files can be listed by using `what`: for more details see `help what`.

Example 2.2 *If we create a MATLAB file called* `power.m` *using the editor it can be saved in the current directory: however the code will not work. The reason for this can be seen by typing* `which power` *which produces the output*

```
>> which power
power is a built-in function.
```

So MATLAB will try to run the built-in function.

2.1.1 Functions

In the previous sections we wrote codes which could be legitimately run at the MATLAB prompt. We now discuss the important class of codes which actually act as functions. These codes take inputs and return outputs. We shall start with a very simple example:

```
function [output] = xsq(input)
output = input.^2;
```

which we will save as `xsq.m`. It is important that we get the syntax for functions correct so we'll go through this example in some detail.

- The first line of `xsq.m` tells us this is a function called `xsq` which takes an input called `input` and returns a value called `output`. The input is contained in round brackets, whereas the output is contained within square brackets. It is crucial for good practice that the name of the function `xsq` corresponds to the name of the file `xsq.m` (without the `.m` extension), although there is some flexibility here.
- The second line of this function actually performs the calculation, in this case squaring the value of the input, and storing this result in the variable `output`. Notice that the function uses dot arithmetic `.^` so that this function

will work with both vector and matrix inputs (performing the operation element by element). Notice also that we have suppressed the output of the calculation by using a semicolon; in general all communication between a function subroutine and the main calling program should be done through the input and output.

Having written our function we can now access it from within MATLAB. Consider the following:

```
x = 1:10;
y = xsq(x)
```

Here we have initialised the vector `x` to contain the first ten integers. We call the function in the second line and assign the result, in this case `x.^2`, to the variable `y`. Notice that we have called the function with the argument `x` and not something called `input`. The variables `input` and `output` are local variables that are used by the function; they are not accessible to the general MATLAB workspace[3]. When the function is run it merely looks at what is given as the argument. It is therefore important the function has the correct input; in our example scalar, vector or matrix inputs are allowed. In other cases, if the function expects an argument of a certain type then it must be given one otherwise an error will occur (which will not always be reported by MATLAB). Of course it is not possible to call the function just using `xsq` since the code cannot possibly know what the input is. MATLAB will return an error stating that the `Input argument 'input' is undefined`. As noted above this function can also be used for scalars, for instance `xsq(2)` returns the value 4, and for vectors

```
>> A = [1 2 3 4 5 6];
>> y = xsq(A)

y =

     1     4     9    16    25    36
```

Functions can also take multiple inputs and give multiple outputs. Consider the following examples:

Example 2.3 *Suppose we want to plot contours of a function of two variables* $z = x^2 + y^2$. *We can use the code*

[3] You can find out what variables are in use by MATLAB by typing `who`, which lists all variables in use, or `whos`, which lists all variables along with their size and type.

```
function [output] = func(x,y)
output = x.^2 + y.^2;
```

which should be saved in the file **func.m**. *The first line indicates this is a function which has two inputs* **x** *and* **y**, *and returns a single output* **output**. *The next line calculates the function* $x^2 + y^2$; *again we have used dot arithmetic to allow for the possibility of vector or matrix arguments. For the calculation to be valid the vectors* **x** *and* **y** *must have the same size. To plot the contours (that is the level curves) of the function* $z = x^2 + y^2$ *we would proceed as follows:*

```
x = 0.0:pi/10:pi;
y = x;
[X,Y] = meshgrid(x,y);
f = func(X,Y);
contour(X,Y,f)
axis([0 pi 0 pi])
axis equal
```

This gives us the plot

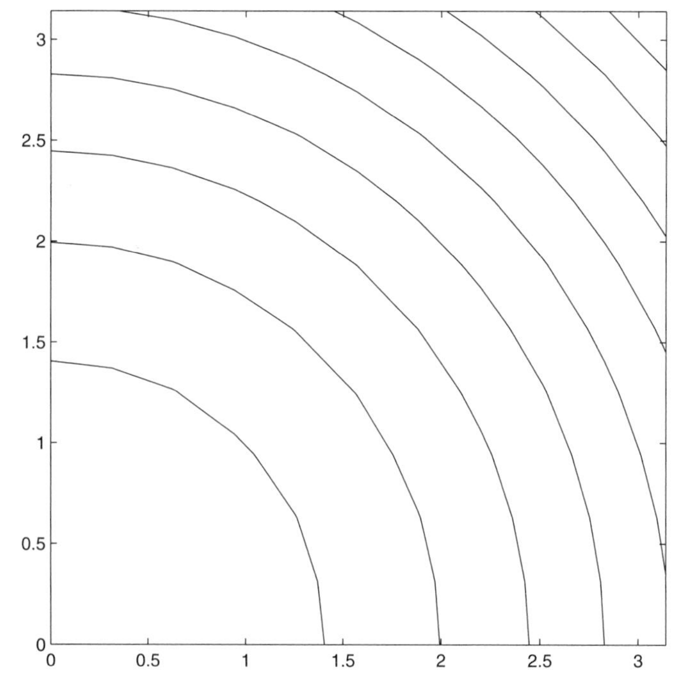

For the moment we do not need to worry about the plotting commands used to display the curves: we will return to plotting in more detail later in this chapter.

Important Point

> It is just as important that the function receives the correct number of inputs as it is that they are of the correct type. For example, we cannot call our function using `func` or `func(1)` (or with arguments of different size). In all these cases MATLAB will give an error message.

Example 2.4 *Suppose we now want to construct the squares and cubes of the elements of a vector. We can use the code*

```
function [sq,cub] = xpowers(input)
sq = input.^2;
cub = input.^3;
```

Again, the first line defines the function `xpowers`*, which has two outputs* `sq` *and* `cub` *and one input. The second and third lines calculates the values of* `input` *squared and cubed. This function file must be saved as* `xpowers.m` *and it can be called as follows:*

```
x = 1:10;
[xsq,xcub] = xpowers(x);
```

This gives

```
>> xsq

xsq =

     1     4     9    16    25    36    49    64    81   100

>> xcub

xcub =

  Columns 1 through 6

           1           8          27          64         125         216

  Columns 7 through 10

         343         512         729        1000
```

The output is two row vectors, one containing the values of the first ten integers squared and the second the values of the first ten integers cubed. Notice that when the function is called we must know what form of output we expect, whether it be a scalar, a vector or a matrix. The expected outputs should be placed within square brackets.

Example 2.5 *As you might expect a function can have multiple inputs and outputs:*

```
function [out1,out2] = multi(in1,in2,in3)
out1 = in1 + max(in2,in3);
out2 = (in1 + in2 + in3)/3;
```

which should be saved as `multi.m`*. This gives us a function called* `multi` *that takes three inputs* `in1`*,* `in2` *and* `in3` *and returns two outputs* `out1` *and* `out2`*. The first output is the sum of the first input and the maximum of the latter two, calculated using the MATLAB intrinsic function* `max`*. The second output is simply the arithmetic mean (the average) of the inputs. We can call this function in the following way*

```
x1 = 2; x2 = 3; x3 = 5;
[y1,y2] = multi(x1,x2,x3);
y1, y2
```

For this example we obtain `y1=7` *and* `y2=3.3333`*.*

The input and output of a function do not have to be the same size (although in most cases they will be). Consider the following example:

Example 2.6 *Consider a code which returns a scalar result from a vector input. For example*

```
function [output] = sumsq(x)
output = sum(x.^2);
```

Our function `sumsq` *takes a vector (or potentially a scalar) as an input and returns the sum of the squares of the elements of the vector. The MATLAB intrinsic function* `sum` *calculates the sum of its vector argument. For instance*

```
x = [1 2 4 5 6];
y = sumsq(x)
```

sets `y` *equal to the scalar* $1^2 + 2^2 + 4^2 + 5^2 + 6^2 = 82$.

2.1.2 Brief Aside

For those of you familiar with matrices we pause here and note that the command in the previous example will also work with matrices:

```
>> A=[1 2 3; 4 5 6];
>> sumsq(A)

ans =

    17    29    45
```

The command has squared (and summed) the elements of the matrix `A`, which is two-by-three. This has exploited the property that the `sum` command sums the columns of a matrix. If we want to sum the rows of a matrix we use `sum(A,2)`, so that we have

```
>> sum(A,1)  % which is equivalent to sum(A)

ans =

     5     7     9

>> sum(A,2)

ans =

     6
    15
```

Notice that the answers are the shape we would expect: the first is a row vector whereas the second is a column vector. Many of the MATLAB commands we shall meet in this text, in general those commands which reduce the dimension of the input object by one, can operate along the rows or the columns; which is specified using an additional argument.

Many of MATLAB's intrinsic commands work in the same way and so care is needed to ensure that the correct number and form of inputs to functions are used.

2.2 Plotting Simple Functions

One of the most powerful elements of MATLAB is its excellent plotting facilities which allow us to easily and rapidly visualise the results of calculations. We have already met some examples of plotting (the line graph plotting of the functions and the contour plot on page 32). We pause here to try some examples of the plotting facilities available within MATLAB. We start with the simplest command `plot` and use this as an opportunity to revisit the ways in which functions can be initialised. We start with initialising an array, in this case `x`

```
x = 0:pi/20:pi;
```

which as we know sets up a vector whose elements are

$$\left(0, \frac{\pi}{20}, 2\frac{\pi}{20}, \cdots, 19\frac{\pi}{20}, \pi\right),$$

(that is, a vector whose elements range from 0 to π in steps of $\pi/20$). This array is of size one-by-twenty one, which can be confirmed by using the command `size(x)` or, if we know it is a one-dimensional array as is the case here, by using the command `length(x)` (in general `length` gives the maximum of the dimensions of a matrix). We can plot simple functions, for instance `plot(x,sin(x))` or more complicated examples such as

`plot(x,sin(3*x),x,x.^2.*sin(3*x)+cos(4*x))`

(this plots $\sin 3x$ and $x^2 \sin 3x + \cos 4x$). Try these examples out for yourself.

Now we have an array `x` we begin using it as an argument to other functions. We start with calculating the point on a straight line $3x - 1$ using

```
y = 3*x-1;
```

Again `size(y)`, or `length(y)`, confirms that this array has the same dimensions as the vector `x`. We can plot y versus x using the command `plot(x,y)` to produce a straight line (in the default colour blue). You can change the colour or style of the line, or force the individual data points to be plotted, using a third argument for the plot command; more details will be given later on page 45.

Proceeding to polynomials, the code

```
y = x.^2+3;
```

produces a vector `y` whose elements are given by the quadratic $x^2 + 3$. Notice that by using the dot before the operator (here exponentiation ^) we are performing the operation element by element on the array `x`.

Example 2.7 *To plot the quadratic* $x^2 + 7x - 3$ *from* x *equals* -3 *to* 3 *in steps of* 0.2 *we use the code*

```
x = -3:0.2:3;
y = x.^2+7*x-3;
grid on
plot(x,y)
```

The resulting plot is given below

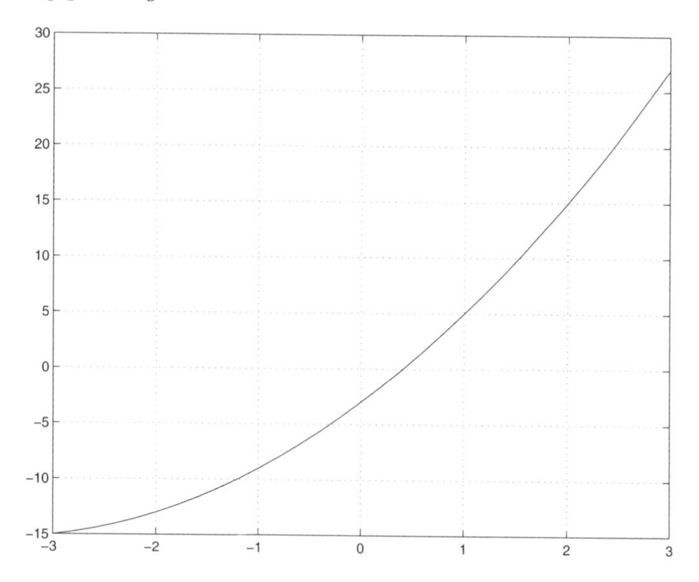

We have given the plot a grid by using the command `grid on`; this can be removed using the command `grid off`.

MATLAB provides an excellent computing environment for producing results which can be viewed quickly and easily. This is essential when we come to analyse the results of our calculations, a task that is usually necessary in order to obtain useful information from an otherwise mathematical calculation. MATLAB is capable of producing very intricate and clear plots, as the following example illustrates.

Example 2.8 *Consider the following code:*

```
x = 0:pi/20:pi;
n = length(x);
r = 1:n/7:n;
y = x.^2+3;
plot(x,y,'b',x(r),y(r),'r*')
axis([-pi/3 pi+pi/3 -1 15])
xlabel('x values')
ylabel('Function values')
title('Demonstration plot')
text(pi/10,0,'\alpha=\beta^2')
```

If we dissect this piece of code we see the first line initializes the vector `x`. *The second and third lines pick out every third integer value between 1 and* `length(x)`. *The fourth line computes the values of the quadratic* $y = x^2 + 3$ *at the points in the vector* `x` *(using dot arithmetic to achieve this). We then reach the* `plot` *command. Here we are telling MATLAB to plot the curve* y *versus* x *and to colour the plot blue (using the flag* `'b'`*). The second part of the* `plot` *command tells MATLAB to plot every third point on the curve (here represented by* `x(r)` *and* `y(r)`*) as points which are labelled with a red asterisk using* `'r*'`. *The commands directly following the* `plot` *command are used to manipulate the final look of the plot. The command* `axis` *sets the start and end points of the horizontal (from* $-\pi/3$ *to* $\pi + \pi/3$*) and then the vertical axis (from* -1 *to* 15*). The commands* `xlabel` *and* `ylabel` *add labels to the horizontal and vertical axes and the command* `title`*, not surprisingly adds the title to the figure. The command* `text` *allows the user to add text to the figure at a coordinate specified within the units of the plot. The arguments of these commands include a string which starts and ends with a quotation mark. The text can include many characters but here we have included Greek letters, using the LaTeX[4] construction \alpha for* α *and \beta for* β. *The properties of a figure can be edited using the drop down menus on its window. This includes being able to increase the size of the characters used in labels. By changing the line which sets the title to*

```
title('Demonstration plot','FontSize',24)
```

we see quite a dramatic change in the size of the characters in the title.

[4] LaTeX is a language that is almost universally used for typesetting mathematics. It is available for almost all computing platforms and operating systems. Further details can be found by consulting any good LaTeX text.

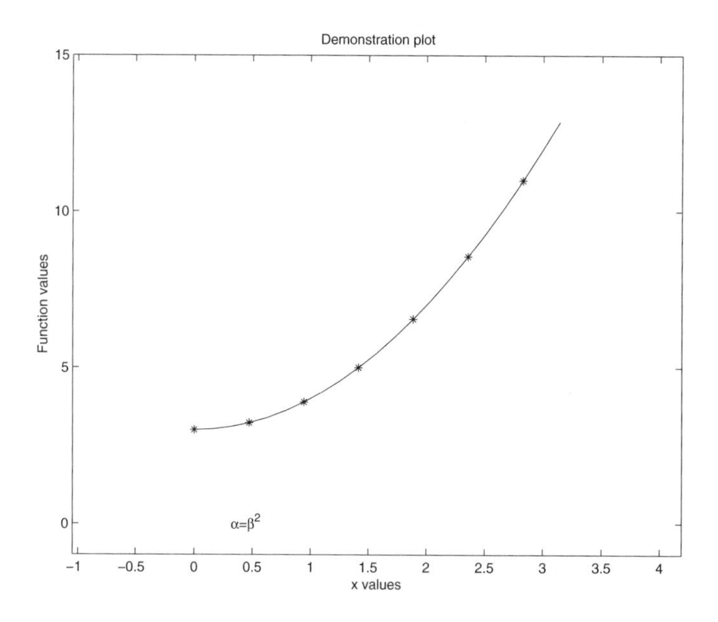

There are a wide variety of other plotting options available. For instance `loglog(x,y)` produces a log-log plot. Similarly `semilogx` and `semilogy` produces a log plot for the x and y-axis, respectively. You should also be aware of the commands `clf` which clears the current figure and `hold` which holds the current figure. We will explore the use of these commands in the tasks at the end of this section.

One of the excellent features of MATLAB is the way in which it handles two and three-dimensional graphics. Although we will have little need to exploit the power of MATLAB's graphical rendering we should be aware of the basic commands. Examples serve to highlight some of the many possibilities:

```
x = linspace(-pi/2,pi/2,40);
y = x;
[X,Y] = meshgrid(x,y);
f = sin(X.^2-Y.^2);
figure(1)
contour(X,Y,f)
figure(2)
contourf(X,Y,f,20)
figure(3)
surf(X,Y,f)
```

This gives the three figures

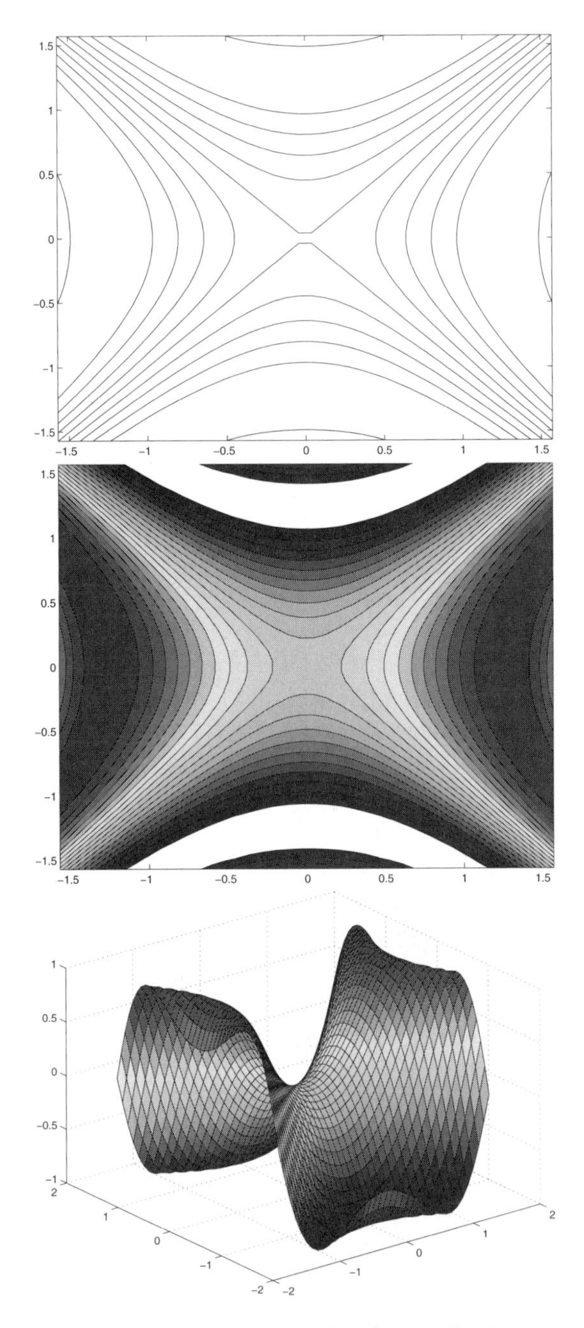

These figures are, respectively, a contour plot (created using `contour(X,Y,f)`), a filled contour plot with 20 contour levels (created using `contourf(X,Y,f,20)` and a surface plot (created using `surf(X,Y,f)`). The function `f` is plotted on a grid generated using the command `meshgrid(x,y)`. As the name suggests `meshgrid` sets up a grid of points by generating copies of the `x` and `y` values; see `help meshgrid` for details of this command. To further explore the potential of MATLAB's plotting facilities see the MATLAB `demo` topics `Visualization` and `Language/Graphics`.

2.2.1 Evaluating Polynomials and Plotting Curves

We return to the topic of functions by first constructing a simple program to evaluate a quadratic. We shall start with a code to generate the value of a specific quadratic at a specific point:

```
x = 3;
y = x^2+x+1;
disp(y)
```

This code sets `x` equal to 3, calculates $x^2 + x + 1$ and then displays the answer. We now expand the above code so that the user can enter the point and the coefficients of the quadratic. In general suppose we have the general quadratic

$$y = a_2x^2 + a_1x + a_0.$$

Firstly we need to define the quadratic and this is done by fixing the three coefficients a_0, a_1 and a_2. This can be done using the following script, which we will call `quadratic.m`

```
% quadratic.m
% This program evaluates a quadratic
% at a certain value of x
% The coefficients are stored in a2, a1 and a0.
%                                                SRO & JPD
%
str = 'Please enter the ';
a2 = input([str 'coefficient of x squared: ']);
a1 = input([str 'coefficient of x: ']);
a0 = input([str 'constant term: ']);
x = input([str 'value of x: ']);
y = a2*x*x+a1*x+a0;
% Now display the result
disp(['Polynomial value is: ' num2str(y)])
```

Again we have a code where the first few lines of the code start with a percent sign `%`. MATLAB treats anything coming after a `%` sign as a comment.

Comments at the start of a code have a special significance in that they are used by MATLAB to provide the entry for the help manual for that particular script. The manual entry can then be accessed by typing `help quadratic` at the MATLAB prompt to produce

```
quadratic.m
This programme evaluates a quadratic
at a certain value of x
The coefficients are stored in a2, a1 and a0.
                                        SRO & JPD
```

Notice the manual entry terminates once MATLAB reaches a line in the file `quadratic.m` in which the first character is **not** a % sign. The line later on in the code which starts with a percent sign is again a comment. In this case we have inserted this line to provide information to the user about the calculation which is to be performed; in this case display the result of our calculation. Judicious use of comments throughout a code makes it readable by a person who may not be fully conversant with the precise details of the calculations to be performed. It is good programming practice to insert text as a manual entry at the start of any code as well as including comments on particular aspects of the calculation which may not be transparent to the user of the code. Note it is also possible to obtain a complete listing of the code from within MATLAB by typing `type quadratic` at the prompt.

Back to our code. The first three non-comment lines use the `input` command to prompt the user to enter the values of the constants `a2`, `a1` and `a0`. As we saw earlier the `input` command takes an argument in the form of a character string, contained between quotes, that acts as the prompt which appears on the screen. Here we have introduced a variable `str` which is equal to `'Please enter the '`, a phrase which is common to each of the input statements. The argument to `input` is then a vector containing strings. Notice in this example we have appended to the prompt a colon followed by a space so that when the user types the values, rather than them being flush against the final character, there is a space. This is not necessary but simply a style convention we will usually adopt.

For the moment we shall assume the user responds to the prompts by entering values which are reasonable. The user is then prompted for the value of x at which they wish to evaluate the polynomial. The following line actually does the calculation of evaluating the polynomial; at this stage we have used the fact that $x^2 = x \times x$, but of course we could use `a2*x^2+a1*x+a0`. There are many ways of writing polynomials, for instance this could have been written recursively as

$$a_0 + x(a_1 + x a_2). \tag{2.1}$$

This general recursive form goes under the name of Horner's method. With the function evaluated the answer is displayed, using the function `num2str` ('number-to-string') which takes as input a number and returns a character string which is then appended to the phrase `'Polynomial value is: '`. The concatenation (joining together) of strings is achieved by writing them as elements in a row vector and then using the `disp` function to write them to the screen.

There is no obvious reason why we couldn't type the commands given above at the MATLAB prompt and enter all the values required for the calculation to proceed. However, it is preferable to carry out the calculation by simply typing the command `quadratic`; this also allows us to have access to a general program which calculates the value of a quadratic. An even better approach is to write a user callable function. We will use this example to build up the level of complexity of the function until we have one which is as general as we can possibly make it.

Firstly let's suppose the values of the coefficients in the quadratic are known *a priori*[5]. For example, suppose we want to evaluate the quadratic $y = 3x^2 + 2x + 1$. We could then use the function

```
%
% evaluate_poly.m
%
function [value] = evaluate_poly(x)
value = 3*x.^2+2*x+1;
```

Here we have again used dot arithmetic in the construction `x.^2`, thus making this function able to take both a scalar or vector (or even a matrix) argument. As we have seen earlier if this function is called with a vector, then a vector is returned as output.

Now we can use the function `evaluate_poly`, in the form `evaluate_poly(2)` or `x = 2;y = evaluate_poly(x)`. Note within MATLAB we are able to call a function with a variety of different inputs; whether this is valid depends upon the structure of the function[6]. With this in mind we can now use our function to generate a vector containing the values of the polynomial at a number of x points and plot the result using

```
>> x = -5:0.5:5;
```

[5] A very similar command already exists in MATLAB, namely `polyval`, but this function is developed to demonstrate the construction.

[6] This is similar to the idea of overloading which is intrinsic to object orientated languages like C++ and Java. However, in those languages a different function is called depending on the type of input, whereas here it is one function which apparently deals with all the cases.

```
>> y = evaluate_poly(x);
>> plot(x,y)
```

which gives

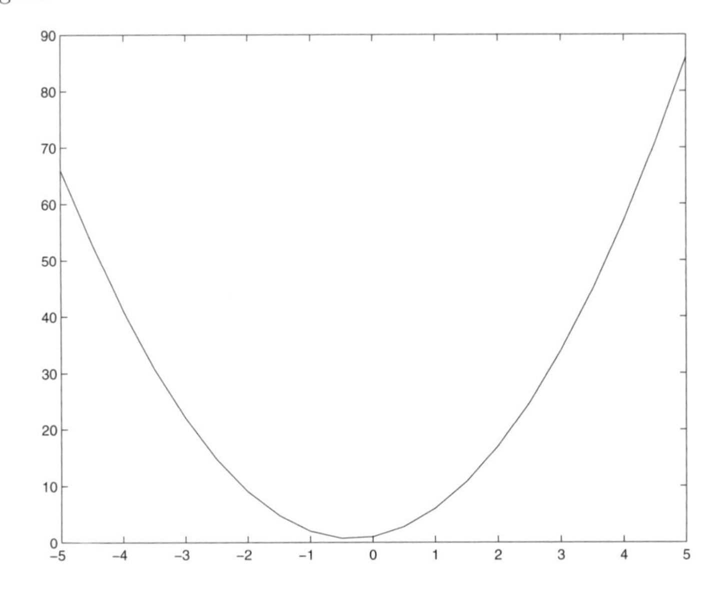

2.2.2 More on Plotting

At this point we briefly return to the problem of generating plots using MATLAB. The simple code given above first sets up a vector which runs from $x = -5$ to $x = 5$ in steps of 0.5. The value of the quadratic at each point is then calculated, using our newly defined function. The final `plot(x,y)` command 'simply' draws a line through these points. The way the plot is constructed can best be seen by making the x-grid coarser (that is, with less points); we then find the plot is constructed by drawing a straight line through each pair of points. MATLAB is slightly more subtle than this in that it checks whether there is a window open ready for plotting, and if not it opens one. The axes are automatically rescaled and tick marks are drawn. The default settings are generally fine, although at some points you may want to look at certain regions of the plot. At the moment we will not worry about the details of how this is achieved[7]. In general we shall introduce the features of the plotting package as we need them. Here we extend our plotting capability by considering the impact of a

[7] Try the commands `h = gca`. This returns what MATLAB calls a 'handle' to the current axis, and then the function `get(h)` lists all the properties of the axis. There is also a command `gcf` which returns a handle to the entire figure.

third argument of the `plot` command, such as in `plot(x,y,'r.')`. The string which is passed as the third argument conveys information as to how the plot should be presented: here the first character `r` is the colour (red) and the second is the symbol (in this case a dot `.`) to be used at each point (as opposed to a line joining the points). The colour options are

y	yellow	m	magenta
c	cyan	r	red
g	green	b	blue
w	white	k	black

and the choice of symbols are

.	point	v	triangle (down)
o	circle	X	triangle (up)
x	x-mark	<	triangle (left)
+	plus	>	triangle (right)
*	star	p	pentagram
s	square	h	hexagram
d	diamond		

It is also possible to control the line style. Instead of using a symbol, as in the previous command, we can draw the line using one of the following options:

-	solid
:	dotted
-.	dashdot
- -	dashed

As we saw earlier (on page 38) we can also plot more than one curve on the same graph. One simple way of achieving this is through the following code

```
>>x = -pi:pi/20:pi;
>>plot(x,sin(x),'r-',x,cos(x),'b:')
```

This generates the plot

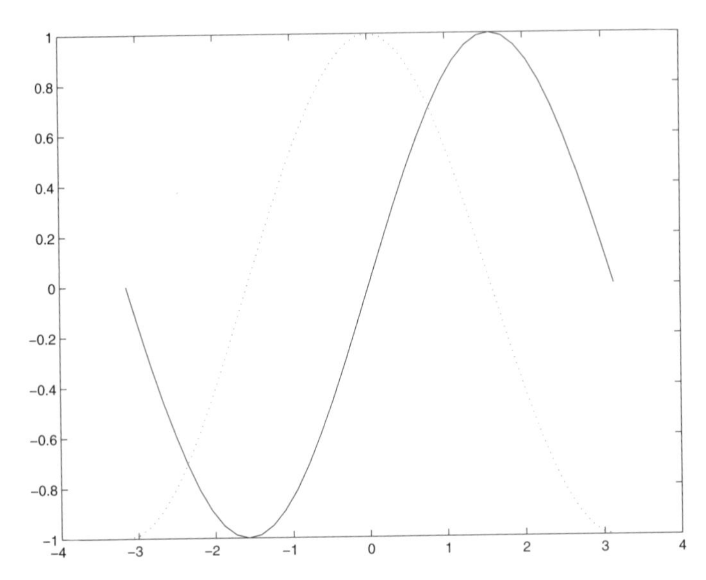

from which the command syntax can be deduced; a plot of $\sin(x)$ versus x using a solid red line and a plot of $\cos(x)$ versus x using a blue dotted line (if the plot were to be viewed in colour). This could have been achieved in another way

```
x = -pi:pi/20:pi;
clf
plot(x,sin(x),'r-')
hold on
plot(x,cos(x),'b:')
hold off
```

As you might infer from this, using the command `plot` clears the current plot and replaces it with the new figure, unless the figure is being "held". So here we plot $\sin(x)$, hold this current figure using `hold on` and then plot $\cos(x)$. The current figure is released using `hold off`.

We could add a legend to this plot to help us identify which line is which. This is accomplished with the command

```
legend('sine','cosine')
```

which now gives

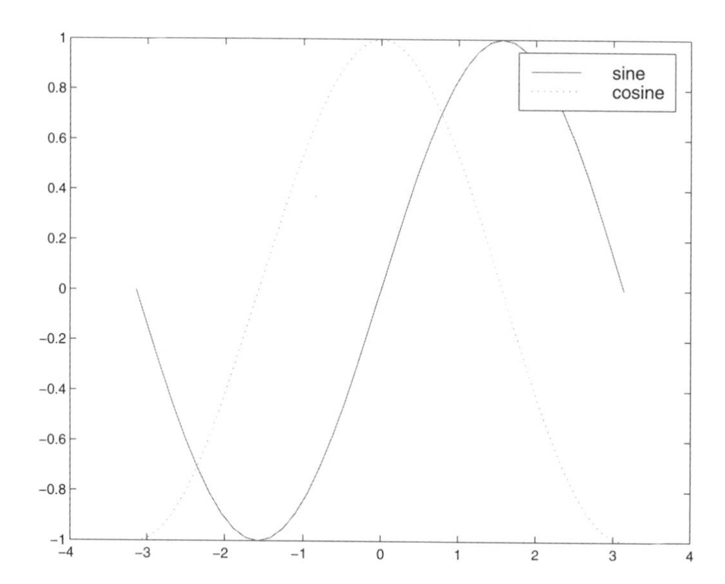

In this case the legend has been placed in the top right corner, however its location can be changed, see **help legend** for details.

We now return to our discussion of functions and consider in more detail functions which can be called with more than one argument and return as many arguments as we wish. In constructing functions we do need to take some care with how our outputs are returned; we have already seen that if the function is called with a vector then the "output" is, in general, a vector. However, there is no reason why there should not be two (or more) outputs returned. Consider the following modification to the function `evaluate_poly` from page 43

```
%
% evaluate_poly2.m
%
function [f, fprime] = evaluate_poly2(x)
f = 3*x.^2+2*x+1;
fprime = 6*x+2;
```

This MATLAB function calculates the values of the polynomial and its derivative. This could be called using the sequence of commands

```
x = -5:0.5:5;
[func,dfunc] = evaluate_poly2(x);
```

We can further generalise our code by passing the coefficients of the quadratic to the function, either as individual values or a vector. The code then becomes

```
%
% evaluate_poly3.m
%
function [f,fprime] = evaluate_poly3(a,x)
f = a(1)*x.^2+a(2)*x+a(3);
fprime = 2*a(1)*x+a(2);
```

which can be called using

```
x = -5:0.5:5;
a = [1 2 1];
[f,fp] = evaluate_poly3(a,x);
```

Before proceeding, we make one final comment regarding the general form for function files. The only communication between the function and the calling program is through the input and output variables; the function should not prompt a user for input nor should it echo the result of any calculation. This is not a necessary constraint on MATLAB functions but more a useful programming convention which is worth following.

This fairly simple routine `evaluate_poly3(x)` was chosen to demonstrate some of the particulars of MATLAB function files because MATLAB has a built-in function, called `polyval`, which also evaluates polynomials. The extract from `help polyval` yields the information

```
Y = POLYVAL(P,X), when P is a vector of length N+1 whose elements
    are the coefficients of a polynomial, is the value of the
    polynomial evaluated at X.

        Y = P(1)*X^N + P(2)*X^(N-1) + ... + P(N)*X + P(N+1)
```

Care needs to be taken with the order in which the coefficients of the polynomial are presented. In fact these are in the same order as in our routine. This example provides a good opportunity to investigate the full code for this built-in function; the code listing can be obtained using `type polyval`. Compare this with our three-line function `evaluate_poly3(x)`.

Example 2.9 *We can use* `polyval` *to evaluate polynomials. We will use it to test a hypothesis that the values of* $n^2 + n + 41$*, where* n *is an integer, are prime (at least for the first few integer values of* n*: obviously this is not so for* $n = 41$*). Consider*

```
p = [1 1 41];
x = 1:40;
f = polyval(p,x);
isprime(f)
```

the function `isprime(f)` *returns a value of* 1 *if the element of* f *is a prime and* 0 *if it is not. The result of our calculation is a string of* 39 *ones demonstrating that all but the value for* $n = 40$ *are prime* $(40^2+40+41 = 40\times(40+1)+41 = 41\times(40+1) = 41^2)$*. We note that the value of quadratic* $n^2 - n + 41$ *is also prime for all integers up to and including* $n = 40$.

2.3 Functions of Functions

One MATLAB command which provides us with considerable freedom in writing versatile code is the command `feval` which loosely translates as *function evaluation*. The simplest use for this command is

```
y= feval('sin',x);
```

evaluate the function `sin` at `x`. This is equivalent to `sin(x)`. In general the arguments for `feval` are the name of the function, which must be either a MATLAB built-in function or a user defined function, contained between quotes, and the value (or values in the case of a vector) at which the function is to be evaluated. The utility of `feval` is that it allows us to use function names as arguments. By way of introduction we provide an example. Consider the function

$$h(x) = 2\sin^2 x + 3\sin x - 1.$$

One way of writing code that would evaluate this function is

```
function [h] = fnc(x)
h = 2*sin(x).*sin(x)+3*sin(x)-1
```

However we can recognise this function as a composition of two functions $g(x) = \sin(x)$ and $f(x) = 2x^2 + 3x - 1$ so that $h(x) = f(g(x))$. We can easily write a function file `f.m` to evaluate $f(x)$

```
function [y] = f(x)
y = 2*x.^2+3*x-1;
```

(using dot arithmetic to allow for the possibility of vector arguments x). To use this in calculating the value of the composite function $h(x)$ we need to be able to pass the function name to `f` as an argument. We can do this with the following modification to our code

```
function [y] = f(fname,x)
z = feval(fname,x);
y = 2*z.^2+3*z-1;
```

Calculating the function $h(x)$ is now as simple as

```
x = -pi:pi/20:pi;
y = f('sin',x);
plot(x,y)
```

This gives a plot of the function $h(x)$ as

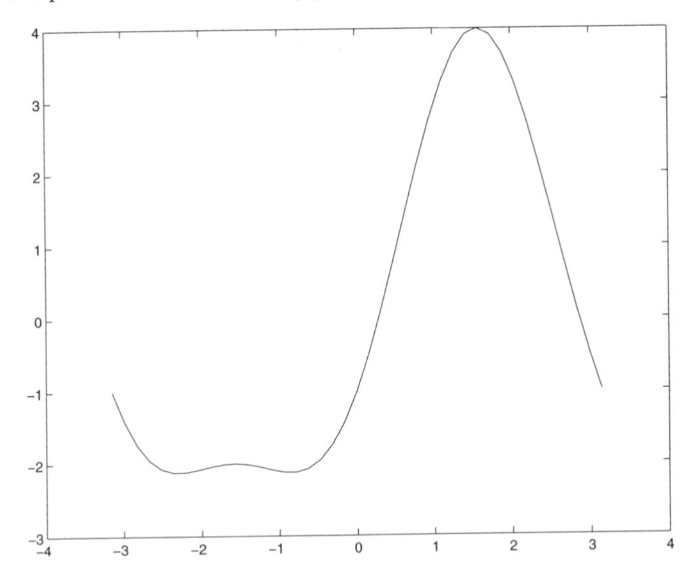

A more useful example is given by the function `plotf`

```
function plotf(fname,x)
y = feval(fname,x)
plot(x,y)
grid on
axis([min(x) max(x) min(y) max(y)])
xlabel('x')
ylabel([fname '(x)'])
```

This function takes as input a function name and a vector `x` and produces a labelled plot. Notice although there are two inputs for this function there are no values output; the figure is the output.

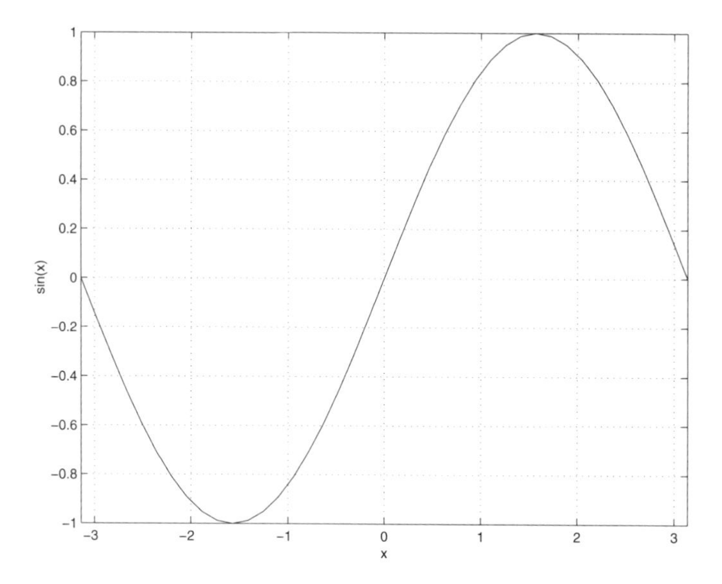

The command `feval` has a wide range of uses, some of which we will exploit later throughout the text.

2.4 Errors

2.4.1 Numerical Errors

We mentioned right at the start of this text that numerical methods are usually not exact (that is, they are approximate methods). In fact it is very hard to get computers to perform exact calculations. If we add (or subtract) integers then a computer can be expected to get the exact answer, but even this operation has its limits. Once we try to perform the division operation we run into trouble. We are happy with the fact that one divided by three is a third, which we can write as $1/3$, but if we need to store this on a computer we run into difficulties.

For the moment let's suppose we have a virtual computer that works in decimal. By this we mean that any number which can be written as a terminating decimal (that is, one which stops), can be stored 'exactly' (provided we do not require too many digits). We know that a third does not have a finite representation, so there is no way we can store this number in our imaginary computer other than by truncating the sequence. We would use the notation $1/3 \approx 0.3333$. Obviously the more three's we retain the more accurate the answer.

Almost all numerical schemes are prone to some kind of error. It is important to bear this in mind and understand the possible extent of the error. Intrinsic to this has to be the knowledge of how much we can trust the results which are produced. Errors can be expressed as two basic types:

Absolute error: This is the difference between the exact answer and the numerical answer.

Relative error: This is the absolute error divided by the size of the answer (either the numerical one or the exact one), which is often converted to a percentage.

(In each case we take the modulus of the quantity.)

Example 2.10 *Suppose an error of £1 is made in a financial calculation of interest on £5 and on £1,000,000. In each case the absolute error is £1, whereas the relative errors are 20% and 0.0001% respectively.*

Example 2.11 *Suppose a relative error of 20% is made in the above interest calculations on £5 and £1,000,000. The corresponding absolute errors are £1 and £200,000.*

Example 2.12 *Estimate the error associated with taking 1.6 to be a root of the equation* $x^2 - x - 1 = 0$.

The exact values for the roots are $(1 \pm \sqrt{5})/2$ *(let us take the positive root). As such the absolute error is*

$$\left| \frac{1+\sqrt{5}}{2} - 1.6 \right| \approx 0.01803398874989$$

and the relative error is the absolute error divided by the value 1.6 *(or alternatively the exact root) which is approximately equal to* 0.01127124296868 *or 1.127%.*

We could also substitute $x = 1.6$ *into the equation to see how wrong it is:* $1.6^2 - 1.6 - 1 = -0.04$. *Although it is difficult to understand how this can be used, it is often the only option (particularly if the exact answer cannot be found).*

This method can be used to determine roots of a function. We now discuss an example which uses this technique:

Example 2.13 *Determine a value of x such that*

$$f(x) = x^2 + 4x = 40.$$

We start by guessing that $x = 6$ is the root we require:

$x = 6,\ f(6) = 60 > 40$ *which is too big, try $x = 5$.*

$x = 5,\ f(5) = 45 > 40$ *which is still too big, try $x = 4$.*

$x = 4,\ f(4) = 32 < 40$ *now this is too small, so we shall try $x = 4.5$.*

$x = 4.5,\ f(4.5) = 38.25 < 40$ *a bit too small, try $x = 4.75$*

$x = 4.75,\ f(4.75) = 41.5625 > 40$ *a bit too big, back down again to $x = 4.625$*

$x = 4.625,\ f(4.625) = 39.890625 < 40$ *a bit too small, back up again to $x = (4.75 + 4.625)/2$*

$x = 4.6875,\ f(4.6875) = 40.72265625 > 40$ *and we can continue this process.*

Here we have just moved around to try to find the value of x such that $f(x) = 40$, but we could have done this in a systematic manner (actually using the size of the errors).

It appears that in order to work out the error one needs to know the exact answer in which case the whole idea of an error would seem to be superfluous. In fact when we focus on errors we are merely trying to work out their size (or magnitude) rather than their actual numerical value.

Before we move on we revisit the MATLAB variable `eps`. This is defined as the smallest positive number such that `1+eps` is different from `1`. Consider the calculations:

```
x1 = 1+eps;
y1 = x1-1
x2 = 1+eps/2;
y2 = (x2-1)*2
```

In both cases using simple algebra you would expect to get the same answer, namely `eps`; but in fact `y2` is equal to zero. This is because MATLAB cannot distinguish between `1` and `1+eps/2`. The quantity `eps` is very useful, especially when it comes to testing routines.

Example 2.14 *Calculate the absolute errors associated with the following calculations:*

```
sin(15*pi)
(sqrt(2))^2
1000*0.001
1e10*1e-10
```

To calculate the absolute errors we need to know the exact answers which are 0, 2, 1 *and* 1 *respectively. We can use the code:*

```
abs(sin(15*pi))
abs((sqrt(2))^2-2)
abs(1000*0.001-1)
abs(1e10*1e-10-1)
```

(notice in the last case the exponent form of the number takes precedence in the calculation: we could of course make sure of this using brackets). The errors are 10^{-15}, 10^{-16} *and zero (in the last two cases).*

2.4.2 User Error

The elimination of user error is critical in achieving accurate results. In practice user error is usually more critical than numerical errors, and after some practise their identification and elimination becomes easier. User error is avoidable, whereas many numerical errors are intrinsic to the numerical techniques or the computational machinery being employed in the calculation.

The most severe user errors will often cause MATLAB to generate error messages; these can help us to understand and identify where in our code we have made a mistake. Once all the syntax errors have been eliminated within a code (that is MATLAB is prepared to run the code), the next level of errors are harder to find. These are usually due to:

1. *An incorrect use of variable names.*
   This may be due to a typographical error which has resulted in changes during the coding.

2. *An incorrect use of operators.*
   The most common instance of this error occurs with dot arithmetic.

3. *Syntactical errors which still produce feasible MATLAB code.*
   For instance in order to evaluate the expression $\cos x$, we should use `cos(x)`: unfortunately the expression `cos x` also yields an answer (which is incor-

rect); somewhat bizarrely `cos pi` yields a row vector (which has the cosines of the ASCII values of the letters `p` and `i` as elements).

4. *Mathematical errors incorporated into the numerical scheme the code seeks to implement.*
   These usually occur where the requested calculation is viable but incorrect (see the example on page 181).

5. *Logical errors in the algorithm.*
   This is where an error has occurred during the coding and we find we are simply working out what is a wrong answer.

Avoiding all of these errors is something which only comes with practice and usually after quite a lot of frustration. The debugging of any program is difficult, fortunately in MATLAB we have the luxury of being able to access variables straight away. This allows us to check results at any stage. Simply removing the semicolons from a code will cause the results to be printed. Although this is not recommended for codes which will produce a deluge of results, it can be very useful.

In Task 2.10 at the end of this chapter and similarly in other chapters[8], we have included some codes with deliberate errors. They also contain commands from future chapters: however most of the errors do not rely on understanding the nuances of these unfamiliar commands. Hopefully by trying to locate the errors in these you should start to develop the debugging skills which will eventually ensure you become a competent programmer.

We shall start with a couple of examples which demonstrate some of the most common sources of problems. Of course it is impossible to predict all possible errors which could be made, but the more you are aware of the better!

Example 2.15 *This code purports to obtain three numbers a, b and c from a user and then produce the results $a+b+c$, $a/((b+c)(c+a))$ and $a/(bc)$.*

[8] These tasks are marked with **(D)**.

```
a = input(' Please entere a )
b = 1nput(' Please enter b )
a = Input ' Please enter c '

v1 = a+ B+d
v2 = a/((b+c)(c+a)
v3 = a/b*c

disp(' a + b + c= ' int2str(v1)])
disp(['v2 = ' num2str v2 ]
disp(['v3 =  num2str(v4) ]);
```

We shall now work through the lines one-by-one detailing where the errors are and how they can be fixed.

`a = input(' Please entere a )` *Firstly the single left-hand quote mark should be balanced by a corresponding right-hand quote mark. The command should also end with a semicolon. This will stop the value of* `a` *being printed out after it is entered. The misspelling of the word enter is annoying but will not actually affect the running of the code: however it is worth correcting if only for presentation's sake.*

`b = 1nput(' Please enter b )` *Again we have an imbalance in the single quote mark and the lack of a semicolon. We also have the fact that* `input` *has become* `1nput`.

`a = Input ' Please enter c '` *Here we have a capitalised command (which MATLAB will actually point out in its error message – try typing* `Input` *in MATLAB). Again we are missing the semicolon, but the more severe errors here are the lack of brackets around the message and the fact that we are resetting the variable* `a`. *This erases the previous value of* `a` *and also does not set the value of* `c` *as we intended.*

`v1 = a+ B+d` *The spacing is unimportant and is merely a matter of style. Here again we should add a semicolon. The main problem here is that the variable* `b` *has been capitalized to* `B` *and since MATLAB is case sensitive these are treated as different variables. Another new variable* `d` *has also slipped in. This means MATLAB will either complain that the variable has not been initialised or will use a previous value which has nothing to do with this calculation. To avoid this it is a good idea to start the code with* `clear all` *to clear all variables from memory.*

`v2 = a/((b+c)(c+a)` *This command presents a number of errors in the evaluation of the denominator of the fraction. The first of these is the missing asterisk between the two factors and secondly we have a unbalanced bracket.*

`v3 = a/b*c` *This has one error but it is one of the most common ones. As mentioned previously this evaluates* $\left(\frac{a}{b}\right) c$. *We need to force the calculation* bc *to be performed first by using brackets. With this (and the previous two commands we might want to elect to add semicolons to stop extra output).*

`disp(' a + b + c= ' int2str(v1)])` *This command is missing a left square bracket to show that we are going to display a vector (with strings as elements). We have used the command* `int2str` *which takes an integer and returns a character string. However at no point have we specified that* a, b *and* c *are integers. We should instead use the command* `num2str` *which converts a general number to a character string.*

`disp(['v2 = ' num2str v2 ]` *This command is missing a round bracket from the end of the expression and a pair of brackets to show that* `num2str` *is being applied to the variable* `v2`.

`disp(['v3 = num2str(v4) ]);` *Here we are missing a single quote to show that the string has ended (after the equals sign). We have also erroneously introduced a new variable* `v4` *which should be* `v3`. *The semicolon on the end of this line is superfluous since the* `disp` *command displays the result.*

The corrected code looks like:

```
clear all
a = input(' Please enter a ');
b = input(' Please enter b ');
c = input(' Please enter c ');

v1 = a+b+c;
v2 = a/((b+c)*(c+a));
v3 = a/(b*c);

disp([' a + b + c= ' num2str(v1)])
disp([' v2 = ' num2str(v2)])
disp([' v3 = ' num2str(v3)])
```

2.5 Tasks

Task 2.1 *Enter and run the code*

```
a = input('Enter a : ');
b = input('Enter b : ');
res = mod(a,b);
str1 = 'The remainder is ';
str2 = '  when ' ;
str3 = ' is divided by ';
disp([str1 num2str(res) str2 ...
  num2str(a) str3 num2str(b)]);
```

which should be saved as `remainders.m`*. Experiment with this code by running it with various values for* `a` *and* `b`*. Make sure you understand how this code works, in particular the* `mod` *command and the* `disp` *command.*

Task 2.2 *Modify the code in Task 2.1 so it returns the answer to a^b and change the character strings* `str1`*,* `str2` *and* `str3` *so that the format of the answer is 'The answer is a^b when a is raised to the power b'.*

Task 2.3 *Write a code which takes a variable x and returns the value of 2^x. Make sure that your code works for variables which are scalars, vectors and matrices.*

Task 2.4 *By modifying the function* `func.m` *(given earlier on page 32) repeat the example for the functions $x^2 - y^2$ and then $\sin(x + y)$ (extending the range to $[0, 2\pi]$ in the latter case).*

Task 2.5 *Modify the code* `xpowers.m` *to simultaneously give the values of the functions $\sin x$, $\cos x$ and $\sin^2 x + \cos^2 x$.*

Task 2.6 *Modify the code* `multi.m` *in Example 2.5 to work out the values of the map:*

$$\begin{aligned} x &\mapsto (x+y)|1 \\ y &\mapsto (x+2y)|1, \end{aligned}$$

where $(a|b)$ is the remainder when a is divided by b and can be calculated using the MATLAB function `mod`*. If b is unity this is the fractional part. This new function will take two inputs and return two outputs.*

Task 2.7 *The code*

```
clear x y
x = -2:0.1:2;
y = 9-x.^2;
plot(x,y)
```

plots the function $y = 9 - x^2$ *for* $x \in [-2, 2]$ *in steps of* $1/10$. *Modify the code so the function* $y = x^3 + 3x$ *is plotted between the same limits and then for* $x \in [-4, 6]$ *in steps of* $1/4$. *(If you can't see the current figure, type* `figure(1)` *which should bring it to the foreground). This code can be run from the prompt or can be entered using* `edit` *and then saved.*

Task 2.8 *Consider the quartic* $y = x^4 + x^2 + a$. *For which values of* a *does the equation have two real roots?*

Task 2.9 *Plot on the same graph the functions* $f(x) = x + 3$, $g(x) = x^2 + 1$, $f(x)g(x)$ *and* $f(x)/g(x)$ *for the range* $x \in [-1, 1]$. *(You need to decide how many points to use to get a smooth curve, and whether to set up the vector* `x` *either using the colon construction or* `linspace`*). You also need to remember to use dot arithmetic since you are operating on vectors.*

Task 2.10 (D) *This task contains codes which are written with deliberate mistakes. You should try to debug the codes so that they actually perform the calculations they are supposed to:*

1. *Perform the calculations*

$$
\begin{aligned}
x &= 4 \\
x + 2 &= y \\
z &= \frac{1}{y^2 \pi}
\end{aligned}
$$

```
x=4
x+2=y
z=1/y^2Pi
```

2. *Calculate the sum*

$$\sum_{i=1}^{N} \frac{1}{i} + \frac{1}{(i+2)(i+3)}$$

where the user inputs N.

```
N=input('Enter N )

for i=1:n
    sum = 1/j + 1/(j+2)*(j+3)
end

disp( ' The answer is ' s])
```

Make sure the code gives the correct answer, for instance for $N = 1$.

3. *Calculate the function*

$$f(x) = \frac{x \cos x}{(x^2 + 1)(x + 2)}$$

for x from 0 to 1 in steps of 0.1.

```
x==0.0:0.1:1.0;
f=xcos x/*([x^2+1]*(x+2)
```

4. *Set up the vector 1 3 3 3 5 3 7 3 9*

```
w = ones(9);

w(1) = 1;

for j = 1:4
    w(2j) = 3:
    w(2j+1) = 2j+1:
```

Make sure that the code returns the correct values of the entries of w.

Task 2.11 *The following codes should be written to produce conversion between speeds in different units.*

(a) *Construct a code which converts a speed in miles per hour to kilometres per hour.*

(b) *Write a code which converts metres per second to miles per hour and use it to determine how fast a sprinter who runs the 100 metres in 10 seconds is travelling in miles per hour (on average).*

(c) *Rewrite your code from part (a) so that it is now a function that takes a single input, the speed in miles per hour, and returns a single output, the speed in kilometres per hour.*

(d) *Now modify the code you wrote in part (b) to determine the sprinter's speed in kilometres per hour by calling the function from the previous part.*

(**NB**: *1 mile = 1760 yards; 1 yard = 36 inches; 1 inch = 2.54 cm; 1 m = 100 cm)*

Task 2.12 *The functions $f(x)$ and $g(x)$ are defined by*

$$f(x) = \frac{x}{1+x^2} \text{ and } g(x) = \tan x.$$

Write MATLAB codes to calculate these functions and plot them on the interval $(-\pi/2, \pi/2)$. Also plot the functions $f(g(x))$ and $g(f(x))$ on this interval.

Task 2.13 *Write a code which enables a user to input the coefficients of a quadratic $q(x) = ax^2 + bx + c$ and plot the function $q(x)$ for $x = \sin y$ where $y \in [0, \pi]$.*

3
Loops and Conditional Statements

3.1 Introduction

We now consider how MATLAB can be used to repeat an operation many times and how decisions are taken. We shall conclude with a description of a conditional loop. The examples we shall use for demonstrating the loop structures are by necessity simplistic and, as we shall see, many of the commands can be reduced to a single line. The true power of computers comes into play when we need to repeat calculations over and over again.

In order to help you to understand the commands in this chapter, it is suggested that you work through the codes on paper. You should play the rôle of the computer and make sure that you only use values which are assigned at that time. Remember that computers usually operate in a serial fashion and that they can only use a variable once it has been defined and given a value. This kind of thought process is very helpful when designing your own codes.

3.2 Loops Structures

The basic MATLAB loop command is `for` and it uses the idea of repeating an operation for all the elements of a vector. A simple example helps to illustrate this:

```
%
% looping.m
%
N = 5;
for ii = 1:N
    disp([int2str(ii) ' squared equals ' int2str(ii^2)])
end
```

This gives the output

```
1 squared equals 1
2 squared equals 4
3 squared equals 9
4 squared equals 16
5 squared equals 25
```

The first three lines start with % indicating that these are merely comments and are ignored by MATLAB. They are included purely for clarity, and here they just tell us the name of the code. The fourth line sets the variable `N` equal to 5 (the answer is suppressed by using the semicolon). The `for` loop will run over the vector `1:N`, which in this case gives `[1 2 3 4 5]`, setting the variable `ii` to be each of these values in turn. The body of the loop is a single line which displays the answer. Note the use of `int2str` to convert the integers to strings so they can be combined with the message " squared equals ". Finally we have the `end` statement which indicates the end of the body of the loop.

We pause here to clarify the syntax associated with the `for` command:

```
for ii = 1:N
    commands
end
```

This repeats the `commands` for each of the values in the vector with `ii`$=1,2,\cdots,N$. If instead we had `for ii = 1:2:5` then the commands would be repeated with `ii` equal to 1, 3 and 5. Notice in the code above we have indented the `disp` command. This is to help the reading of the code and is also useful when you are debugging. The spaces are not required by MATLAB. If you use MATLAB built-in editor (using the command `edit`) then this indentation is done automatically.

Example 3.1 *The following code writes out the seven times table up to ten seven's.*

```
str = ' times seven is ';

for j = 1:10
    x = 7 * j ;
    disp([int2str(j) str int2str(x)])
end
```

The first line sets the variable `str` *to be the string " times seven is " and this phrase will be used in printing out the answer. In the code this character string is contained within single quotes. It also has a space at the start and end (inside the quotes); this ensures the answer is padded out.*

The start of the `for` *loop on the third line tells us the variable* j *is to run from 1 to 10 (in steps of the default value of 1), and the commands in the* `for` *loop are to be repeated for these values. The command on the fourth line sets the variable* x *to be equal to seven times the current value of* j*. The fifth line constructs a vector consisting of the value of* j *then the string* `str` *and finally the answer* x*. Again we have used the command* `int2str` *to change the variables* j *and* x *into character strings, which are then combined with the message* `str`*.*

Example 3.2 *The following code prints out the value of the integers from 1 to 20 (inclusive) and their prime factors.*

To calculate the prime factors of an integer we use the MATLAB command `factor`

```
for i = 1:20
  disp([i factor(i)])
end
```

This loop runs from `i` *equals 1 to 20 (in unit steps) and displays the integer and its prime factors. There is no need to use* `int2str` *(or* `num2str`*) here since all of the elements of the vector are integers.*

The values for which the `for` loop is evaluated do not need to be specified inline, instead they could be set before the actual `for` statement. For example

```
r = 1:3:19;
for ii = r
    disp(ii)
end
```

displays the elements of the vector `r` one at a time, that is 1, 4, 7, 10, 13, 16 and 19; of course we could have used more complicated expressions in the loop.

The following simple example shows how loops can be used not only to repeat instructions but also to operate on the same quantity.

Example 3.3 *Suppose we want to calculate the quantity six factorial* ($6! = 6 \times 5 \times 4 \times 3 \times 2 \times 1$) *using MATLAB.*

One possible way is

```
fact = 1;
for i = 2:6
    fact = fact * i;
end
```

To understand this example it is helpful to unwind the loop and see what code has actually executed (we shall put two commands on the same line for ease)

```
fact = 1;
i=2; fact = fact * i; At this point fact is equal to 2
i=3; fact = fact * i; At this point fact is equal to 6
i=4; fact = fact * i; At this point fact is equal to 24
i=5; fact = fact * i; At this point fact is equal to 120
i=6; fact = fact * i; At this point fact is equal to 720
```

The same calculation could be done using the MATLAB command `factorial(6)`.

Example 3.4 *Calculate the expression* ${}^{n}C_{m}$ *for a variety of values of* n *and* m. *This is read as 'n choose m' and is the number of ways of choosing* m *objects from* n. *The mathematical expression for it is*

$$ {}^{n}C_{m} = \frac{n!}{m!(n-m)!}. $$

We could rush in and work out the three factorials in the expression, or we could try to be a little more elegant. Let's consider $n!/(n-m)!$ *is equal to* $n \times (n-1) \times (n-2) \times \cdots \times (n-m+1)$. *We can therefore use the loop structure. Note also that there are* m *terms in this product, as there are in* $m!$, *so we can do both calculations within one loop*

```
prod = 1;
mfact = 1;
for i = 0:(m-1)
    mfact = mfact * (i+1);
    prod = prod * (n-i);
end
soln = prod/mfact;
```

Breaking up the calculation like this can lead to problems for large values of m and so it is often best to work out the answer directly:

```
soln = 1;
for i = 0:(m-1)
    soln = soln * (n-i) / (i+1);
end
```

This product could also be written as:

```
soln = 1;
for i = 0:(m-1)
    soln = soln * (n-i) / (m-i);
end
```

This version may have a slight computational advantage since the terms appearing in the fractions are closer in magnitude. That these two versions are identical can be seen by rewriting the product as

$$t = \left(\frac{n}{1}\right)\left(\frac{n-1}{2}\right)\cdots\left(\frac{n-m+2}{m-1}\right)\left(\frac{n-m+1}{m}\right)$$

or as

$$s = \left(\frac{n}{m}\right)\left(\frac{n-1}{m-1}\right)\cdots\left(\frac{n-m+2}{2}\right)\left(\frac{n-m+1}{1}\right).$$

This example could also be done using the MATLAB command `prod`, which calculates the product of the elements of a vector; in fact MATLAB has a command `nchoosek` designed for just this calculation. In order to use `prod` we could use:

```
M = 1:m;
t = (n-1+M)./M; prod(t)
s = (n-1+M)./(m-M+1); prod(s)
```

where t and s correspond to the two expressions above.

So far we have dealt with products: in the next example we shall consider a simple summation.

Example 3.5 *Determine the sum of the geometric progression*

$$\sum_{n=1}^{6} 2^n.$$

This is accomplished using the code:

```
total = 0
for n = 1:6
    total = total + 2^n;
end
```

which gives the answer 126, as you might expect from the formula for the sum of a geometric progression

$$S = a\frac{1-r^n}{1-r}$$

where a is the first term, r is the ratio of the terms and n is the number of terms. In our case $a = 2$, $r = 2$ and $n = 6$, which gives $S = 126$.

We now go through the ideas of summing series more thoroughly since it provides important information on how loop structures work.

3.3 Summing Series

In Example 3.5 we have summed a series: we now describe this topic in more detail. We start by constructing a code to evaluate

$$\sum_{i=1}^{N} i^2.$$

Firstly, we note that we do not necessarily know N so this will need to be entered by the user or on a command line. We will first do one case by "hand". Let us consider $N = 4$, so we wish to determine

$$\sum_{i=1}^{4} i^2.$$

If we were to do this on paper we would probably write down all the terms in the summation and then add them up: evaluating the terms in the series

$$\sum_{i=1}^{4} i^2 = 1 + 4 + 9 + 16$$

and adding them up we obtain

$$30.$$

At this stage we can add the terms up in our heads: however, what would happen if N was larger, for instance 10. In this case note the answer is given by

$$\sum_{i=1}^{10} i^2 = 1 + 4 + 9 + 16 + 25 + 36 + 49 + 64 + 81 + 100.$$

But in fact we may actually do it like this:

- The first term corresponds to $i = 1$, for which $i^2 = 1$.
- The second term, that is $i = 2$, has $i^2 = 4$ and adding to the previous answer gives $1 + 4 = 5$.
- The third term, that is $i = 3$, has $i^2 = 9$ and adding to our previous answer gives $5 + 9 = 14$, etc.

We can automate this process by using the MATLAB code:

```
N = input('Enter the number of terms required: ');
s = 0;

for i = 1:N
    s = s + i^2;
end

disp(['Sum of the first ' int2str(N) ...
      ' squares is ' int2str(s)])
```

As we mentioned earlier the command `int2str` converts an integer to a string. This means we can concatenate these strings into a sentence. If we tried

```
apple = 8;disp(['I have ' apple ' apples'])
```

MATLAB misses out the number and gives "I have apples", whereas the command

```
disp(['I have ' int2str(apple) ' apples'])
```

gives "I have 8 apples", as required: for more information on this command see page 185.

We can break down our summation code as follows:

- The first line simply asks the user to enter the value of `N` when prompted with the string contained in the `input` statement.
- The second line sets a variable `s` to be zero: this will be used to store the cumulative sum. The blank line is included to make our code more readable.
- The next three lines form a loop. The loop variable is `i` which runs from 1 to `N`. This means the command in the loop will be repeated for each of these values. The value of `i^2` will be added to the previous value of `s`.
- After the loop there is another blank line (again purely for readability).
- Finally, we display our results using the `disp` command. We have constructed a row vector with four elements: the first and third elements are merely the strings `'Sum of the first '` and `' squares is '` whereas the second and fourth are the values `N` and `s` converted from integers to strings using the MATLAB command `int2str`.

We can now modify the above code to determine the value of the summation

$$\sum_{i=3}^{7} i^3.$$

Previously we were asked to calculate the sum of square terms and now we need the sum of cubes so the line `s = s + i^2;` needs to be modified to `s = s + i^3;`. We also need to change the range which the loop runs over from `1:N` to `3:7`, which is accomplished by changing the argument of the `for` loop to `i = 3:7`. The modified code will now work, however it will ask for a value of `N` which is no longer relevant. We can comment out this line by prepending it with a percentage sign. This instructs MATLAB to ignore everything else on this line. This can be very useful when you want to change a code but do not want to just delete lines. It can also be used to include comments so you can understand the code. Finally, we need to change the line which outputs the results, the amended code looks like this:

```
% N = input('Enter the number of terms required: ');
s = 0;
for i = 3:7
    s = s + i^3;
end
disp(['Required value of the summation is ' int2str(s)])
```

Again we have used the command `int2str` to change the variable `s` into a string. This only works correctly if `s` is an integer. If `s` is not an integer we need to use the command `num2str` which changes a general number to a string.

In all the cases so far we have only been interested in the final answer. Let's consider instead the problem of determining the values of

$$I_N = \sum_{i=1}^{N} f(i),$$

where the `f(i)` are defined by the problem at hand. At the moment we will stick to `f(i) = i^2`. Again we shall start by doing an example by hand, let us determine the values of `I_N` for `N` equals one to four. Firstly `N = 1`

$$I_1 = \sum_{i=1}^{1} i^2 = 1,$$

and now `N = 2`, which is

$$I_2 = \sum_{i=1}^{2} i^2 = 1 + 4 = 5,$$

and `N = 3`,

$$I_3 = \sum_{i=1}^{3} i^2 = 1 + 4 + 9 = 14,$$

and finally `N = 4`, which gives

$$I_4 = \sum_{i=1}^{4} i^2 = 1 + 4 + 9 + 16 = 30.$$

You should have noticed that `I_2 = I_1+2^2`, `I_3 = I_2+3^2` and `I_4 = I_3+4^2`: These are just like the elements of an array (or vector). In general

$$I_N = I_{N-1} + N^2. \tag{3.1}$$

It is exactly this kind of recursion relation which a computer can take advantage of. In order to calculate these values we can use the code:

```
maxN = input('Enter the maximum value of N required: ');
I(1) = 1^2;

for N = 2:maxN
    I(N) = I(N-1) + N^2;
end

disp(['Values of I_N'])
disp([1:N; I])
```

This code uses a vector `I` to store the results of the calculation. It first prompts the user to input the value of `N`, which it stores in the variable `maxN`. The first value of the vector is then set as 1^2. Equation (3.1) now tells us the relation between the first and second elements of `I`, which we exploit in the loop structure. The final two lines display the string `Values of I_N` and the actual values. The answers are given as a matrix, the first row of which contains the numbers 1 to N and the second row gives the corresponding values of `I_N`.

This produces the results:

```
Enter the maximum value of N required: 10
Values of I_N
     1     2     3     4     5     6     7     8     9    10
     1     5    14    30    55    91   140   204   285   385
```

(for $N = 10$). This can be compared with the analytical solution, see equation (3.4). Up until now we have considered very simple summations. Let's now see how we can sum the series where the terms are defined by some function `f(i)`. We use a separate MATLAB program to construct the series. Let's consider

$$I_N = \sum_{i=1}^{N} i \sin \frac{i\pi}{4}.$$

We start by using a code `f.m` which will return the coefficients we are going to sum. You should enter this code and save it as `f.m`.

```
function [value] = f(inp)
value = inp * sin(inp*pi/4);
```

– The first line tells the computer this program is a function, which should take as input a value `inp` and return the variable `value` (if `value` is undefined MATLAB will complain).

– The second line simply works out the required value. Notice that MATLAB already has a variable `pi`.

The above code is now modified to:

```
maxN = input('Enter the maximum value of N required: ');
I(1) = f(1);

for N=2:maxN
    I(N) = I(N-1) + f(N);
end

disp(['Values of I_N'])
disp([1:N; I])
```

This produces the result

```
Enter the maximum value of N required: 5
Values of I_N
    1.0000    2.0000    3.0000    4.0000    5.0000
    0.7071    2.7071    4.8284    4.8284    1.2929
```

(for $N = 5$). We note that here we have a good example of the differing meanings of the command `y(n)`. In the case of `I(N)`, this refers to the Nth entry of the array `I`, whereas for `f(N)`, it refers to the function `f` evaluated at the point `N`. This subtle difference is critical!

3.3.1 Sums of Series of the Form $\sum_{j=1}^{N} j^p$, $p \in \mathbb{N}$

Let's consider the simple code which works out the sum of the first N integers raised to the power p:

```
% Summing series

N = input('Please enter the number of terms required ');
p = input('Please enter the power ');

sums = 0;
for j = 1:N
    sums = sums + j^p;
end

disp(['Sum of the first ' int2str(N) ...
      ' integers raised to the power ' ...
      int2str(p) ' is ' int2str(sums)])
```

so this produces

$$S_N = \sum_{j=1}^{N} j^p.$$

We note the formula when $p = 1$ is given by $N(N+1)/2$. We pause here to think how we could work this out. If we substitute in the values $N = 1$, $N = 2$ and $N = 3$ (all for $p = 1$) we would obtain three points on the 'curve' and these will uniquely determine its coefficients (assuming it is a quadratic). We assume that $S_N = aN^2 + bN + c$ and use the three values above to give three simultaneous equations from which we can determine the coefficients a, b and c, these equations are:

$$N = 1 \qquad a + b + c = S_1 = 1, \tag{3.2a}$$

$$N = 2 \qquad 4a + 2b + c = S_2 = 1 + 2 = 3, \tag{3.2b}$$

$$N = 3 \qquad 9a + 3b + c = S_3 = 1 + 2 + 3 = 6. \tag{3.2c}$$

From (3.2a) we see that $c = 1 - a - b$ and this can be substituted into the other two equations to yield

$$3a + b = 2, \tag{3.3a}$$

$$8a + 2b = 5. \tag{3.3b}$$

Now using (3.3a) in (3.3b) we find $a = 1/2$ and then in (3.3a), $b = 1/2$ and finally using (3.2a) we have $c = 0$. Hence

$$S_N = \frac{1}{2}N^2 + \frac{1}{2}N, \quad \text{for} \quad N = 1, 2, 3.$$

In order to prove that this is true for all N we can use proof by induction.

It seems reasonable to expect that the sum for a certain power of p will be of degree $p+1$. In order to determine the coefficients of a polynomial of degree $p+1$ we require $p+2$ points. Consider this example:

```
clear all
format rat
p = input('Please enter the power you require ');

points = p+2;
n = 1:points;
for N=n
    sums(N)=0;
    for j = 1:N
        sums(N) = sums(N) + j^p;
    end
end

[coe] = polyfit(n,sums,p+1)

format
```

This code is worth dissecting, since it is our first example of nested loops.

- The first line ensures all values of variables have been cleared. This is good programming practice since it means undeclared variables will be noticed rather than used incorrectly.
- The second line sets the format for the duration of this run to be `rat`, forcing MATLAB to try to print the answers as rationals (notice this is coupled with the final line which merely resets the formatting to the default).
- The next line asks the user for the value of p. As noted above we need to determine $p+2$ coefficients and as such we need $p+2$ equations. A vector `n` is set up running over these values.
- We now have a nested `for` loop structure which sets up the points (from 1 to $p+2$).
- The next line sets the initial value of the element of the array `sums` to be zero, ready to be set to the required cumulative sum. This sum is calculated within the inner loop. We note again that each `for` statement needs to be balanced with an `end`.
- Now we use the MATLAB intrinsic command `polyfit` which gives the coefficients of the required polynomial. The syntax of this command is

`polyfit(x,y,n)` which returns the coefficients of order n through the points in (x, y).

For the examples $p = 2$ and $p = 3$ we have

```
>> sumser2
Please enter the power you require 2

coe =

     1/3          1/2          1/6           *

>> sumser2
Please enter the power you require 3

coe =

     1/4          1/2          1/4           *           *
```

Notice here we have asked the code to try to give us rational answers. The asterisks are merely where MATLAB cannot express zero (or something close to it) as a rational. In the first case we have

$$\sum_{j=1}^{N} j^2 = \frac{N^3}{3} + \frac{N^2}{2} + \frac{N}{6} = \frac{N}{6}(2N+1)(N+1), \tag{3.4}$$

(where we have factorised our answer) and in the second case

$$\sum_{j=1}^{N} j^3 = \frac{N^4}{4} + \frac{N^3}{2} + \frac{N^2}{4} = \frac{N^2}{4}(N+1)^2.$$

We could use this procedure for any integer value of p: however a rigorous proof requires mathematical induction (or another similar process).

3.3.2 Summing Infinite Series

We now consider examples where we need to truncate the series.

Example 3.6 *The Taylor series for a function $f(x)$ about a point $x = a$ is given by*

$$f(x) = f(a) + \sum_{n=0}^{\infty} \frac{(x-a)^n}{n!} f^{(n)}(a).$$

We can use this to approximate $\sin x$ *by an infinite series in* x *as*

$$\sin x = \lim_{N\to\infty} \sum_{n=0}^{N} (-1)^n \frac{x^{2n+1}}{(2n+1)!}.$$

Of course in using a computer we cannot actually take N *to be infinity, but we shall take it to be large in the hope that the error will be small. We can write the above series (taking* $N = 10$*) and evaluate the series at the points* 0, 0.1, 0.2, 0.3, 0.4 *and* 0.5 *using*

```
v = 0.0:0.1:0.5;
sinx = zeros(size(v));
N = 10; range = 0:N;
ints = 2*range+1;
for n = range
     sinx = sinx+(-1)^n*v.^ints(n+1)...
        /(factorial(ints(n+1)));
end
```

This is perhaps the most complicated code we have produced so far, so we shall pause and break it down.

`v = 0.0:0.1:0.5;` *sets up the vector* `v` *running from zero to a half in steps of a tenth.*

`sinx = zeros(size(v));` *In order to understand this command it is best to start with what's inside the brackets. The command* `size(v)` *gives the size of the vector* `v` *and then the command* `zero` *sets up* `sinx` *as an array of zeros of the same size.*

`N = 10; range = 0:N;` *sets a variable N to be equal to 10 and then sets up a vector* `range` *which runs from zero to N.*

`ints = 2*range+1;` *This gives the mathematical expression* $2n+1$ *for all the elements of the vector* `range` *and puts them in* `ints`*.*

`for n = range ..... end` *This loop structure repeats its arguments for* `n` *equal to all the elements of range.*

`sinx = sinx+(-1)^n*v.^ints(n+1)/(factorial(ints(n+1)));` *This mathematical expression evaluates the subject of the summation as a function of* n*. (Notice the necessary offset when referring to the elements of* `ints`*, for instance* `ints(1)` *refers to the value when* $n = 0$*). Also note the use of* `.^` *when operating on* `v` *as this is a vector.*

Running this gives

```
>> sinx

sinx =

         0    0.0998    0.1987    0.2955    0.3894    0.4794
```

These values can be compared to those calculated directly by MATLAB to give

```
>> sin(v)

ans =

         0    0.0998    0.1987    0.2955    0.3894    0.4794

>> sinx-sin(v)

ans =

   1.0e-16 *

         0    0.1388    0.2776         0   -0.5551         0
```

Since all the elements of the vector `sinx-sin(v)` *share a factor of* `1.0e-16` *this has been extracted. This results shows that our approximation has worked very well (for relatively few terms). For larger values of x more terms might be needed. For a value of $x = \pi$ we find the same number of terms as used above produces a value of $\sim 10^{-11}$, which is still a good approximation (since the actual value of $\sin \pi$ is zero), but not quite as good as those shown above. The infinite series given above actually converges for all values of x.*

Example 3.7 *Calculate the sum*

$$\sum_{n=0}^{\infty} \mathrm{e}^{-n}.$$

This poses a problem for us since we can obviously not add up an infinite number of terms, we need to truncate the calculation. In this case we do not need very many terms since e^{-x} gets very small very quickly. We can use the code:

```
N = 10;
total = 0;
for n = 0:N
    total = total + exp(-n);
end
```

To test for convergence of our results we should compare our answer for different values of N. *Fortunately, in this case, we are able to work out the value of the truncated series, since it is merely a geometric progression (the formula for the sum of a geometric progression is given on page 68). The first term is* $e^0 = 1$ *and the ratio between successive terms is* $1/e$*, hence the sum is*

$$S_N = \frac{1 - e^{-(N+1)}}{1 - e^{-1}}.$$

In the limit as $N \to \infty$ *this tends to*

$$S_\infty = \frac{1}{1 - e^{-1}} = \frac{e}{e - 1} \approx 1.581977.$$

If we compare the results for $N = 10$ *the error is already* $\sim 10^{-5}$*. To find a formula for the error we need to calculate* $|S_N - S_\infty|$*, which is*

$$|S_N - S_\infty| = \left| \frac{1 - e^{-(N+1)}}{1 - e^{-1}} - \frac{1}{1 - e^{-1}} \right| = \left| \frac{e^{-N}}{e - 1} \right|,$$

which decreases (exponentially fast) as N increases.

3.3.3 Summing Series Using MATLAB Specific Commands

So far we have used commands which are common to many programming languages (or at least similar to) and have not exploited the power of MATLAB. In the first example, given above, we mentioned the idea of adding things up in our head. MATLAB is very good at this type of exercise. Consider the previous example

$$\sum_{i=1}^{10} i^2.$$

Firstly we set up a vector running from one to ten:

```
i = 1:10;
```

and now a vector which contains the values in `i` squared:

```
i_squared = i.^2;
```

Now we use the MATLAB command `sum` to evaluate this:

```
value = sum(i_squared)
```

(*notice here we have left off the semicolon so the result is displayed automatically*). The full code for this example is

```
i = 1:10;
i_squared = i.^2;
value = sum(i_squared)
```

This can all be contracted on to one line `sum((1:10).^2)`: *you should make sure you know how this works!*

Using the command `sum` allows us to simplify our codes: however it is essential we understand exactly what it is doing. Consider a problem where we have a set of values y_1, y_2 up to y_N, where for the sake of argument N is taken to be odd. The problem is to work out the sum of the series

$$\sum_{i=1}^{N} y_i f(i),$$

where $f(i) = 1$ when i is odd and $f(i) = 2$ when i is even. We can set up these values of f using the commands:

```
N = 11;
iodd = 1:2:N;
ieven = 2:2:(N-1);
f(iodd) = 1;
f(ieven) = 2;
```

which gives

```
f =

     1     2     1     2     1     2     1     2     1     2     1
```

This can be achieved in one command, namely `2-mod(1:11,2)`. We can now use a similar code to those we wrote earlier to sum the series. For example if $y = x^2$, we have

```
x = 1:11;
y = x.^2;
sum(y.*f)
```

This kind of expression will prove to be very useful when we come to consider numerically evaluating integrals in a later chapter.

Example 3.8 *Evaluate the expression*

$$\prod_{n=1}^{N}\left(1+\frac{2}{n}\right)$$

for $N = 10$ *(the symbol* $\prod$ *means the product of the terms, much in the same way* $\sum$ *means summation). This can be done using the code:*

```
n = 1:10;
f = 1+(2)./n;
pr = prod(f)
```

The first line sets up a vector with elements running from 1 to 10; the second line sets up a vector **f** *whose elements have the values* $1 + 2/n$*; notice* **f** *is the same shape as* **n**. *We have divided 2 by a vector and so we have used "./" rather than just "/". If you want to see what the code does, just leave off the semicolons; this will show you the vectors which are generated. Finally, the last line gives the product of all the elements of the vector* **f**.

This gives the answer 66. This can be seen by a slight rearrangement of the expression

$$\prod_{n=1}^{N}\left(1+\frac{2}{n}\right)=\prod_{n=1}^{N}\left(\frac{n+2}{n}\right)=\frac{\prod_{n=1}^{N} n+2}{\prod_{n=1}^{N} n},$$

where we have used the fact $\prod a_n b_n = (\prod a_n)(\prod b_n)$. *Now actually evaluating the products:*

$$\begin{aligned}&=\frac{(N+2)(N+1)\cdots 3}{N(N-1)\cdots 1}\\&=\frac{(N+2)!/(2.1)}{N!}=\frac{(N+2)(N+1)}{2}.\end{aligned}$$

When $N = 10$ *this gives a value of 66.*

3.3.4 Loops Within Loops (Nested)

Many algorithms require us to use nested loops (loops within loops), as in the example of summing series on page 75. We illustrate this using a simple example of constructing an array of numbers:

```
for ii = 1:3
    for jj = 1:3
        a(ii,jj) = ii+jj;
    end
end
```

Notice that the inner loop (that is, the one in terms of the variable `jj`) is executed three times with `ii` equal to 1, 2 and then 3. These structures can be extended to have further levels. Notice each `for` command must be paired within an `end`, and for the sake of readability these have been included at the same level of indentation as their corresponding `for` statement.

Example 3.9 *Calculate the summations*

$$\sum_{j=1}^{N} j^p$$

for p equal to one, two and three for $N = 6$.

We could perform each of these calculations separately but since they are so similar it is better to perform them within a loop structure:

```
N = 6;
for p = 1:3
    sums(p) = 0.0;
    for j = 1:N
        sums(p) = sums(p)+j^p;
    end
end
disp(sums)
```

The order in which the loops occur should be obvious for each problem but in many examples the outer loop and inner loop could be reversed.

3.4 Conditional Statements

MATLAB has a very rich vocabulary when it comes to conditional operations but we shall start with the one which is common to many programming languages (even though the syntax may vary slightly). This is the `if` command which takes the form:

```
if (expression)
        commands
        ...
end
```

As you might expect, if the `expression` is true then the commands are executed, otherwise the programme continues with the next command immediately beyond the `end` statement. There are more involved forms of the command, but before proceeding with these let's first discuss the construction of the `expression`. Firstly there are the simple mathematical comparisons

`a < b`	True if a is less than b
`a <= b`	True if a is less than or equal to b
`a > b`	True if a is greater than or equal to b
`a >= b`	True if a is greater than or equal to b
`a == b`	True if a is equal to b
`a ~= b`	True if a is not equal to b

More often than not we will need to form compound statements, comprising more than one condition. This is done by using logical expressions, these are:

`and(a,b)`	`a & b`	Logical AND
`or(a,b)`	`a	b`	Logical OR
`not(a)`	`~a`	NOT
`xor(a,b)`		Logical exclusive OR

The effect of each of these commands is perhaps best illustrated by using tables

AND	false	true
false	false	false
true	false	true

OR	false	true
false	false	true
true	true	true

XOR	false	true
false	false	true
true	true	false

NOT (~) This simply changes the state so ~(true)=false and ~(false)=true.

We pause and just run through these logical operators:

a AND b This is true if both a and b are true

a OR b This is true if one of a and b is true (or both).

a XOR b This is true if one of a and b is true, but not both.

In many languages you can define Boolean (named after George Boole) variables and these will have actual logical meaning. For instance in Excel you can set values to be `True` or `False`. In MATLAB zero represents False and any other value implies True. We shall adopt the convention that `1` (one) is true and `0` (zero) is false. In fact sometimes it is useful to define variables such that `false=0; true= (~false)`. This statement is read as `set true to be equal to not false`; we have added the brackets for clarification.

Example 3.10 *In the following examples it may be helpful to draw pictures, but we should at least teach you to read these statements. We recall that in this context a square bracket means a closed set (that is, including the end point) whereas a round bracket indicates an open set (that is, not including the end point).*

Determine the sets for which these statements are true:

1. `x>1 & x<2`
   We can read this command as "x is greater than one and less than two". This is obviously true only for values between one and two (not inclusive). This is the open set $(1, 2)$.

2. `x<0 | x>=1`
   This command can be read as "x less than zero or greater than (or equal to) one". This set has two parts: x strictly negative or x greater than (or equal to one). Hence the set is $(-\infty, 0) \cup [1, \infty)$.

3. `x>1 | x<2`
   This command can be read as "x greater than 1 or x less than 2". In fact all values of x are greater than one or less than 2, hence the answer is $(-\infty, \infty)$.

4. `x<=1 | x>=1`
   This one says "x is less than (or equal to) 1 or x is greater than (or equal to 1)". This again is true for all values of x, hence the answer is $(-\infty, \infty)$.

5. `x<=1 & x>=1`
   This one is similar to the previous one with one word changed so it now reads:"x is less than (or equal to) 1 and x is greater than (or equal to 1)".

The only value which is less than (or equal to) one and greater than (or equal to) one is one itself. The answer is the single value one, written as $\{1\}$.

6. `~(x>2)`
   This is our first example of negation, which is read as "x is not greater than 2", which is true for values of x less than (or equal to) 2, that is the set $(-\infty, 2]$.

7. `(x>1) & (~(x<2))`
   Here we have "x is greater than 1 and x is not less than 2". The second part of this expression means that x is greater than (or equal to 2) (and the first part is always true for this range), so the solution is $[2, \infty)$.

8. `abs(x-1) < 2`
   Here we have an expression involving a mathematical function: this one is read as: "the modulus of x minus 1 is less than two". This means points which are within 2 units of the point 1, that is $(-1, 3)$.

9. `rem(n,4) == 1`
   This command calculates the remainder when divided by 4 and checks to see if this is equal to 1. The values for which this is true are $\{4n+1 : n \in \mathbb{Z}\}$.

3.4.1 Constructing Logical Statements

We shall now actually construct logical arguments which can be used in `if` statements.

Example 3.11 *Let us consider a command which is only executed if a value x lies between 1 and 2 or it is greater than or equal to 4. We shall try to describe the thought processes involved:*

- *In order that a value lies between one and two, it has to be greater than one* **and** *less than two, so this component is written as:*

```
(x>1) & (x<2)
```

- *If x is greater than or equal to 4, which is written simply as* `x>=4`.
- *Finally, we need to combine these conditions and this is done using the logical operation* `or`, *since the value of x could lie in one or the other of the regions. Hence we have*

```
((x>1) & (x<2)) | (x>=4)
```

We could use the commands in a different form

```
a = and(x>1,x<2);
b = (x>=4);
c = or(a,b)
```

by making use of the `and` *and* `or` *commands. Notice here we have actually set "Boolean" variables* `a`*,* `b` *and* `c` *(in fact they are only normal variables which take the values zero or one).*

We can also use these commands for vectors and matrices (comparing them element-wise). Consider the following

```
>> x = [1 2 0 3];
>> and(x>1,x<4)

ans =

     0     1     0     1
>> A = eye(2);
>> and(A==1,A>0)

ans =

     1     0
     0     1
```

If two sets do not intersect then by definition $A \cap B = \emptyset$ (the null, or empty, set). In terms of logical statements we say that the conditions are mutually exclusive. For instance, the condition `(x>1)&(x<0)` is never true, since x cannot be greater than one **and** less than zero.

Notice that we have used brackets to group the terms in the previous example. This is not always necessary but it is very good practice, especially since we cannot guarantee that the precedence in other languages will be the same. In fact although the expression `(x>=4) | ((x>1) & (x<2))` is the same as that given above, removing the brackets from both of them yields different results. In the former case the expression still works, whereas in the latter case

MATLAB complains about precedence and invites the user to find out more by using `help precedence`.

It is convenient at this stage to introduce some of the other commands which are available to us when constructing conditional statements, namely `else` and `elseif`. The general form of these is given by:

```
if (expression)
          commands ...
elseif (expression)
          commands ...
else
          commands ...
end
```

Notice that each `if` statement must be paired with an `end` statement, but we can use as many `elseif` statements as we like, but only one `else` (within a certain level). We are also able to nest `if` statements: again it is recommended that the new levels are indented so as to aid readability.

Example 3.12 *Consider the following piece of code which determines which numbers between 2 and 9 go into a specified integer exactly:*

```
str = 'Divisible by ';
x = input('Number to test: ');

for j = 2:9
    if rem(x,j) == 0
      disp([str int2str(j)])
    end
end
```

The results of this code can be checked using the `factor` *command. We have used the command* `rem` *to obtain the remainder.*

Example 3.13 *Here we construct a conditional statement which evaluates the function:*

$$f(x) = \begin{cases} 0 & x < 0 \\ x & 0 \leqslant x \leqslant 1 \\ 2 - x & 1 < x \leqslant 2 \\ 0 & x > 2 \end{cases}$$

One of the possible solutions to this problem is:

```
if x >= 0 & x <= 1
   f = x;
elseif x > 1 & x <= 2
   f = 2-x;
else
   f = 0;
end
```

Notice that it has not been necessary to treat the end conditions separately; they are both included in the final `else` *clause.*

Example 3.14 (Nested if statements) *The ideas behind nested if statements is made clear by the following example*

```
if raining
    if money_available > 20
       party
    elseif money_available > 10
       cinema
    else
       comedy_night_on_telly
    end
else
    if temperature > 70 & money_available> 40
       beach_bbq
    elseif temperature>70
       beach
    else
       you_must_be_in_the_UK
    end
end
```

3.4.2 The MATLAB Command `switch`

MATLAB has a command called `switch` which is similar to the BASIC command `select` (the syntax is virtually identical). The command takes the form:

```
switch switch_expr
        case case_expr1
          commands ...
        case {case_expr2,case_expr3}
          commands ...
        otherwise
          commands ...
end
```

We shall work through the example given in the manual documentation for this command:

```
switch lower(METHOD)
        case {'linear','bilinear'}
                disp('Method is linear')
        case 'cubic'
                disp('Method is cubic')
        case 'nearest'
                disp('Method is nearest')
        otherwise
                disp('Unknown method.')
end
```

This code assumes that `METHOD` has been set as a string. The first command `lower(METHOD)` changes the string to be lower case (there is also the corresponding command `upper`). This value is then compared to each case in turn and if no matches are found the `otherwise` code is executed.

Example 3.15 *We refer to the old rhyme that allows us to remember the number of days in a particular month.*

Thirty days hath September
April, June and November
All the rest have thirty-one
Except February alone
which has twenty-eight days clear
and twenty-nine on leap year

This is exploited in the code:

```
msg = 'Enter first three letters of the month: ';
month = input(msg,'s');
month = month(1:3); % Just use the first three letters
if lower(month)=='feb'
   leap = input('Is it a leap year (y/n): ','s');
end
switch lower(month)
     case {'sep','apr','jun','nov'}
          days = 30;
     case 'feb'
          switch lower(leap)
               case 'y'
                    days = 29;
               otherwise
                    days = 28;
          end
     otherwise
          days = 31;
end
```

Before we proceed it is worth discussing a couple of the commands used above. The MATLAB command `lower` forces the characters in the string to be lower case: for instance if the user typed 'Feb' the command `lower` would return 'feb'. We have also included a command which makes sure the code only considers the first three characters of what the user inputs. The command `input` is used here with a second argument `'s'` which tells MATLAB that a string is expected as an input; without this the user would need to add single quotes at the start and end of the string.

You might want to consider what you would need to do to check the user has input a valid month. The next command is very useful for this kind of problem.

3.5 Conditional loops

Suppose we now want to repeat a loop until a certain condition is satisfied. This is achieved by making use of the MATLAB command `while`, which has the syntax

```
while (condition)
     commands...
end
```

This translates into English as: while `condition` holds continue executing the `commands....` The true power of this method will be found when tackling real examples, but here we give a couple of simple examples.

Example 3.16 *Write out the values of x^2 for all positive integer values x such that $x^3 < 2000$. To do this we will use the code*

```
x = 1;
while x^3 < 2000
     disp(x^2)
     x = x+1;
end
value = floor((2000)^(1/3))^2;
```

This first sets the variable `x` *to be the smallest positive integer (that is one) and then checks whether $x^3 < 2000$ (which it is). Since the condition is satisfied it executes the command within the loop, that is it displays the values of x^2 and then increases the value of x by one.*

The final line uses some mathematics to check the answer, with the MATLAB command `floor` (which gives the integer obtained by rounding down; recall the corresponding command to find the integer obtained by rounding up is `ceil`).

Example 3.17 *Consider the one-dimensional map:*

$$x_{n+1} = \frac{x_n}{2} + \frac{3}{2x_n},$$

subject to the initial condition $x_n = 1$. Let's determine what happens as n increases. We note that the fixed points of this map, that is the points where $x_{n+1} = x_n$, are given by the solutions of the equation:

$$x_n = \frac{x_n}{2} + \frac{3}{2x_n},$$

which are $x_n = \pm\sqrt{3}$. We can use the code

```
xold = 2; xnew = 1;

while abs(xnew-xold) > 1e-5
     xold = xnew;
     xnew = xnew/2+3/(2*xnew);
end
```

This checks to see if $x_{n+1} = x_n$ to within a certain tolerance. This procedure gives a reasonable approximation to $\sqrt{3}$ and would improve if the tolerance (`1e-5`) was reduced.

3.5.1 The `break` Command

A command which is of great importance within the context of loops and conditional statements is the break command. This allows loops to stop when certain conditions are met. For instance, consider the loop structure

```
x = 1;
while 1 == 1
     x = x+1;
     if x > 10
        break
     end
end
```

The loop structure `while 1==1 ... end` is very dangerous since this can give rise to an infinite loop, which will continue *ad infinitum*; however the break command allows control to jump out of the loop.

3.6 MATLAB Specific Commands

We could have coded the Example 3.16 using the MATLAB command `find`, which returns an array of locations at which a certain condition is satisfied. The condition is formed as above using logical expressions and combinations of them.

Example 3.18 *Find all integers between 1 and 20 for which their sine is negative.*

Firstly we set up the array of integers, then calculate their sines and subsequently test whether these values are negative.

```
ii = 1:20;
f = sin(ii);
il = find(f<0);
disp(ii(il))
```

We briefly dissect this code.

- In the first line set up a vector which runs from one to twenty;
- the second line calculates the sine of the elements in vector `ii` and stores the result in the vector `f`;
- in the third line we use `find` to determine the locations in the vector `f` where the sine is negative. The command returns a list of positions in the vector `f` (and equivalently `ii`) where the condition is true.
- The final command prints out a list of these integers.

Other similar commands are `any` and `all`. Examples of their use are:

```
x = 0.0:0.1:1.0;
v = sin(x);
if any(v<0)
    disp('Found a negative value')
else
    disp('All values zero or positive')
end

if all(v>0)
    disp('All values positive')
else
    disp('One value is zero or negative')
end
```

Some care is needed with the logic at this point and these commands seem to be, in some sense, complements of each other.

There are many other commands which test the properties of a variable, for instance `isempty`, `isreal` etc.

Example 3.19 *Now we evaluate the expression*

$$f(x) = \frac{1}{(x-1)(x-2)(x-3)}$$

on a grid of points $n/10$ for $n = 1$ to $n = 40$ for which the expression is finite, else the function is to be returned as 'NaN'.

This is accomplished using the code:

```
x = (1:40)/10;
g = (x-1).*(x-2).*(x-3);
izero = find(g==0);
ii = find(g~=0);
f(izero) = NaN;
f(ii) = 1./g(ii);
```

This allows us to plot the function whilst missing out the infinite parts. This would be accomplished using the command `plot(x,f)`

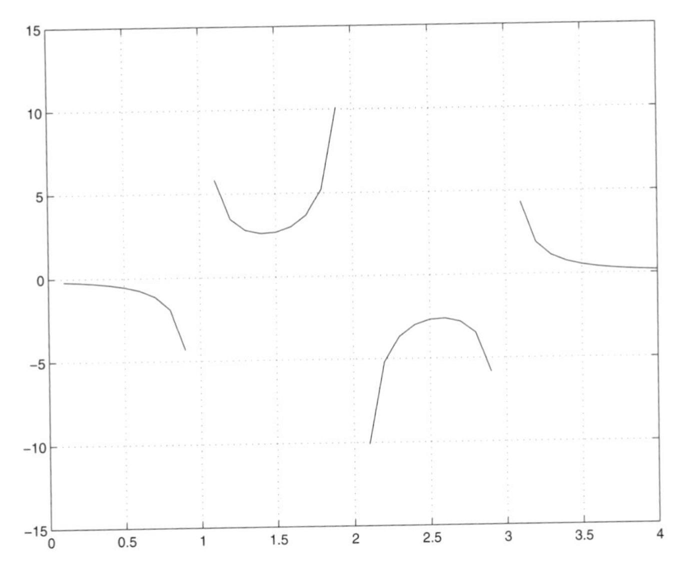

3.7 Error Checking

To this point we have assumed the data made available to a code is suitable; this is generally a very dangerous assumption. We have found MATLAB will object to mathematically impossible tasks: however it will happily perform operations provided they are viable (which unfortunately they are more often than not).

Now that we have seen conditional statements we can make sure our codes are more robust. The ultimate aim would be to make them totally "idiot proof", but this is virtually impossible. No matter how many different scenarios a programmer comes up with, they can never predict everything a user will do.

MATLAB gives us three very useful commands in this context: `break`, `warning` and `error`. The first of these we have already met. The latter two allow us to either warn the user of a problem or actually stop the code because of an irretrievable problem, respectively. Both the commands `warning` and `error` are used with an argument, which is displayed when the command is encountered. The typical structure might be:

```
if code_fails
   error(' Irretrievable error ')
elseif code_problem
   warning(' Results may be suspect ')
end
```

The actual use of the commands is perhaps best demonstrated by a couple of examples:

Example 3.20 *Let's now write a code which asks the user for an integer and returns the prime factors of that integer.*

```
msg = 'Please enter a strictly positive integer: ';
msg0 = 'You entered zero';
msg1 = 'You failed to enter an integer';
msg2 = 'You entered a negative integer';
x = input(msg);
if x==0
   error(msg0)
end
if round(x)~= x
   error(msg1)
end
if sign(x)==-1
   warning(msg2)
   x = -x;
end
disp(factor(x))
```

Fortunately the command ***input*** *has its own error checking routine and will only let you input numbers (if you try inputting a string it will complain). We now check to see whether zero is entered, in which case the code stops after having informed the user that they entered a zero. Similarly, if the user enters a non integer the code will stop. If the user enters a negative integer the code warns the user they have done so but simply makes it equal to the corresponding positive value.*

The command **`input`** *allows us to input a matrix, which would cause our program to fail (in fact the routine* **`factor`** *issues an error message). We could incorporate a check to ensure that what is entered is a scalar. This could be accomplished by using the condition that the* **`length`** *of a scalar is unity. To this end we could add (below the* **`input`** *statement):*

```
msg3 = 'Code needs a scalar integer';
if length(x) ~= 1 | x ~= floor(x)
   error(msg3)
end
```

When MATLAB encounters these commands it produces a message which indicates on which line they occurred and in which code. This is very helpful when debugging codes.

Example 3.21 *Now consider a code which compels the user to enter a four character string and until they have the code will not proceed.*

```
msg = ['Please enter a ' ...
     'four character string '];
msg0 = ' is not valid, please re-enter';

str = 'X'; % Clearly not valid
first = 1;
while length(str) ~= 4
   if ~(first)
      warning([str msg0])
   end
   first = 0;
   str = input(msg,'s');
end
```

In order to avoid the warning message during the first run, we have used the flag **`first`** *which basically says if this is the first time through don't issue the warning. This uses a* **`while`** *loop to repeat until the user enters a string of length 4.*

As we can see this additional code can make the program cumbersome, but sometimes it is necessary, especially if other users are going to read and use your codes. Again, it also helps when it comes to debugging.

Another very useful command in this context is `exist`. This can be used to check whether the requisite variables exist. This is particularly useful if input is required from another code.

Example 3.22 *Let us consider this header for a code:*

```
msg = ['The variable z does not exist ' ...
       'and will be required for this code'];
if ~(exist('z'))
   error(msg)
end
```

We note this code will not work at the prompt, since the command `error` *will only run during the execution of an m-file.*

3.8 Tasks

Task 3.1 *Modify the code on page 71 to calculate the value of the summation*

$$\sum_{i=1}^{100} \frac{1}{i^2}.$$

(Notice the answer will probably not be an integer, so you will have to modify the display line to use `num2str`*).*

Task 3.2 *Modify the code from Task 3.1 to only sum the values corresponding to odd values of i. (Hint: This only involves a very minor change to the* `for` *loop. You should check the answer you get is smaller than the one from the previous task.)*

Task 3.3 *Modify the function* `f.m` *(on page 72) to calculate the values of:*

$$I_N = \sum_{i=1}^{N} \frac{\sin \frac{i\pi}{2}}{i^2 + 1},$$

for N up to `N=20`*.*

Task 3.4 *Construct a program to display the values of the function $f(x) = x^2 + 1$ for $x = 0$ to $x = \pi$ in steps of $\pi/4$. The key here is to set up an array of points from a to b in steps of h (you can either use the colon command or* `linspace`*). Make sure you remember to use the dot operator.*

Task 3.5 *Change the code given in Example 3.6, used to evaluate* $\sin x$*, to one that calculates* $\cos x$ *whose series expansion is given by*

$$\cos x = \lim_{N \to \infty} \sum_{n=0}^{N} (-1)^n \frac{x^{2n}}{(2n)!}.$$

Compare the approximations to the values calculated directly from MATLAB for $x = 0$, $1/4$, $1/2$ and $3/4$. As for Example 3.6 you will need to truncate the summation, choosing N to be suitably large.

Task 3.6 *Calculate the sum of the series S_N, where*

$$S_N = \sum_{n=1}^{N} \frac{1}{n^2}$$

for different values of N. Given that as $N \to \infty$, $S_N \to \pi^2/c$ where c is a constant. Determine the value of c. (Here you should modify one of the codes for summation: the best one is probably the solution to Task 3.1.)

Task 3.7 *Calculate the summations*

$$\sum_{j=1}^{p+1} j^p$$

for p equal to one, two, three and four (using a nested loop structure).

Task 3.8 *Show that, to within the accuracy permitted by MATLAB,*

$$\lim_{N \to \infty} \sum_{n=1}^{N} \frac{(-1)^n}{n} = -\ln 2 \text{ and } \lim_{N \to \infty} \sum_{n=1}^{N} \frac{1}{n(n+1)} = 2.$$

Task 3.9 *Write down (on paper) a logic expression which is true for values of x such that $2 < x < 4$. Make sure your expression works for $x = 1$, 3 and 5. Repeat the exercise for the union of the sets $x > 3$ and $x < -1$. (You need to write the answers in the form of the combination of x is greater than something and/or x is less than something else).*

Task 3.10 *Write down a logical expression which is true for even values of n. Extend this to only be true for even values of n greater than 20. (Hint: Try using the MATLAB command* `mod`*: for information on the command type* `help mod`*.)*

Task 3.11 *Try to work out what value of x the following code returns (initially without running it).*

```
x = 1;
if tan(73*pi*x/4) >= 0
   x = 2;
else
   x = pi;
end
if floor(x) == x
   x = 10;
else
   x = 7;
end
if isprime(x)
   x = 'True';
else
   x = 'False';
end
```

Is this result true whichever value of `x` *we start with? (You can find out about the command* `isprime` *by typing* `help isprime`*.)*

Task 3.12 (*) *Write a loop structure which iterates $x_{n+1} = 5x_n|1$ for $x_n = 1/7$ until it returns to the same value (use* `while`*).*

Task 3.13 (*) *Determine all integers between 1 and 50 for which $n^3 - n^2 + 40$ is greater than 1000 and n is not divisible by 3. Are* **any** *integers between 1 and 50 perfect (that is, are they equal to the sum of their factors)?*

Task 3.14 *Write a code which only allows the user to input an integer n between 1 and 10 (inclusive) and then prints out a string of the first n letters of the alphabet. (Hint: You could start with a string of the form "abcdefghij".)*

Task 3.15 (*) *Write a code which allows the user to input a two character string, the first being a letter and the second being a digit. (Note that to check if a character is a letter we can use inequalities on strings.)*

Task 3.16 (*) *By modifying the code on page 88 use the* ***find*** *command to plot the function*

$$f(x) = \begin{cases} 0 & x < -1 \\ x^2 & -1 \leqslant x \leqslant 1 \\ 1 & 1 < x < 4 \\ 0 & x \geqslant 4 \end{cases}$$

for the range $x \in [-3, 5]$ using one hundred points (try using the command ***linspace*** *to set up the array of points).*

Task 3.17 (*) *Using the command* ***find****, and modifying the code on page 93, plot the function*

$$f(x) = \frac{1}{\cos \pi x}$$

for $x \in [-3, 3]$. You will need to change the code to determine when the denominator becomes small, rather than when it is identically zero.

Task 3.18 (D) *The following code is supposed to evaluate the function*

$$f(x) = \begin{cases} 0 & x < 0 \\ x & 0 \leqslant x \leqslant 1 \\ 2 - x & 1 \leqslant x \leqslant 2 \\ 0 & x > 2 \end{cases}$$

```
x=lnspace(-4,4);
N = length x

for j = 1;N

if x(j)>=0 and x(i)<=1
   f(j) = x(j);
elseif x(j)>1 or x(i)<2
   f(j) = 2 - x
else
   f(j) = zero;
end
```

Correct the code so that it accomplishes this. This can be checked by using the command `plot(x,f)` *and comparing the figure to what you expect the function to look like.*

4
Root Finding

4.1 Introduction

In many problems we are required to determine when a function is zero. It should have become clear even for relatively simple examples there were many subcases. It is our intention to discuss methods which are in some sense robust and so the form of the actual function we are studying is not important, at least in the construction of the method.

We now re-introduce the MATLAB function `feval` which may seem a little like an extra step, but it will enable us to write more general codes. This function has the syntax `feval(f,x1,...,xn)` and as you might expect from the name evaluates the function `f` using the arguments `x1,...,xn`. It is normally used within other functions which require a function name to be passed to them. This function can be used as

```
feval('sin',0.3)
feval('mycode',0.2,0.3)
```

In the first example this gives `sin(0.3)` and in the second it returns the value of `mycode(0.2,0.3)`.

4.2 Initial Estimates

In order to determine a root it is usually essential to have an initial estimate of its value. In some cases you may have more than one root (or none) and you wish to identify which one you are concerned with. The method we will describe now involves user interaction and is used as an illustration. As we are developing the ideas in this chapter we could consider how they could be generalised to include root selection. In general the methods we will consider will require either an initial guess for the root or a bracketing interval containing a root.

We will use the graphical capabilities of MATLAB to identify the root (or this bracketing interval). Firstly, we shall set up a small m-file called `userfn.m`, which will used to specify an example function.

```
function [value] = userfn(x,par1,par2)
value = x-par1*sin(x.^par2);
```

This gives $f(x) = x - \alpha_1 \sin x^{\alpha_2}$, where α_1 and α_2 are parameters, which the user will specify. Now in order to plot the function we select a range and use the `plot` command. The commands are

```
x = -2.0:0.01:2.0;
y = feval('userfn',x,2,2);
plot(x,y)
grid on
```

This code sets up a vector `x` running from -2 to 2 in steps of 0.01. The next line calculates the value of the function $x - 2\sin x^2$ at these points and puts the answer in `y`. Note we could also have used the command `y = userfn(x,2,2);` for this purpose. We have used the `feval` format so that we could potentially vary the function name. The next two lines plot the graph and add the grid lines.

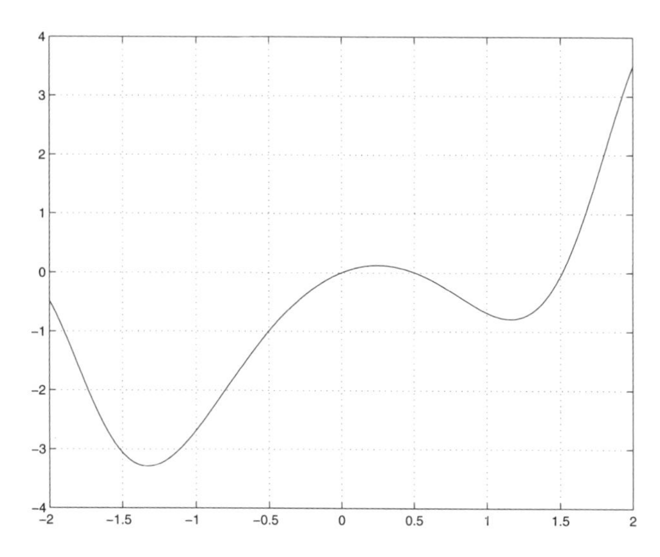

From this initial figure it is not that clear where the zeros lie. There appears to be one at the origin, which we can see straight away from consideration of the function, $f(x) = x - 2\sin x^2$. By adding the command `grid on` (either to the programme `testplot.m` or at the command line) we can see two additional zeros (one near 0.5 and another near 1.5: there is also a hint of an extra one to the left of the figure, which could be determined by extending the range leftwards). In order to investigate further we use the `zoom on` command. By clicking the left-hand mouse button we can enlarge areas of the figure, and using the right-hand button we can pull back. Typing `zoom off` disables this feature.

This can give us a good idea of where the roots lie. Notice this command is only working using the original data and if we zoom too close we will be able to see the straight line segments used for the plotting. Here we have exploited 401 points, so this should not be a problem.

We can now use another command `ginput` to actually return a value rather than just using our eyes. The syntax we will exploit is

```
% Modified testplotm.m
x = -2.0:0.01:2.0;
y = feval('userfn',x,2,2);
disp('Click the mouse near the zero')
disp('and when you have finished press')
disp('the return key')
plot(x,y)
grid on
[xvalues,yvalues] = ginput
[yy,ii] = min(abs(yvalues));
disp(xvalues(ii))
```

The first few lines are the same as `testplot.m` and the next line uses the command `ginput`. The way `ginput` works is to allow the user to select points in the window (using a cross hair) by clicking one of the mouse buttons. This is terminated by pressing the return button. The results are stored in the arrays `xvalues` and `yvalues`; these essentially contain the x and y coordinates of all the points which the user clicks. The next command finds the minimum value of the function and its integer location within the array. Finally the value of x at this location is displayed. This gave a guess of `0.4977` for the zero near 0.5 and `1.4931` for the zero near 1.5.

Using these initial guesses we can now proceed with our description of some of the techniques available to use. We shall also comment on the likely failings of the methods and we should be mindful of how these failings could manifest themselves and how we might monitor them.

Example 4.1 *Determine initial estimates for the zeros of the function* $f(x) = x\sin x - \sqrt{x}$ *between 0 and 10.*

Firstly we construct the MATLAB code:

```
function [value] = userf1(x)
value = x.*sin(x)-sqrt(x);
```

and now run the commands

```
x = linspace(0,10);
y = feval('userf1',x);
plot(x,y)
grid on
```

which yield

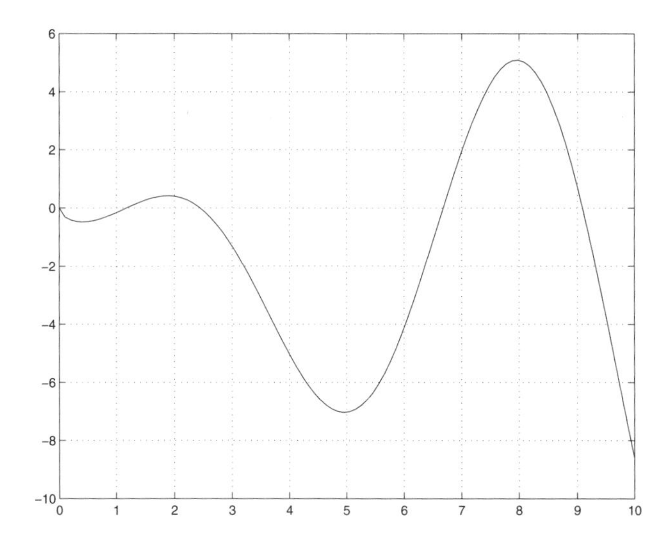

We can see that there are zeros at 0 and near 1.2, 2.5, 6.7 and 9.3. We could use the above data set but it would seem sensible to consider each point separately. For this purpose we use the code:

```
a = input('Start of interval ');
b = input('End of  interval  ');
x = linspace(a,b);
y = feval('userf1',x);
clf
plot(x,y)
grid on
zoom on
```

Running this code with the inputs 1 *and* 2, *we obtain:*

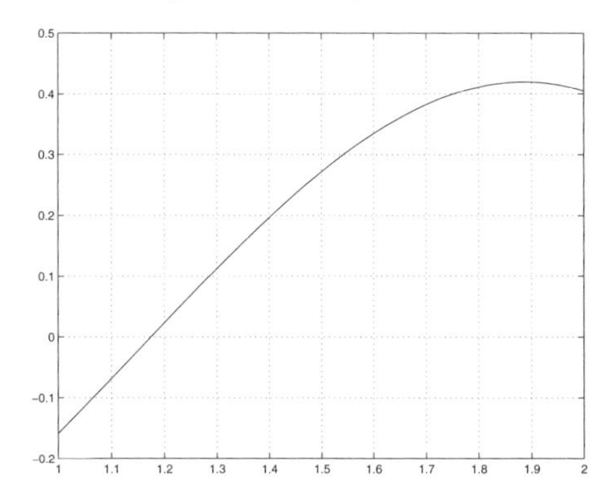

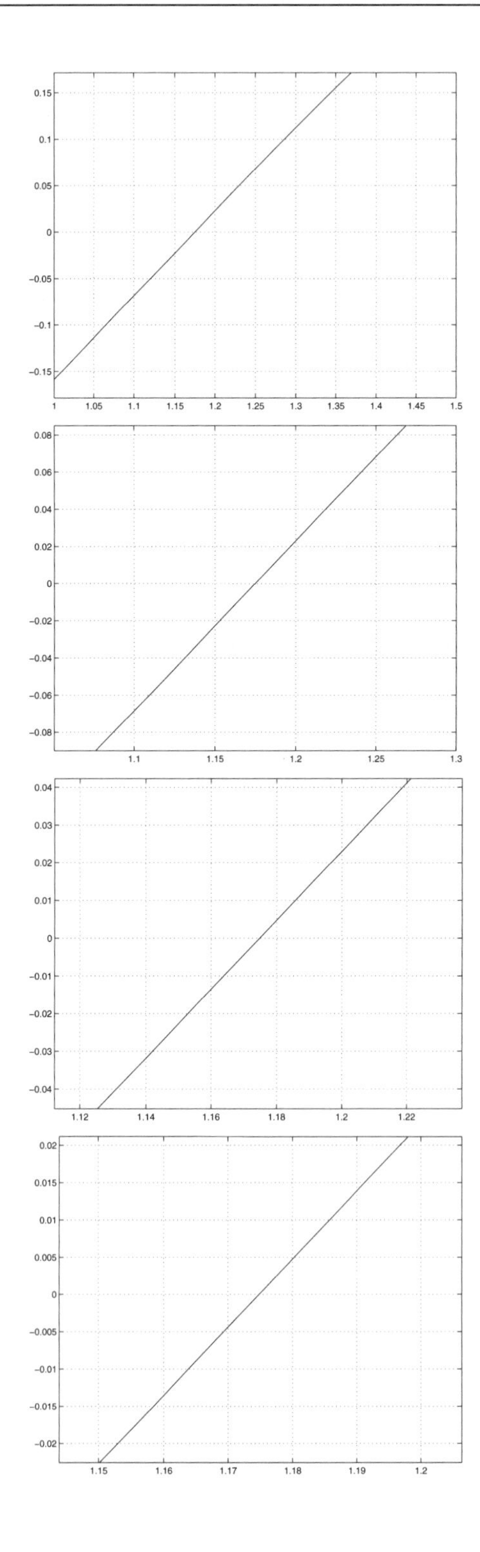
0.15
0.1
0.05
0
−0.05
−0.1
−0.15
1
1.05
1.1
1.15
1.2
1.25
1.3
1.35
1.4
1.45
1.5
0.08
0.06
0.04
0.02
0
−0.02
−0.04
−0.06
−0.08
1.1
1.15
1.2
1.25
1.3
0.04
0.03
0.02
0.01
0
−0.01
−0.02
−0.03
−0.04
1.12
1.14
1.16
1.18
1.2
1.22
0.02
0.015
0.01
0.005
0
−0.005
−0.01
−0.015
−0.02
1.15
1.16
1.17
1.18
1.19
1.2

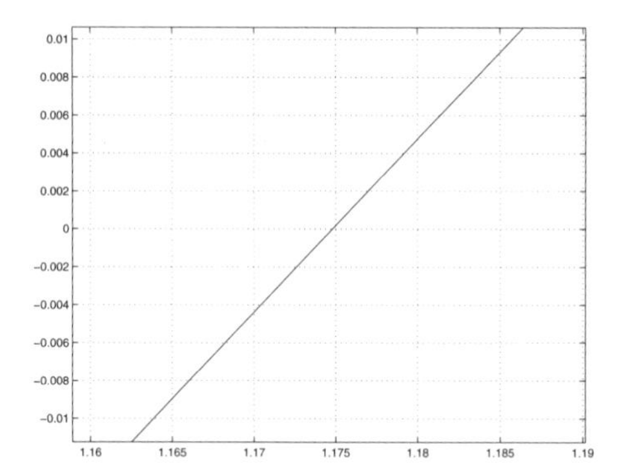

after successive clicks of the left mouse button. This allows us to obtain a better estimate of the root, namely 1.175. *You should try this interval and repeat the exercise for the other roots.*

In reality you should plot the function and then use a combination of the `grid` and `zoom` commands to obtain estimates for the root or intervals which contain them. This also helps to ensure that the answers you obtain are sensible.

4.3 Fixed Point Iteration

Most of the techniques we will discuss are iterative in nature and the first one is called the fixed point iteration scheme. Instead of looking for a zero of the function $f(x)$, it determines a fixed point of an equivalent equation. The equation is rewritten in the form $x = g(x)$, so when x is substituted in the function $g(x)$ it returns the value x, hence the nomenclature fixed point. The conversion of the equation $f(x) = 0$ into one of the form $x = g(x)$ is not always straightforward and definitely can be done in many ways. For instance the previous example, for which $f(x) = x - 2\sin x^2$ could be rewritten as $x = 2\sin x^2$ (where $g(x) = 2\sin x^2$) or $x = \sqrt{\sin^{-1}\frac{x}{2}}$. Although the former version might be more intuitive, there are many examples in which this choice is not so. For instance consider the quadratic $f(x) = x^2 + 2x - 3$, which could be manipulated to give $x = (3 - x^2)/2$ or $x = \sqrt{3 - 2x}$. With these two forms neither of them seems to have any advantages over the other. Further the equation could be written as $x = 3/(x + 2)$ and so on. Before we try to resolve this issue let us say how this is then implemented as a numerical scheme.

We rewrite the equation as

$$x_{n+1} = g(x_n) \qquad n = 0, 1, \cdots,$$

which is a recursion formula. It starts with an initial guess, namely x_0 (which may be determined graphically or by another means).

Just about the simplest code for this purpose would be

```
x0 = 1;
for j=1:10
    x0 = g(x0);
end
```

where we have a routine `g.m` which defines the function $g(x)$ and `x0 = 1` is a suitable initial guess. This runs through the iterative process ten times. This kind of code presupposes that it is going to work and will converge in ten steps (or an appropriate number). You will quickly learn that you can't rely on this kind of thing.

We need to pause here to think what we mean by convergence. We could work out the value of the function $f(x)$ at the current iterate as a check. We would require this to be less than a certain tolerance (the accuracy to which we would expect to know the answer). Notice that this is slightly different to knowing the root to within a certain tolerance.

Alternatively in this case the code may be deemed to be successful when the difference between x_{n+1} and x_n is less than a certain tolerance. In this case we have $x_{n+1} \approx x_n \Rightarrow x_n \approx g(x_n) \Rightarrow f(x_n) \approx 0$. This tolerance reflects how well we want to know the answer and the parameter **`maxits`** is how many times we are prepared to perform the iterations. This is to eliminate problems with cases which don't converge and hence cause infinite loops. The code we shall use for this is:

```
%
% fixed.m
%
function [answer,iflag] = fixed(g,xinit)
global tolerance maxits
iflag = 0;
iterations = 0 ;
xnext = feval(g,xinit);
while (iterations<maxits) & abs(xnext-xinit)>tolerance
        iterations = iterations + 1;
        xinit = xnext;
        xnext = feval(g, xinit);
end
if iterations == maxits
        iflag = -1;
        answer = NaN;
else
        iflag = iterations;
        answer = xnext;
end
```

```
%
% eqn.m
%
function [g] = eqn(x)
g = 2*sin(x.^2);
```

and the main function

```
%
% mfixed.m
%
global tolerance maxits
tolerance = 1e-4;
maxits = 30;
[root,iflag] = fixed('eqn',0.2);
switch iflag
case -1
      disp('Root finding failed')
otherwise
      disp([' Root = ' num2str(root) ...
          ' found in ' num2str(iflag) ' iterations'])
end
```

We have exploited the new command **global** which allows variables to be used by any routine which contains a global statement referring to the same variables, or a subset of them. This is useful when a quantity (or group of quantities) is unlikely to change and it saves on long argument lists. We have also used the number `1e-4` which is 1×10^{-4} (refer to page 7).

The code above is run and gives the result of $\approx 1 \times 10^{-14}$ after only four iterations. In order to identify the other roots (near 1/2 and 3/2) we must use guesses close to these values. Starting around 0.5 (for instance guesses of 0.4 and 0.6 lead to the code finding the root at zero again). Starting close to the other root has a variety of outcomes: starting below it produces the zero root, whereas above leads to the iterations diverging. The reason for this is linked to the derivative of $g(x)$. In fact the errors are multiplied by $|g'(x)|$ at each iteration and if this value is greater than one, the method will fail. In this case we recall that $g(x) = 2\sin x^2$ so that $g'(x) = 4x\cos x^2$ which when $x = \frac{1}{2}$ equals 1.9378 and when $x = \frac{3}{2}$ gives -3.7690. Near the origin the derivative is very small and consequently the technique works well.

It appears this method is only able to find one root and is thus not very useful. However, at this stage we could change the code `eqn.m` to

```
%
% eqn.m
%
function [g] = eqn(x)
g = sqrt(abs(asin(x/2)));
```

This form of the code is a simple alternative way of writing the equation $f(x) = 0$ as given above. We use the function `abs` to ensure that the square root function will not lead to imaginary quantities[1]. In this case the code finds the root near $x = 1/2$ but again fails to obtain the other one. There may well be a form of the function which gets this root but any method which requires this amount of manipulation and user interaction is not suitable for detailed calculations. There are many flaws with this method and consequently it is not widely used for one-dimensional problems. However, there are some higher-dimensional (and more complex) problems for which it is the only feasible option.

We shall discuss whereby this method is improved upon and instead of using substitution into a function it uses the fact that a zero corresponds to a change in sign.

4.4 Bisection

In the previous example we used an initial estimate for a root: here we shall use a bracketing interval. Our initial assumption is that this interval contains a single root. In order to check this hypothesis we shall build in checks at each stage.

If the initial interval is between $x = a$ and $x = b$ we know that $f(a)$ and $f(b)$ are of different signs (for the interval to contain a single root): in which case their product must be negative. As the name of the method suggests we bisect the interval and define the point $c = (b + a)/2$. We can now evaluate the function to obtain $f(c)$.

[1] To evaluate the arcsine we have used the MATLAB command `asin`, as you might expect there are also `acos` and `atan`, as mentioned previously.

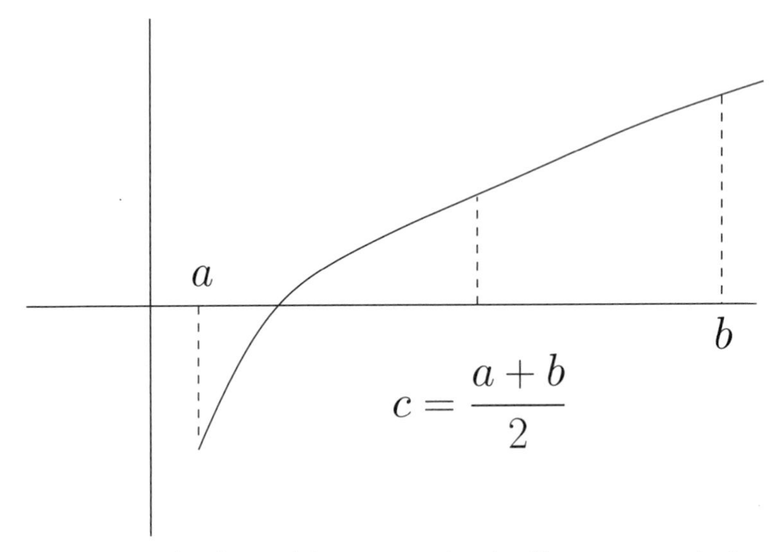

This must either be positive or negative (or if we are very lucky zero). In which case $f(c)$ will have the same sign as either $f(a)$ or $f(b)$. The new interval can then be defined as c and whichever end of the previous interval represents a change in sign of the function.

A simple version of a code to use this method in an interval $[a, b]$ would be

```
a = 1; b = 5;
for j = 1:10
   c = (a+b)/2;
   if f(c)*f(a) > 0
      a = c;
   else
      b = c;
   end
end
```

where in this case $a = 1$, $b = 5$ and the function $f(x)$ is defined in a routine `f.m`. This method is perfectly adequate if you know that the method will converge in ten steps, the interval $[a, b]$ actually contains a root and no future iterations actually coincide with a root (see Task 4.8).

If we wish to be slightly more careful we would use the codes

```
%
% bisect.m
%
function [answer,iflag] = bisect(fun,a,b)
global tolerance maxits
iflag = 0;
iterations = 0 ;
f_a = feval(fun,a);
f_b = feval(fun,b);
while ((f_a*f_b<0) & iterations<maxits) & (b-a)>tolerance
        iterations = iterations + 1 ;
        c = (b+a)/2;
        f_c = feval(fun,c);
        if f_c*f_a<0
                b=c; f_b = f_c;
        elseif f_b*f_c<0
                a=c; f_a = f_c;
        else
                iflag = 1; answer = c;
        end
end
switch iterations
case maxits
        iflag = -1; answer = NaN;
case 0
        iflag = -2; answer = NaN;
otherwise
        iflag = iterations; answer = c;
end
```

```
%
% func.m
%
function [f] = func(x)
f = x-2*sin(x.^2);
```

and the main controlling function

```
%
% mbisect.m
%
global tolerance maxits
tolerance = 1e-4;
maxits = 30;
xlower = 0.4;
xupper = 0.6;

[root,iflag] = bisect('func',xlower,xupper);

switch iflag
case -1
      disp('Root finding failed')
case -2
      disp('Initial range does not only contain one root')
otherwise
      disp([' Root = ' num2str(root) ...
          ' found in ' num2str(iflag) ' iterations'])
end
```

This method is guaranteed to work provided only one root is in the relevant interval and the function is continuous. It may work if there are three roots but this is not recommended: in fact it appears to work provided there are an odd number of roots, because each iteration may remove a number of roots which is guaranteed to be even. We note that the length of the interval after n iterations is $(b-a)/2^n$. Hence if we wish to know the root to within a given tolerance we can work out how many iterations we need to perform. For instance if the required tolerance is ϵ, we find that we need

$$n > \frac{1}{\ln 2} \ln \left(\frac{b-a}{\epsilon} \right).$$

Example 4.2 *To determine a root of a continuous function $f(x)$ between zero and one, given that $f(0)f(1) < 0$ to within 1×10^{-4} requires $n > \ln(10^4)/\ln 2 \approx 13.28$: in other words fourteen iterations.*

We shall now discuss other methods for solving equations of the form $f(x) = 0$ which exploit the form of the derivative (or at least an approximation to it) as well as the function.

4.5 Newton–Raphson and Secant Methods

These methods are related and often confused. We shall derive them in tandem, since in reality they only differ in the last stage. The central premise for both methods is that the function is locally linear and the next iteration for the required value can be attained via linear extrapolation (or interpolation).

4.5.1 Derivation of the Newton–Raphson Method

We will start using a Taylor series to derive the Newton–Raphson technique[2]. We assume the current guess is x and this is incorrect by an amount h, so that $x + h$ is the required value. It now remains for us to determine h or at least find an approximation to it. The Taylor expansion for the function $f(x)$ at the point $x + h$ is given by

$$f(x + h) = f(x) + hf'(x) + O(h^2).$$

This can be interpreted as: the value of the function at $x + h$ is equal to the value of the function at x plus the gradient times the distance between the points. This can be considered to include further terms; at the moment we are fitting a straight line.

In this expression we have used the term $O(h^2)$: loosely this means something the same size as h^2 and also the prime means differentiated with respect to the argument of the function. We now note that $x + h$ is supposedly the actual root so $f(x+h) = 0$, and discarding the higher-order terms we find that

$$h \approx -\frac{f(x)}{f'(x)}. \tag{4.1}$$

This presumes we are close to the actual root, and consequently we can discard the terms proportional to h^2 since these should be smaller than those proportional to h.

This allows us to construct the iterative scheme

$$x_{n+1} = x_n - \frac{f(x_n)}{f'(x_n)}, \qquad n = 0, 1, 2, \cdots .$$

This method can also be derived using geometric arguments. In these derivations the function is taken to be approximated by a straight line in order to determine the next point.

[2] If you haven't seen Taylor series refer to the geometric argument on the next page.

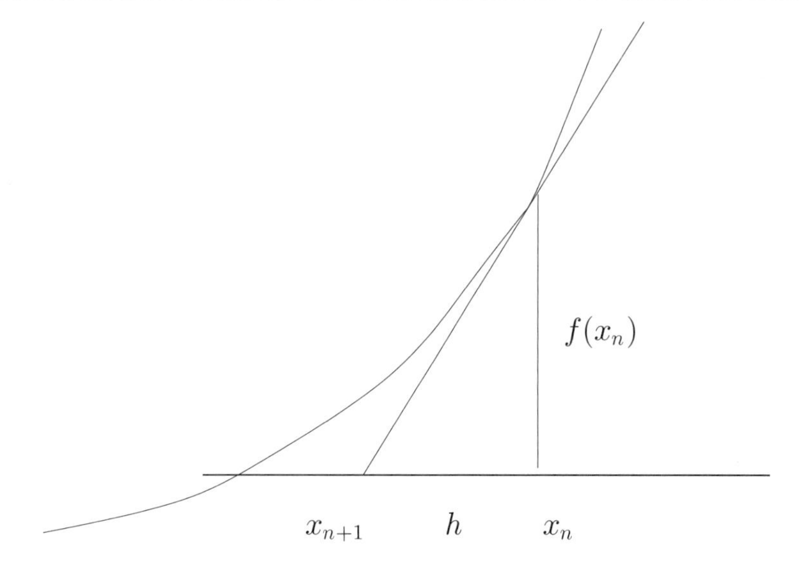

The value of h is determined by using the fact that the ratio of the two sides h and $f(x_n)$ has to be equal to $f'(x_n)$.

At the moment we shall presume that we have two routines `func.m` and `func_prime.m` which give us the function and its derivative. For ease let us consider the function $f(x) = x - 2\sin x^2$, so $f'(x) = 1 - 4x\cos x^2$ (using the chain rule). The code `func.m` is given on page 115 and the other one is:

```
%
% func_prime.m
%
function [value] = func_prime(x)
value = 1 - 4*x.*cos(x.^2);
```

Notice that we have used the dot operators, even though this routine is only ever likely to be called in this context using a scalar. This permits the routine to be used from other codes in a portable fashion.

This can be coded simply using

```
x = 1;
for j = 1:10
    x = x - func(x)/func_prime(x);
end
```

where we have set the initial guess to be $x = 1$ and supposed that the method will converge in ten iterations. We now give a more robust code to perform the iterations

```
%
% Newton_Raphson.m
%
x = input('Starting guess :');
tolerance = 1e-8;
iterations = 0;
while (iterations<30) & (abs(func(x))>tolerance)
       x = x-func(x)/func_prime(x);
       iterations = iterations + 1;
end
if iterations==30
       disp('No root found')
else
       disp([' Root = ' num2str(x,10) ' found in ' ...
         int2str(iterations) ' iterations.'])
end
```

Hopefully you can see the difference between the two codes and see that ultimately the second version is more useful. By entering the values 0.1, 0.6 and 1.5 we can obtain the three roots we are concerned with. Notice we have increased the number of digits printed in the answer to 10 using the form of the MATLAB command `num2str` which accepts two arguments.

We shall discuss the possibilities open to us when we do not know the derivative of the function, but we are able to evaluate the function. This relies on the classical definition of the derivative, which we shall write as

$$f'(x) = \lim_{\delta \to 0} \frac{f(x+\delta) - f(x)}{\delta}.$$

This allows the previous routine to be rewritten without using `func_prime.m`, but we need to define a practical value of δ for this purpose. In general a value of around 10^{-6} is suitable, but this may need to be changed depending on the problem. There is also a version of the method whereby δ is changed during the iteration process (dependent on the current value of h, see Equation (4.1)). The previous code is now changed to be

```
%
% Newton_Raphson_S.m
%
x = input('Starting guess :');
tolerance = 1e-8;
iterations = 0;
delta = 1e-6;
while (iterations<30) & (abs(func(x))>tolerance)
        f0 = func(x);
        f1 = func(x+delta);
        x = x-f0*delta/(f1-f0);
        iterations = iterations + 1;
end
if iterations==30
        disp('No root found')
else
        disp([' Root = ' num2str(x,10) ' found in ' ...
          num2str(iterations) ' iterations.'])
end
```

or alternatively the short version would be

```
del = 1e-6;
x = 1;
for j = 1:10
    x = x-del*func(x)/(func(x+del)-func(x));
end
```

We have changed the name of this method to `Newton_Raphson_S.m` since it is similar to the next method we will discuss, namely the secant technique.

This method starts with two values of x, namely $x = x_0$ and $x = x_1$. A straight line is drawn between the points $(x_0, f(x_0))$ and $(x_1, f(x_1))$, which has the equation

$$\frac{y - f(x_0)}{f(x_1) - f(x_0)} = \frac{x - x_0}{x_1 - x_0}.$$

We wish to find the value of x (x_2 say) for which $y = 0$ which is given by

$$x_2 = x_0 - f(x_0)\frac{x_1 - x_0}{f(x_1) - f(x_0)}.$$

We then move on so that we are going to use $x = x_1$ and $x = x_2$ as the next two points. This is coded to give:

```
%
% Secant.m
%
x0 = input('Starting guess point 1 :');
x1 = input('Starting guess point 2 :');
tolerance = 1e-8;
iterations = 0;
while (iterations<30) & (abs(func(x1))>tolerance)
         iterations = iterations + 1 ;
         f0 = func(x0);
         f1 = func(x1);
         x2 = x0-f0*(x1-x0)/(f1-f0);
         x0 = x1;
         x1 = x2;
end
if iterations==30
       disp('No root found')
else
       disp([' Root = ' num2str(x1,10) ' found in ' ...
         num2str(iterations) ' iterations.'])
end
```

This method works far better if the two initial points are on opposite sides of the root. In the above method we have merely chosen to proceed to use x_1 and the newly attained point x_2: however we could equally have chosen x_0 and x_2. In order to determine which we should use we require that the function changes sign between the two ends of the interval. This is done by changing the lines where the next interval is chosen.

```
%
% False_Position.m
%
x0 = input('Starting guess point 1 :');
x1 = input('Starting guess point 2 :');
x2 = x0;
tol = 1e-8;
iters = 0;
while ((iters<30) & (abs(func(x2))>tol))|(iters==0)
          iterations = iterations + 1 ;
          f0 = func(x0);
          f1 = func(x1);
          x2 = x0-f0*(x1-x0)/(f1-f0);
          if func(x2)*f0 < 0
             x1 = x2;
          else
             x0 = x2;
          end
end
if iters==30
        disp('No root found')
else
        disp([' Root = ' num2str(x2,10) ' found in ' ...
          num2str(iters) ' iters.'])
end
```

Example 4.3 *We consider the function $f(x) = x^m - a$ where $a > 0$. The roots of this equation are $\sqrt[m]{a}$ or, written another way, $a^{1/m}$. For this equation we can write the Newton–Raphson scheme as:*

$$x_{n+1} = x_n - \frac{x_n^m - a}{m x_n^{m-1}},$$

which can be simplified to yield:

$$x_{n+1} = x_n \left(1 - \frac{1}{m}\right) + \frac{a}{m x_n^{m-1}}.$$

Notice this is exactly the map solved in the example on page 91, with $a = 3$ and $m = 2$.

Example 4.4 *Using the Newton–Raphson routine determine the zero of the function* $f(x) = \mathrm{e}^x - \mathrm{e}^{-2x} + 1$. *This can be done by setting up the functions*

```
function [value] = func(x)
value = exp(x)-exp(-2*x)+1;
```

```
function [value] = func_prime(x)
value = exp(x)+2*exp(-2*x);
```

This yields the result

```
>> Newton_Raphson
Starting guess :-2
 Root = -0.2811995743 found in 8 iterations.
```

(with a tolerance of 1×10^{-10}*). In fact we can solve this equation by introducing the variable* $X = \mathrm{e}^x$*, and noting that* X *is never zero. We can rewrite* $f(x)$ *as*

$$f(x) = \frac{1}{X^2}\left(X^3 - 1 + X^2\right).$$

This equation can be solved directly and the real root is found to be

$$X = \frac{\sqrt[3]{100+12\sqrt{69}}}{6} + \frac{2}{3\sqrt[3]{100+12\sqrt{69}}} - \frac{1}{3},$$

or by using the MATLAB code `co=[1 1 0 -1]; roots(co)`. *Then to retrieve the root of the equation we use the fact that* $x = \ln X$.

4.6 Repeated Roots of Functions

Here we shall briefly discuss an additional technique which can be used to determine repeated roots. We shall presume the function is known analytically. We note the bisection method requires the function to change sign and the Newton–Raphson method requires that the first derivative is non-zero. Both of these limitations pose potential problems. Consider a polynomial $p(x)$ with a repeated root (of order $n > 1$) at the point $x = \lambda$. This means that we can write $p(x)$ as

$$(x-\lambda)^n\, q(x),$$

where $q(\lambda) \neq 0$ (since the root is of order n). Let us now consider the derivative of $p(x)$ at $x = \lambda$, so that

$$\frac{\mathrm{d}p}{\mathrm{d}x} = n\,(x-\lambda)^{n-1}\,q(x) + (x-\lambda)^n\,\frac{\mathrm{d}q}{\mathrm{d}x}$$

so evaluating at $x = \lambda$, we find

$$\left.\frac{\mathrm{d}p}{\mathrm{d}x}\right|_{x=\lambda} = 0.$$

This means the methods which use the derivative are bound to fail.

Let us return to the derivation of the techniques using a Taylor series, where we recall that $x + h$ is deigned to be the correct solution, so that

$$0 = f(x+h) = f(x) + hf'(x) + \frac{h^2}{2}f''(x) + O(h^3).$$

We now note that near the required repeated root $f'(x)$ will be almost zero, so we consider the extra term which is proportional to h^2. To find h we solve the quadratic:

$$h^{(\pm)} = \frac{-f'(x) \pm \sqrt{(f'(x))^2 - 2f(x)f''(x)}}{2f''(x)}.$$

And we can define the next point in the sequence as

$$x_{n+1}^{(\pm)} = x_n + h^{(\pm)}.$$

But we do not know whether to take the positive or negative root. In the following code we calculate the value of the function at each of $x_{n+1}^{(\pm)}$ and determine which has the smaller magnitude.

```
x=1.1;
for its=1:20
    [f,fp,fpp] = fun2(x);
    h1 = (-fp+sqrt(fp^2-2*f*fpp))/(2*fpp);
    h2 = (-fp-sqrt(fp^2-2*f*fpp))/(2*fpp);
    [f1,f1p,f1pp] = fun2(x+h1);
    [f2,f2p,f2pp] = fun2(x+h2);
    if abs(f1)<abs(f2)
        x = x+h1
    else
        x = x+h2
    end
end
```

which calls

```
function [f,fp,fpp] = fun2(x);
f = x.^4+3*x.^3-12*x.^2-20*x+48;
fp = 4*x.^3+9*x.^2-24*x-20;
fpp = 12*x.^2+18*x-24;
```

We note that the above code is not limited to real numbers. In this example we have considered the case $p(x) = x^4 + 3x^3 - 12x^2 - 20x + 48$ which has roots at $x = -3$, $x = -4$ and $x = 2$ (twice).

There are other methods of getting round these problems, and indeed these can be applied for higher-order zeros.

4.7 Zeros of Higher-Dimensional Functions(*)

We note that the Newton–Raphson scheme is readily extended to a technique whereby one can locate the zero of a higher-dimensional function. Let us consider the zero to occur at $\mathbf{x} = \mathbf{x}^*$ and the function is $\mathbf{f}(\mathbf{x})$. For instance $\mathbf{f}(x, y) = (\sin(x + y), x^2 - y^2)$, which has zero at the origin $(0, 0)$. We can derive this scheme using the multi-dimensional Taylor series. We presume that we have a current iterate $\mathbf{x}_g$ and that $\mathbf{x}^* = \mathbf{x}_g + \mathbf{h}$. By definition $\mathbf{f}(\mathbf{x}^*) = \mathbf{0}$, hence

$$\mathbf{0} = \mathbf{f}(\mathbf{x}_g + \mathbf{h}) = \mathbf{f}(\mathbf{x}_g) + \nabla \mathbf{f}|_{\mathbf{x}_g} \cdot \mathbf{h} + O(|\mathbf{h}|^2)$$

where we shall explain the notation $\nabla \mathbf{f} \cdot \mathbf{h}$ in due course, but we note that this can be written as a matrix $\mathbf{A}$ times the vector $\mathbf{h}$. We can solve the above equation for the $\mathbf{h}$ (where we assume that $O(|\mathbf{h}|^2)$ is sufficiently small enough to neglect). This gives

$$\mathbf{h} = -\mathbf{A}^{-1}\mathbf{f}(\mathbf{x}_g).$$

We now discuss the entries of the matrix $\mathbf{A}$. Let us consider $\mathbf{f} : \mathbb{R}^n \to \mathbb{R}^n$, in other words

$$\mathbf{f}(\mathbf{x}) = (f_1(x_1, \cdots, x_n), \cdots, f_n(x_1, \cdots, x_n)).$$

The elements of the matrix $\mathbf{A}$ are

$$a_{i,j} = \frac{\partial f_i}{\partial x_j}.$$

The guess for the value of $\mathbf{x}_g$ is then updated.

Example 4.5 *Determine the zeros of the function* $\mathbf{f} : \mathbb{R}^2 \to \mathbb{R}^2$ *such that*

$$\mathbf{f}(\mathbf{x}) = (f, g) = (x^2 - y^2 + 3, (x+2)^2 - y)$$

where we have used $(x_1, x_2) = (x, y)$. *In this case the matrix* $\mathbf{A}$ *is*

$$\begin{pmatrix} \frac{\partial f}{\partial x} & \frac{\partial f}{\partial y} \\ \frac{\partial g}{\partial x} & \frac{\partial g}{\partial y} \end{pmatrix} = \begin{pmatrix} 2x & -2y \\ 2(x+2) & -1 \end{pmatrix}.$$

The code for solving this is:

```
xg = 1; yg = 1;

for its = 1:10
    f = [xg^2-yg^2+3; (xg+2)^2-yg];
    A = [2*xg -2*yg; 2*(xg+2) -1];
    h = -A\f;
    xg = xg+h(1);
    yg = yg+h(2);
end
```

This gives values $(-0.6410, 1.8469)$ *and the value of* $\mathbf{f}$ *of the order of* 10^{-15}. *We can check this by noting that* $y = (x+2)^2$ *and then substituting into* $f(x, y)$ *we have*

$$\begin{aligned} x^2 - (x+2)^4 + 3 &= x^2 - (x^4 + 8x^3 + 24x^2 + 32x + 16) + 3 \\ &= -x^4 - 8x^3 - 23x^2 - 32x - 13. \end{aligned}$$

The roots of which can be found using the commands

```
a = [-1 -8 -23 -32 -13];
x =  roots(a)
y = (x+2).^2;
```

which we shall meet in due course. This gives:

```
>> x

x =

  -4.1124
  -1.6233 + 1.5154i
  -1.6233 - 1.5154i
  -0.6410
```

```
>> y

y =

   4.4623
  -2.1546 + 1.1417i
  -2.1546 - 1.1417i
   1.8469
```

Notice we find another real root and pairs of complex answers.

In this example we have just performed 10 iterations: however in reality we would wish to stop once a certain tolerance is achieved. This could be accomplished by adding an `if` statement:

```
if norm(f) < 1e-12
   break
end
```

The command `norm` calculates the magnitude of the vector (see `help norm`). This code should be added after the line evaluating `f`. We should also check whether the calculation has actually converged or merely exceeded the maximum number of iterations. This can be done by adding the code

```
if norm(f) > 1e-12
   disp('Routine did not converge')
else
   disp('Answer is')
   disp([' x=' num2str(xg)])
   disp([' y=' num2str(yg)])
end
```

In general we will not be able to determine the partial derivatives of the functions analytically and we rely on the local approximations to the derivatives such that

$$\frac{\partial f}{\partial x} = \lim_{\delta \to 0} \frac{f(x+\delta, y) - f(x,y)}{\delta} \text{ and } \frac{\partial f}{\partial y} = \lim_{\delta \to 0} \frac{f(x, y+\delta) - f(x,y)}{\delta}.$$

We note that this method can be used to find turning points of functions, in higher dimensions. For instance consider $f : \mathbb{R}^2 \to \mathbb{R}$ the turning points are such that $\partial f/\partial x = \partial f/\partial y = 0$, so we define $\mathbf{f}$ as $(\partial f/\partial x, \partial f/\partial y)$ and seek zeros of $\mathbf{f}$.

4.8 MATLAB Routines for Finding Zeros

4.8.1 Roots of a Polynomial

As we have seen MATLAB has a specific command for finding the roots of a polynomial, namely `roots`. The coefficients of the polynomial are placed in a vector `c` and the routine returns the corresponding roots. It is very simple to use, but care is need when entering the coefficients.

Example 4.6 *Find the roots of the quintic equation* $f(x) = x^5 - x^4 + x^3 + 2x^2 - 1$.

This is accomplished using the code:

```
c = [1 -1 1 2 0 -1];
roots(c)
```

There are a couple of things to note from this example:

- The polynomial's coefficients are listed starting with the one corresponding to the largest power.
- It is crucial that zeros are included in the sequence where necessary (in the above we have included the zero times x term).
- As a simple check, a polynomial of order p has $p + 1$ coefficients, i.e. a quadratic has three coefficients and a polynomial of degree p will have p roots. So as long as the coefficients of the polynomial are real, the roots will be real or occur in complex conjugate pairs.

4.8.2 The Command `fzero`

The MATLAB command `fzero` is very powerful. It actually chooses which scheme should be used to solve a given problem. Let us try the form of the command

```
fzero('func', 0.4, optimset('disp','iter'))
```

This produces

```
 Func-count       x             f(x)          Procedure
     1           0.4        0.0813636          initial
     2      0.388686        0.0876803          search
     3      0.411314        0.0745675          search
     4         0.384        0.0901556          search
     5         0.416         0.071613          search
     6      0.377373        0.0935142          search
     7      0.422627         0.067296          search
     8         0.368        0.0979791          search
     9         0.432        0.0609148          search
    10      0.354745         0.103721          search
    11      0.445255        0.0513434          search
    12         0.336         0.110687          search
    13         0.464        0.0367268          search
    14       0.30949         0.118215          search
    15       0.49051        0.0139394          search
    16         0.272         0.124167          search
    17         0.528       -0.0223736          search

  Looking for a zero in the interval [0.272, 0.528]

    18      0.488914        0.0153797          interpolation
    19      0.504837      0.000616579          interpolation
    20      0.505484      -1.5963e-06          interpolation
    21      0.505482      1.80934e-09          interpolation
    22      0.505482      5.32907e-15          interpolation
    23      0.505482                0          interpolation
Zero found in the interval: [0.272, 0.528].

ans =

   0.50548227233930
```

The command uses many different forms but here we are looking for a root of the function defined in `func.m` near the value $x = 0.4$ and the options are set to `display iterations`.

Example 4.7 *Determine the zero of the function* $f(x) = x - x^2 \sin x$ *nearest to* $x = 1$. *We need to set up the code*

```
function [val] = myfunc(x)
val = x-x.^2.*sin(x);
```

and then use the inline command `fzero('myfunc',1)`*. This gives*

```
>> fzero('myfunc',1)
Zero found in the interval: [0.84, 1.16].

ans =

    1.1142
```

4.9 Tasks

Task 4.1 *Modify the codes* `userfn.m` *and* `testplotm.m` *to determine the locations of the roots of the function $f(x) = x \sin x$ between 0 and 10. Use the* `ginput` *command to make sure that you have reasonable estimates of the locations of the zeros. Notice the code* `userfn.m` *as it stands takes two input parameters, whereas this function does not need any parameters and as such the function can be changed to only have one input, namely* `x`*. You should also make sure you use a dot in the calculation, since the function will be applied to vectors.*

Task 4.2 *Given that $e^x > 0 \; \forall x$ show that $\cosh x$ has no zeros and $\sinh x$ has only one. Consequently comment on the zeros of the function $f(x) = \cosh^m x \sinh^n x$ where $n, m \in \mathbb{N}$.*

Task 4.3 *Determine the location of the roots of the polynomial $f(x) = x^2 + bx + 1$, $b \in \mathbb{R}$, giving details of when you expect there to be no real roots, one roots or two real roots. (Attempt this mathematically but then use MATLAB's plotting routine to verify your findings.)*

Task 4.4 *Consider the function $f(x) = \sin x + \beta \cos x$ $\beta \in \mathbb{R}$ and derive an expression for the value of the zeros of $f(x)$ mathematically. Use MATLAB to verify your answer for certain values of β (perhaps using zero and one as a start).*

Task 4.5 *Modify the code* `eqn.m` *(page 112) to determine the zeros of the function* $f(x) = 2\cos x - \sin x$ *between* -2π *and* 2π *using the fixed point method. (You may need to use a plot of the function to identify initial guesses for the values.)*

Task 4.6 *Consider, using the fixed point method, the zeros of the function* $f(x) = x^2 + bx + 1$ *and the two ways of rewriting the function in the required form as* $g(x) = -(x^2 + 1)/b$ *and* $g(x) = -\sqrt{-(bx + 1)}$*. Using the results from Task 4.3 state which method is appropriate (if either) for a given value of* b*. (Recall we require* $|g'(x)| < 1$ *for the method to work.)*

Task 4.7 *Use the bisection method to determine the zero of the function* $f(x) = 2x^2 - x^3 + \sin x$ *between one and three. You will need to enter the codes* `bisect.m`*,* `mbisect.m` *and* `func.m`*. The code* `bisect.m` *does not need modification: however you need to change* `func.m` *to evaluate the function* $f(x)$ *at a point* x *(again remember to use the dots in the correct place). You also need to change* `mbisect.m` *to reflect the fact that we are seeking a zero of the function defined in* `func.m` *in the range 1 to 3.*

Task 4.8 *Consider the function* $f(x) = \cos 3x$ *between 0 and* π*. Comment on the bisection method using the ranges* $[0, \pi]$*,* $[0, 7\pi/8]$ *and finally* $[\pi/8, \pi]$*. It is not necessary to write a code to solve this problem; you should sketch the curve and work out what is likely to happen.*

Task 4.9 *Using the Newton–Raphson method calculate the roots of the functions* $f(x) = x\cos x - \sin x$ *and* $g(x) = (x^3 - x)\sin x$*. These should be done as two separate calculations. You will need to write two MATLAB functions for each case to evaluate the function and its derivative. These should be called* `func.m` *and* `func_prime.m` *so that they can be called from* `Newton_Raphson.m` *(alternatively you could change the line in* `Newton_Raphson.m` *to reflect that you are using different function names).*

Task 4.10 *Using the method of False Position find the roots of the function* $\sin(x - x^3)$ *between* $x = 0$ *and* $x = 6$*.*

Task 4.11 *Using the knowledge that the polynomial* $p(x) = x^3 - 4x^2 + 5x - 2$ *has a repeated root, determine its location.*

Task 4.12 *Using the command* `roots` *calculate the solutions of the equations*

$$x^3 + x^2 + x + 1 = 0 \text{ and } -2x + x^5 + x^2 = 4.$$

Here you need to be careful that the coefficients of the polynomials are entered correctly as vectors.

Task 4.13 *Using the command* `fzero` *investigate the roots of the functions* $x \sin x + \cos x$, $(\sin x)^3 - \sin x$ *and* $(x^2 - 4x + 5)/(x - 1)$.

Task 4.14 (*) *The solutions of the differential equation*

$$x^2 \frac{\mathrm{d}^2 y}{\mathrm{d}x^2} + x\frac{\mathrm{d}y}{\mathrm{d}x} + \left(x^2 - \frac{1}{4}\right) y = 0$$

are given by $J_{1/2}(x)$ *and* $Y_{1/2}(x)$, *which are called Bessel functions. Using the MATLAB command* `besselj(nu,x)` *(where* `nu` *is equal to* $1/2$ *in this case) and* `fzero` *determine the zeros of the function* $J_{1/2}(x)$ *in the interval* $(0, 20]$.

Task 4.15 (D) *The following codes are designed to find a root of the function* $f(x) = x \sin x - x^2 \cos x - 1$ *using the Newton–Raphson method*

f.m

```
function [out] = f(in)
f = x*sin x - x^2 cos x:
```

fp.m

```
function [out] = fp(in)
fp = (x+1)*cos x - 2*x sin x:
```

```
x = 0;

for j = 1:1
    x = x - fp(x)/f(x);
end
```

5
Interpolation and Extrapolation

5.1 Introduction

We shall now consider the problem where we know the values of a function at a certain set of predetermined points, (x_1, f_1), (x_2, f_2), $\cdots$ through to (x_N, f_N). The question to be answered in this chapter is what values does the function take at intermediate values of x (which is called interpolation), or alternatively what values does the function take external to this range (which is called extrapolation).

It is worth pausing here to think how we would determine an intermediate data value in practice:

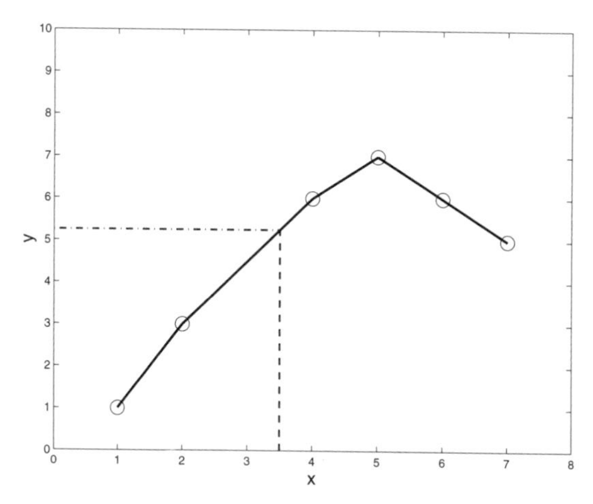

If we presume we are given the original data set, shown here as circles, we might join them together with straight lines (as shown). In order to work out the value at $x = 3.5$ say, we would draw a vertical line and see where it crossed the appropriate segment. Then a horizontal line would be drawn and the value read off the y-axis. In the main this is exactly the technique used: however differing degrees of approximations are used to create the lines between the points. We shall start by discussing how we might obtain the original data.

5.2 Saving and Reading Data

In this Chapter we will generally need to read data from a file. Within the context of MATLAB there are two routes open to us for this procedure and it will depend on how the data was initially saved. For illustration we shall presume we have a MATLAB session which has created a set of variables:

```
>> whos
  Name       Size             Bytes  Class

  a          3x4                 96  double array
  a1         2x2                 32  double array
  b          6x6                288  double array
  c          1x5                 40  double array
  z          1x1                 16  double array (complex)

Grand total is 58 elements using 472 bytes
```

Here we have used the command **whos** to list all the current variables, and it gives information concerning: their size (both in terms of dimensions and in memory, given in bytes) and their class. Most of these are classed as "double array" which means they are real matrices. The final variable **z** is a one-by-one matrix which is complex. The command also gives information concerning the total number of elements (that is the sum of the product of the dimensions of the matrices) and the total memory. If we wished to only list those variables starting with a we could use:

```
>> whos a*
  Name       Size             Bytes  Class

  a          3x4                 96  double array
  a1         2x2                 32  double array
```

```
Grand total is 16 elements using 128 bytes
```

or if we just wanted the names of the variables we could use:

```
who

Your variables are:

a          a1         b          c          z
```

If we only wanted information about one variable we could just use the construction `whos a1`, for instance.

In order to save the data we use the command `save`. The help command is very helpful here (`help save`). However, we shall give a synopsis here of the main points.

To save all the variables which are currently in use:

```
% Save all current variables in a file
% session_vars.mat
save session_vars
```

This creates a file called `session_vars.mat`, where the first part of this filename is user-defined to reflect the contents of the file and the second part `.mat` defines it as a MATLAB file containing variables. Note this suffix may also be used by another package and as such it will be difficult to use the `Open` menu. This file now contains the actual variables, which means both their names and values are stored.

They can be reloaded using the command

```
% Loads the variables stored in the
% file session_vars.mat
load session_vars
```

We note the default file name for files of this type end with `.mat` and as such we are asking MATLAB to load the file `session_vars.mat`. This file is very special since it actually contains information concerning the variables as well. If we wish to save only certain variables we can list them:

```
% Save the variables a & a1 to the
% file session_vars.mat
save session_vars a a1

% Save all variables starting with b
% to session_vars.mat
save session_vars b*

% Append all variables starting with ca
% to the file session_vars.mat
save session_vars ca* -append
```

This last command is particularly useful since it adds the variables to the file whereas the previous ones erase all the other variables currently stored in the files.

If we merely want to save the data from variables (rather than their names) we can again use `save` but with a different syntax. This time we shall specify the entire filename, which needs to be enclosed within single quotes.

```
save 'session_vals.dat' a a1 -ascii
```

This produces an ASCII (American Standard Code for Information Interchange – basically human readable) file containing the values of the variables `a` and `a1`. This creates a file whose contents looks like:

```
1.0000000e+000  1.0000000e+000  1.0000000e+000
1.0000000e+000  1.0000000e+000  1.0000000e+000
1.0000000e+000  1.0000000e+000  1.0000000e+000
1.0000000e+000  1.0000000e+000  1.0000000e+000  1.0000000e+000
1.0000000e+000  1.0000000e+000  1.0000000e+000  1.0000000e+000
1.0000000e+000  1.0000000e+000  1.0000000e+000  1.0000000e+000
1.0000000e+000  1.0000000e+000  1.0000000e+000  1.0000000e+000
```

where the variables were defined as `a = ones(3); a1 = ones(4)`. The problem with this technique is we cannot read this back into MATLAB directly using `load` (it complains there are not the same number of pieces of data in each line). It cannot know this is a three-by-three matrix followed by a four-by-four one. There are other options for this command which allow extra precision, for instance `-double`.

The problem here is that we need to have knowledge about the data in the file to understand the form we wish to read it in. For example consider a file containing the data

```
1.0000000e+000  2.0000000e+000  3.0000000e+000  4.0000000e+000
1.0000000e+000  4.0000000e+000  9.0000000e+000  1.6000000e+001
```

(this was created using the code: `x = 1:4; y = x.^2; save 'bob1.dat' x y -ascii`). This contains two rows vectors and we can load this using:

```
load 'bob1.dat'
```

Now using the command `whos` we can see that we have a two-by-four array called `bob1`. We can now extract the data using `a = bob1(1,:); b = bob1(2,:); clear bob1`. These commands give us the first row and the second row in the row vectors `a` and `b`; finally we clear the array `bob1` since we have extracted the requisite data. Whilst this last stage is not necessary it is good practice to keep track of the variables you have. Notice there is no requirement that the variables are called the same as when they were saved (since usually you won't know what they were called).

This is probably the best way of reading in data from another source, provided the data is rectangular in shape. However, there is a further method which can be used. This involves the commands `fopen`, `fprintf`, `fscanf` and `fclose`. These commands are very powerful and as you might expect quite complicated to use:

```
x = 0:.4:2;
y = [x; exp(x).*cos(2*x)];

fid = fopen('data.dat','w');
fprintf(fid,'%6.2f  %12.8f\n',y);
fclose(fid);
```

In the first two lines we set up a vector `x` running from 0 to 2 in steps of 0.4 and then a matrix `y` comprising two columns (first `x` and then the corresponding values of the function $e^x \cos 2x$). The next line opens a file `data.dat` using the command `fopen` (note the fact that the filename must be enclosed in single quotes). The first argument is the filename and the second one reflects what we are going to do: in this case `write` to the file. The options for this second argument are:

`'r'`	read
`'w'`	write (create if necessary)
`'a'`	append (create if necessary)
`'r+'`	read and write (do not create)
`'w+'`	truncate or create for read and write
`'a+'`	read and append (create if necessary)
`'W'`	write without automatic flushing
`'A'`	append without automatic flushing

(this list can be obtained by typing `help fopen`). The variable `fid` is an identifier or handle for the file and as such will be used in all commands related to that file. If the command `fopen` is successful `fid` will be a scalar integer (it will have value minus one if it is unsuccessful). We can now refer to this file merely using this variable (you can actually use the integer value but this is not advisable). The next line uses the command `fprintf` which has three arguments: the file identifier (`fid`), the format and the variable to be written.

We need to ensure the format matches the data. The format here prints two numbers per line, the first one '`%6.2f`' corresponds to a fixed point number with 2 places after the decimal point using six characters in total; the second one '`%12.8f`' has eight places after the decimal point and uses 12 characters in total. There are several other formats and these are based on those from the C language. The other option of the most use is `e`, so the line would be changed to:

```
fprintf(fid,'%6.2e %12.8e \n',y)
```

which produces floating point versions of the above. The command `\n` gives a new line after printing the two numbers, whilst we could place a tab between the characters using `\t`.

Finally we use the command `fclose(fid)` to close the file corresponding to the the file identifier `fid`. It is very good practice to close the files we are using as soon as we have finished with them.

This produces a file called `data.dat` which contains:

```
0.00    1.00000000
0.40    1.03936428
0.80   -0.06498473
1.20   -2.44823335
1.60   -4.94458639
2.00   -4.82980938
```

We could read the data back into MATLAB using the command `fscanf`: this is similar syntactically to `fprintf` except we now specify the shape of what is to be read in. The simplest way of reading in the above data would be:

```
clear all
load 'data.dat'
x = data(:,1);
f = data(:,2);
```

since it is rectangular. We shall now presume we do not know the functional form of our data. The following sections will use the data contained within `data.dat` as an example. It is suggested that you type the above code to create the file (or you can directly enter the data).

5.3 Which Points to Use?

We pause here to discuss the problem of which points to use for the process, in general we apply the following method.

IMPORTANT POINT

Pick those closest!

This can be done by hand, or can be automated. For this purpose we use the routine `findrange.m`:

```
function [ibot,itop] = findrange(x,z,N)
if mod(N+1,2)~=0
   disp('Must use odd N')
   break
end
points = (N+1)/2; % Half number of points in stencil
[ii] = find(x>z);
if isempty(ii)
   itop = length(x);
else
   i = ii(1);
end

itop = i+points-1;
ibot = i-points;
if itop>length(x)
      itop = length(x);
      ibot = length(x)-2*points+1;
elseif ibot<1
      itop = 2*points;
      ibot = 1;
end
```

This takes three arguments: `x` (the data points); `z` (the point at which the function is required) and `N` (the order of the interpolating polynomial; for linear interpolation `N` would be 1 and 3 for cubic interpolation). First we check that the interpolating polynomial is of odd order. This is accomplished by checking that one plus `N` is even. The number of points in half the stencil is then calculated as the variable `points`. We use the command `[ii] = find(x>z);` to list the points contained within `x` which are greater than `z`. If there are no points for which this is true (in which case `ii` is empty), we can infer that `z` is greater than all the values of `x` and we need to extrapolate using the last few points. Else we use the first value for which `x` is greater than `z` (stored in the first element of the array `ii`). We now have a marker `i` and we work out the top and bottom of the stencil as `itop` and `ibot` respectively. We now need to deal with the cases for which the stencil extends beyond the limits of `x`, that is if `itop` is greater than `length(x)` or if `ibot` is less than 1 (the first element of `x`). We now use the end `2*points`.

This routine can now be used in other programs to give the extent of the stencil.

5.4 Newton Forward Differences and Lagrange Polynomials

We shall now discuss the operation of fitting an Nth order polynomial through $N+1$ points and remark again this will yield a unique answer. We consider the set of data points: (x_0, f_0), (x_1, f_1), $\cdots$, (x_N, f_N). We now introduce the difference operator Δ such that

$$\Delta f_0 = f_1 - f_0 \text{ or in general } \Delta f_j = f_{j+1} - f_j.$$

These are called *forward differences*, since points forward of the current value are used (there are also backward differences and central differences, which we shall meet in our discussion of the solution of differential equations). We consider the composition of the operator whereby

$$\begin{aligned} \Delta^2 f_0 = \Delta(\Delta f_0) = \Delta(f_1 - f_0) &= \Delta f_1 - \Delta f_0 \\ &= (f_2 - f_1) - (f_1 - f_0) \\ &= f_2 - 2f_1 + f_0. \end{aligned}$$

Of course we can now proceed to define $\Delta^n f_0$ in an iterative manner.

We introduce the polynomial

$$f(x) = f_0 + (x - x_0)\frac{\Delta f_0}{h} + \frac{(x-x_0)(x-x_1)}{2!}\frac{\Delta^2 f_0}{h^2} + \cdots \quad ,$$

which will ultimately terminate after N terms. Alternatively this can be written as:

$$f(x) = f_0 + \sum_{n=1}^{N-1} \frac{\Delta^n f_0}{h^n n!} \prod_{j=0}^{n-1} (x - x_j). \tag{5.1}$$

We consider the data values to be equally spaced, so the value of Δx_j is $h \; \forall j$. This analysis can be extended to irregularly spaced points, but we shall not attempt that here.

We are now able to construct the polynomial (5.1) for a set of points. We start with a simple example.

Example 5.1 *Let us consider the points* $(0, 1)$, $(1, 3)$ *and* $(2, 4)$. *Here we have three points and as such we would expect to obtain a quadratic. We use a tabular form, which gives:*

x	f	Δf	$\Delta^2 f$
0	1	2	-1
1	3	1	
2	4		

Thus we have

$$f(x) = 1 + (x-0)\frac{2}{1} + \frac{(x-0)(x-1)}{2!}\frac{(-1)}{1^2},$$
$$f(x) = 1 + 2x - \frac{x(x-1)}{2},$$

where $h = 1$, $f_0 = 1$, $\Delta f_0 = 2$ *and* $\Delta^2 f_0 = -1$ *(reading from the top row of the table).*

It is possible to prove that the polynomial (5.1) goes through each of the set of points, but we shall not include that proof here. We remark that various conclusions can be drawn from the data by using the forward differences; for instance whether it was generated using a polynomial (in which case the differences truncate).

We now construct a polynomial which goes through a set of points which are not necessarily evenly spaced. Let us consider the polynomial

$$\begin{aligned} f(z) \quad = \quad & \frac{(z-x_2)(z-x_3)(z-x_4)}{(x_1-x_2)(x_1-x_3)(x_1-x_4)}f_1 + \frac{(z-x_1)(z-x_3)(z-x_4)}{(x_2-x_1)(x_2-x_3)(x_2-x_4)}f_2 \\ & + \frac{(z-x_1)(z-x_2)(z-x_4)}{(x_3-x_1)(x_3-x_2)(x_3-x_4)}f_3 \\ & + \frac{(z-x_1)(z-x_2)(z-x_3)}{(x_4-x_1)(x_4-x_2)(x_4-x_3)}f_4. \end{aligned} \tag{5.2}$$

It is worth pausing at this point and checking that this curve goes through each of the points (x_j, f_j). For example for $j = 3$, we set $z = x_3$ and only the third term is non-zero and we have

$$f(x_3) = \frac{(x_3-x_1)(x_3-x_2)(x_3-x_4)}{(x_3-x_1)(x_3-x_2)(x_3-x_4)}f_3 = 1 \times f_3 = f_3.$$

Hence the value of the polynomial at $z = x_3$ is f_3 (as we would hope). This is an example of a **Lagrange polynomial**; which could have equally been written as

$$f(z) = \sum_{i=1}^{4} f_i \prod_{\substack{j=1 \\ j\neq i}}^{4} \frac{z - x_j}{x_i - x_j}.$$

This is a convenient way to write out the cubic we require (as it is relatively easily extended to higher-order cases) and in order to evaluate it we can use the MATLAB code:

```
ip = 1:4;
f_z = 0.0;
for ii = 1:4
    jj = find(ip~=ii);
    prod = f(ip(ii));
    for kk = jj
       prod = prod *(z-x(ip(kk)))/(x(ip(ii))-x(ip(kk)));
    end
    f_z = f_z + prod;
end
```

This code probably needs some explanation.

- Firstly we set up a vector `ip=1:4` which runs from one to four,
- and then we have a new variable `f_z` which will ultimately contain our estimate of the function at the point z.
- We now run through the terms in Equation (5.2).
- Inside the loop we construct a vector `jj = find(ip~=ii)`, which now contains the indices of the locations in vector `ip` which are not equal to the current value of `ii`. The initial value of the variable `prod` is set equal to the function evaluated at the corresponding grid point `ip(1)` etc.
- We then go through the individual terms in the product
- and then finally add this onto the value `f_z` until we have gone through the four terms.

This code has been written in this way so it could be extended to as many points as we want.

For $z = 0.6$ we find a value of `f_z=0.6386`, which is shown on the figure as an asterisk:

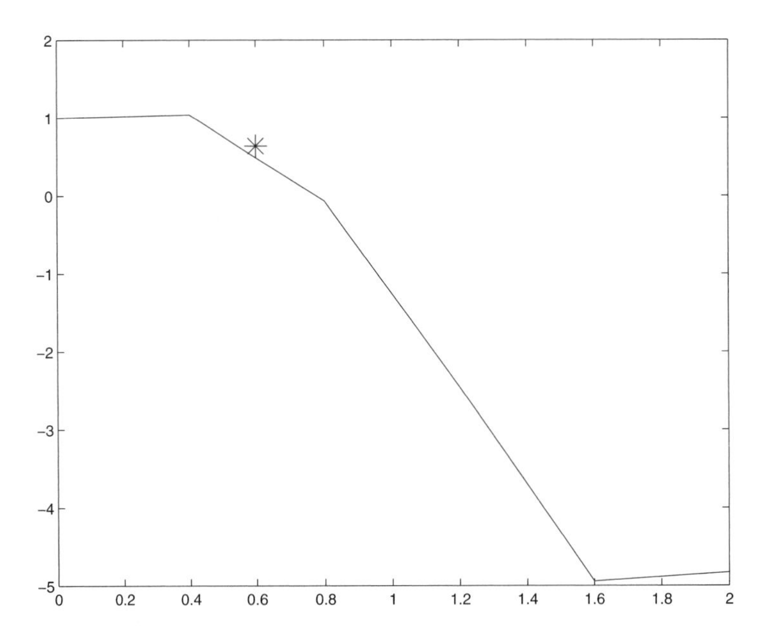

As mentioned we can write this expression as a summation of products and this is easily extended to N points as

$$f(z) \approx \sum_{i=1}^{N} f_i \prod_{\substack{j=1 \\ j \neq i}}^{N} \frac{z - x_j}{x_i - x_j}.$$

Consider the code:

```
function [value] = poly_int(z,N)
global x f
imax = length(x);
if mod(N,2) ~= 0
       disp(' N should be even ')
       break
elseif N >= imax;
       disp('Too many points used')
       break
end
M = N/2;
[ibottom,itop] = findrange(x,z,N);
ip = ibottom:itop;
il = 1:N;
f_z = 0.0;
for ii = 1:N
    jj = find(il~=ii);
    prod = f(ip(ii));
    for kk = jj
       prod = prod *(z-x(ip(kk))) ...
        /(x(ip(ii))-x(ip(kk)));
    end
    f_z = f_z + prod;
end
value = f_z;
```

And now test the code using[1]

[1] We have re-introduced the command `isempty`, which is true if its argument is empty.

```
global x f
load data.dat
x = data(:,1);
f = data(:,2);
N = length(x);
xmin = x(1); xmax = x(N);
xtest = linspace(xmin,xmax,20);
for ii = 1:20
    [ftest(ii)] = poly_int(xtest(ii),N);
end
plot(x,f,'r',xtest,ftest,'b')
```

This gives the plot

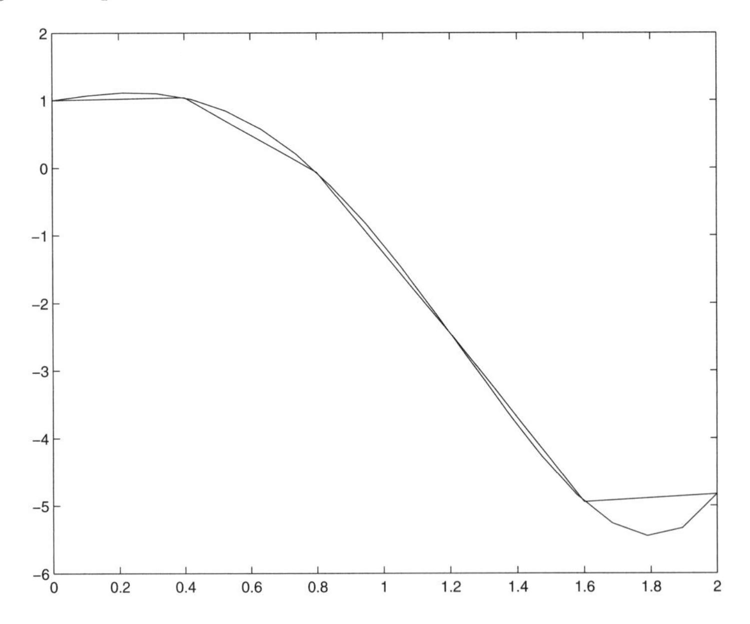

As you can see using six points works quite well at fitting the data. Although as we will see in subsequent examples using high-order polynomials can lead to significant errors.

5.4.1 Linear Interpolation/Extrapolation

We consider the simplest case wherein we have two points (x_1, y_1) and (x_2, y_2). The straight line through these is given by

$$y = \frac{x - x_2}{x_1 - x_2} y_1 + \frac{x - x_1}{x_2 - x_1} y_2,$$

which here we have constructed in a Lagrange polynomial style. We note that the answer is independent of the process here and provided $x_1 \neq x_2$ we will always get a straight line. This can be continued for quadratics and higher order functions.

This is a plot of the data we shall use for this discussion:

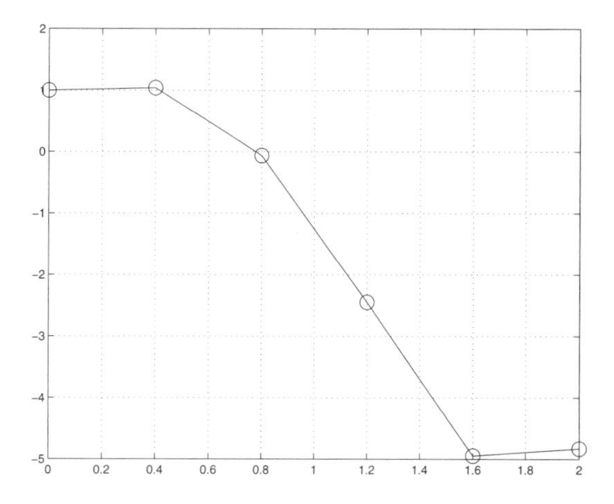

This was obtained using the commands:

```
clear all
load 'data.dat'
plot(data(:,1),data(:,2),'o','MarkerSize',12)
hold on
plot(data(:,1),data(:,2))
hold off
grid on
print -dps2 data.ps
```

The first two commands merely load the data as discussed in the previous section (the code to generate this data is given on page 137). The next command plots the data (the first column against the second) using circles, whose size is specified by `'MarkerSize'` (and set equal to 12). We then `hold` the figure `on` which stops any future plotting commands overwriting the figure. Instead they augment the current image. The next plotting command adds the straight lines. We then turn the `holding off` so the figure will be destroyed if we execute another `plot` command. We also add a `grid` to the figure.

Finally, we print the results to a Postscript file so that it can be included in another document (for instance this text) or sent to a printer. There are many options for this command, for instance we could use `print -djpeg90 data.jpg` to generate a JPEG file.

As mentioned above for convenience we can extract the data from the array `data` using:

```
x = data(:,1);
f = data(:,2);
clear data
```

5.5 Calculating Interpolated and Extrapolated Values

We shall now describe how MATLAB can be used to determine the interpolating polynomial for a set of points. We shall presume that we have the requisite number of points to perform this operation, that is two points for a line, three for a quadratic and four for a cubic, etc. We shall make use of the MATLAB command `polyfit`. The syntax for this command is `polyfit(x,y,N)`, where the points are defined in `x` and `y`, and `N` is the order of the interpolating polynomial. We note that this function actually uses a least squares approach and fits the best polynomial (see page 152). This gives the curve we want provide the number of points represented in `x,y` is $N+1$. Let us consider the interpolation of data points using a straight line.

Example 5.2 *We seek to find the value of the function at* $x = 4.5$ *where the data points are* $(1,-3)$, $(3,4)$, $(5,5)$, $(7,-8)$, $(9,-3)$ *and* $(11,0)$, *using linear interpolation.*

```
x = [1 3 5 7 9 11];
y = [-3 4 5 -8 -3 0];

xi = 4.5;

r = 2:3; % xi lies between the 2nd and 3rd point of x

p = polyfit(x(r),y(r),1);

yi = polyval(p,xi)
```

This gives `p=[0.5000 2.5000]` *(representing the line* $y = x/2 + 5/2$*) and* `yi=4.75`*. We note that here we have determined the range* `r` *by hand but we could employ the function* `findrange` *(see page 140).*

We have used the command `polyval(p,xi)` to calculate the value of the interpolating polynomial (represented by its coefficients `p`) and `xi`.

We can easily extend this method to use higher-order curves.

Example 5.3 *Using the data in the previous example now calculate the value of the interpolating polynomial at* $x = 4.5$ *using cubic interpolation.*

```
x = [1 3 5 7 9 11];
y = [-3 4 5 -8 -3 0];

xi = 4.5;

[ibot,itop] = findrange(x,y,3);
r = ibot:itop;

p = polyfit(x(r),y(r),3);

yi = polyval(p,xi)
```

This gives `p=[-0.1667 0.75 2.667 -6.25]` *and* `yi=5.75`*.*

We note that this is quite different to the answer given by linear interpolation and one might argue that linear interpolation is better here. This of course depends on the underlying function.

5.6 Splines

We now mention another form of interpolation: this fits a curve in an interval where the data at the end of the interval coincides with the data points and also matches the available derivatives with the next interval. If we consider the points (x_i, f_i) and (x_{i+1}, f_{i+1}) and fit a straight line this will match the function values with the intervals on each side (but not necessarily the derivative). We need to use a higher-order curve. In order to satisfy the requirements we need at least a cubic, and this is often the preferred option.

If we consider the points to be (x_i, f_i) for $i = 1$ to N, we have $N - 1$ intervals. Let us consider the cubic valid over the interval $[x_i, x_{i+1}]$ to be

$$y_i(x) = a_i + (x - x_i)b_i + (x - x_i)^2 c_i + (x - x_i)^3 d_i.$$

We note that we have $N - 1$ cubics each with four unknowns; that is $4N - 4$ unknowns in total. We now proceed to give details of the conditions which we can use to specify the values of the coefficients. Firstly we require that the cubic matches the data values at the ends of the interval:

$$y_i(x_i) = f_i \text{ and } y_i(x_{i+1}) = f_{i+1} \text{ for } i = 1, \cdots, N-1;$$

this in effect fixes the a_i and yields $2(N - 1)$ equations. Secondly we require the gradient and the curvature to match at the interior points:

$$y_i'(x_{i+1}) = y_{i+1}'(x_{i+1}) \text{ and } y_i''(x_{i+1}) = y_{i+1}''(x_{i+1}) \text{ for } i = 1, \cdots, N-2.$$

This yields a further $2(N - 2)$ equations. Unfortunately we only have $4N - 6$ equations and as such our system is not totally specified. We elect to consider natural splines, for which the curvature is taken to be zero at the ends of the domain, that is

$$y_1''(x_1) = 0 \text{ and } y_{N-1}''(x_N) = 0.$$

We can construct a matrix system in order to determine the coefficients. This can be coded: however the algorithm is quite involved and again MATLAB comes to our rescue with the command `spline`.

Example 5.4 *Fit a cubic spline to the data* $x = (1, 2, 3, 5)^T$, $f = (4, 2, 0, 3)^T$ *and plot the interpolated function on a grid* $z = 0, 0.1, 0.2, \cdots, 5$.

```
x = [1 2 3 5];
f = [4 2 0 3];
z = 0.:0.1:5;
y = spline(x,f,z);
plot(z,y,x,f,'MarkerSize',12)
```

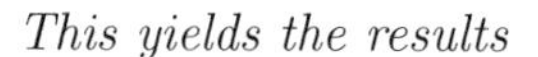

This yields the results

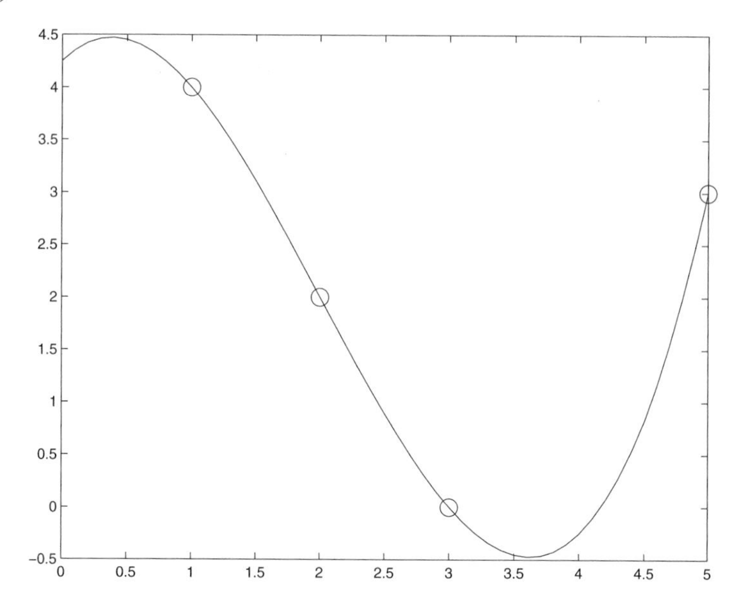

By way of illustration we shall now construct the splines through the points (x_1, y_1), (x_2, y_2), (x_3, y_3).

These curves are taken to have the equations

$$y_j(x) = a_j + b_j(x - x_j) + c_j(x - x_j)^2 + d_j(x - x_j)^3,$$

where $j = 1$, 2 and 3. This gives us 12 undetermined coefficients and we shall now describe the equations which they satisfy.

$y_j(x_j) = f_j$, $j = 1, 2, 3$: This represents the condition that the spline goes through the data point at its left-hand end and these give the equations:

$$a_1 = f_1, \tag{5.3a}$$

$$a_2 = f_2, \tag{5.3b}$$

$$a_3 = f_3. \tag{5.3c}$$

$y_j(x_{j+1}) = f_{j+1}$, $j = 1, 2, 3$: This gives a similar condition at the right-hand ends:

$$a_1 + h_1 b_1 + h_1^2 c_1 + h_1^3 d_1 = f_2, \tag{5.3d}$$

$$a_2 + h_2 b_2 + h_2^2 c_2 + h_2^3 d_2 = f_3, \tag{5.3e}$$

$$a_3 + h_3 b_3 + h_3^2 c_3 + h_3^3 d_3 = f_4. \tag{5.3f}$$

Here we have introduced $h_j = x_{j+1} - x_j$ for convenience.

$y'_j(x_{j+1}) = y'_{j+1}(x_{j+1})$, $j = 1, 2$: This is the matching of the first derivative at the internal points, which gives

$$b_1 + 2h_1c_1 + 3h_1^2d_1 = b_2, \tag{5.3g}$$

$$b_2 + 2h_2c_2 + 3h_2^2d_2 = b_3. \tag{5.3h}$$

$y''_j(x_{j+1}) = y''_{j+1}(x_{j+1})$, $j = 1, 2$: This is a similar condition for the second derivative at the internal points:

$$2c_1 + 6h_1d_1 = c_2, \tag{5.3i}$$

$$2c_2 + 6h_2d_2 = c_3. \tag{5.3j}$$

$y''_1(x_1) = 0$ and $y''_3(x_4) = 0$: These are the conditions that the splines at the ends of the domain are linear at the end points, which gives:

$$c_1 = 0, \tag{5.3k}$$

$$2c_3 + 3h_3d_3 = 0. \tag{5.3l}$$

The solution of the above equations gives the required coefficients. Notice that these coefficients could also have been retrieved using the `spline` command and then using the elements of the answer `coefs`

```
pp = spline(x,f);
pp.coefs
```

Note that the coefficients correspond to a cubic without the $-x_j$ factors. This is an example of an object rather than a variable, that is `pp`.

5.7 Curves of Best Fit

In all of the examples so far we have forced the curves to go through all the data points. We now relax that requirement. We shall start with a straight line and optimise the coefficients used to define it. The straight line is given by $f_L(x)$ which will be of the form $ax + b$. Let us assume that the error at a certain point x_j is given by $(f_L(x_j) - f_j)^2$, so the total error is

$$\begin{aligned} \text{Error: } e = \sum_{i=1}^{N} e_i^2 &= \sum_{i=1}^{N} (f_L(x_i) - f_i)^2 \\ &= \sum_{i=1}^{N} (ax_i + b - f_i)^2 . \end{aligned} \tag{5.4}$$

We wish to minimise this expression by choosing a and b accordingly. In order to determine the actual values of a and b we differentiate with respect to each one and set the result equal to zero.

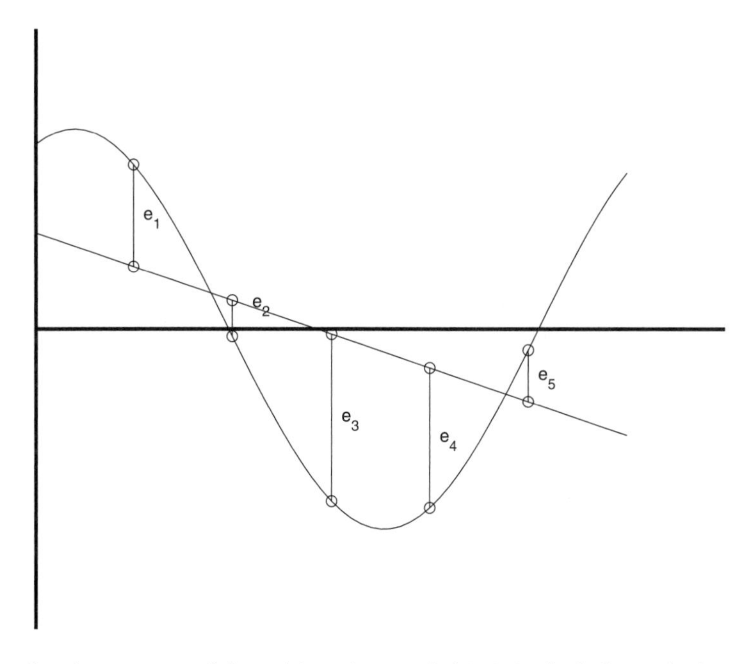

We need to be very careful at this point, and this is included purely for interest:

$$\frac{\partial e}{\partial a} = \sum_{i=1}^{N} 2x_i (ax_i + b - f_i) = 0$$

and

$$\frac{\partial e}{\partial b} = \sum_{i=1}^{N} 2 (ax_i + b - f_i) = 0.$$

These equations can be manipulated to give the simultaneous equations

$$\begin{aligned} a\sum_{i=1}^{N} x_i^2 + b\sum_{i=1}^{N} x_i &= \sum_{i=1}^{N} x_i f_i \\ a\sum_{i=1}^{N} x_i + bN &= \sum_{i=1}^{N} f_i. \end{aligned}$$

We could solve these equations by hand but we shall exploit MATLAB for this purpose, so that we solve the matrix form of the equation:

$$\begin{pmatrix} \sum_{i=1}^{N} x_i^2 & \sum_{i=1}^{N} x_i \\ \sum_{i=1}^{N} x_i & N \end{pmatrix} \begin{pmatrix} a \\ b \end{pmatrix} = \begin{pmatrix} \sum_{i=1}^{N} x_i f_i \\ \sum_{i=1}^{N} f_i \end{pmatrix}$$

This is coded as:

```
load data.dat;
x = data(:,1); f=data(:,2);
N = length(x);
A = ([sum(x.^2) sum(x); sum(x) N]);
rhs = ([sum(x.*f); sum(f)]);
vect = inv(A)*rhs;
a = vect(1); b = vect(2);
fss = a*x+b;
plot(x,f,'r',x,fss,'b')
```

This yields the plot:

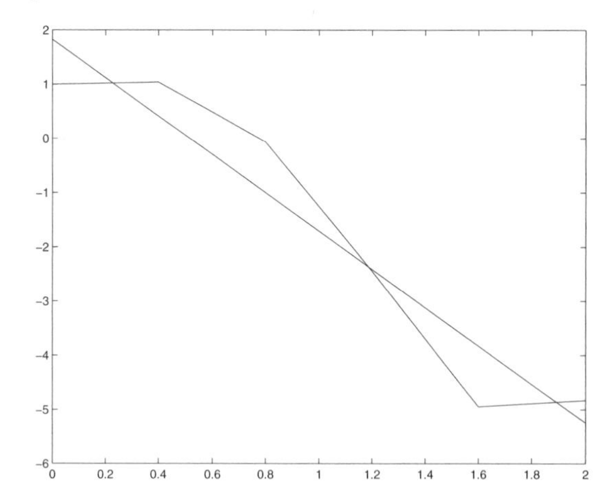

As you can see the curve is a reasonable approximation to the points but does not pass through many (if any) of the actual grid points. This method can be extended to assume other forms of data.

We could have also used the MATLAB command `polyfit` which automates the previous procedure. This is called using `[p,s] = polyfit(x,y,n)` and fits a polynomial of degree n for $y = y(x)$. The coefficients for the polynomial are returned in `p` and the second variable is associated with the structure and is used by other commands to assess the level of the error. We can now use the command `polyval` to determine the value of the fitted polynomial at an intermediate point. This can be called simply as `y = polyval(p,x)` or in a more sophisticated form as `[y,delta] = polyval(p,x,s)`, where `delta` is in some sense representative of the error and the input argument `s` is generated by `polyfit`.

If we know more about a function we can use higher-order approximations, or can use combinations of functions.

Example 5.5 *Given that a set of data is of the form $y = ax + b\mathrm{e}^{-x} + c$ state how one would determine the constants a, b and c.*

As with the linear example we define the sum of the squares of the errors

$$e = \sum_{i=1}^{N} \left(ax_i + b\mathrm{e}^{-x_i} + c - f_i\right)^2$$

and seek the values of the constants which minimises this expression. Again we construct the partial derivatives:

$$\begin{aligned}
\frac{\partial e}{\partial a} &= \sum_{i=1}^{N} 2x_i \left(ax_i + b\mathrm{e}^{-x_i} + c - f_i\right), \\
\frac{\partial e}{\partial b} &= \sum_{i=1}^{N} 2\mathrm{e}^{-x_i} \left(ax_i + b\mathrm{e}^{-x_i} + c - f_i\right), \\
\frac{\partial e}{\partial c} &= \sum_{i=1}^{N} 2 \left(ax_i + b\mathrm{e}^{-x_i} + c - f_i\right).
\end{aligned}$$

These equations can then be combined into a matrix form to give:

$$\begin{pmatrix} \sum x_i^2 & \sum x_i \mathrm{e}^{-x_i} & \sum x_i \\ \sum \mathrm{e}^{-x_i} x_i & \sum \mathrm{e}^{-2x_i} & \sum \mathrm{e}^{-x_i} \\ \sum x_i & \sum \mathrm{e}^{-x_i} & \sum 1 \end{pmatrix} \begin{pmatrix} a \\ b \\ c \end{pmatrix} = \begin{pmatrix} \sum x_i f_i \\ \sum \mathrm{e}^{-x_i} f_i \\ \sum f_i \end{pmatrix}.$$

5.8 Interpolation of Non-Smooth Data

We will demonstrate in this example how one might encounter problems with seemingly simple data. Let us consider

x	f
0	1
1	1
2	1
3	1
4	0
5	0
6	0
7	0

This is a simple step function which has a value of 1 from $x = 0$ to $x = 3$ and a value of 0 from $x = 4$ to $x = 7$. It is tempting to say

$$f(x) = \begin{cases} 1 & x \leqslant 3 \\ 0 & x > 3 \end{cases} .$$

However we are not sure so we shall try to determine the values of the function in a systematic fashion. We shall use the command `poly_int` which was introduced on page 145. This will be supplemented with the code:

```
global x f
str = 'No. of points for interpolation (even): ';
x = 0:1:7;
f = zeros(size(x));
ii = find(x<4); f(ii) = 1;

z = 0:0.2:7;
nz = length(z);
N = input(str);
for j = 1:nz
    f_z(j) = poly_int(z(j),N);
end
figure(1)
t = ['Using ' num2str(N) ' points'];
plot(x,f,'o',z,f_z,'r')
title(t,'FontSize',20)
axis([-1 8 min(f_z)-1 max(f_z)+1])
```

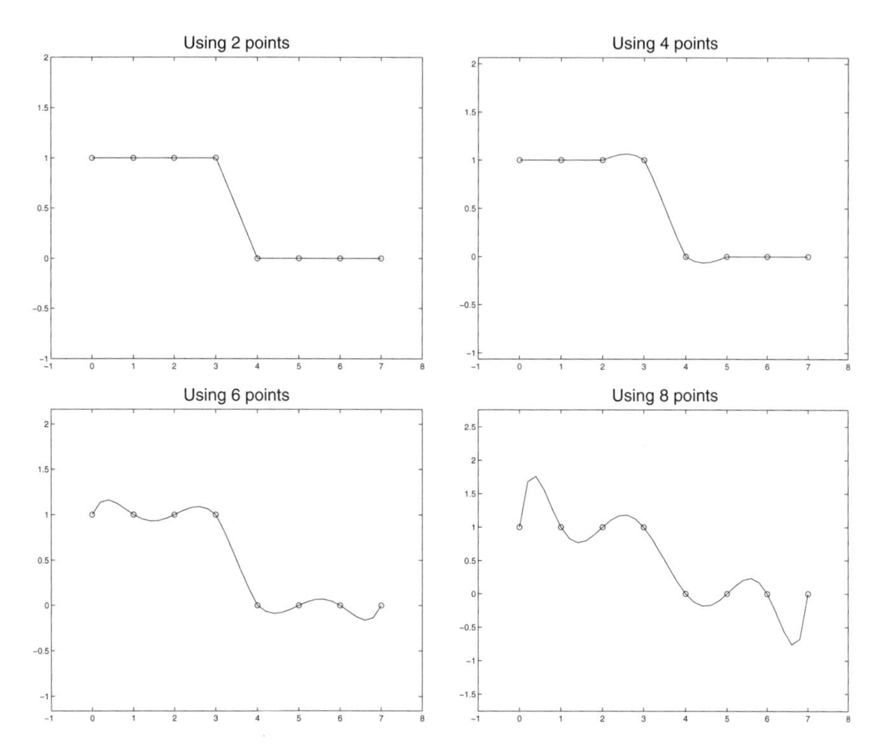

As you can see as more points are used in the interpolation, the approximation gets steadily worse and the effect extends far from the discontinuity (which is causing the problem): in fact less points seem to produce less error. The question remains as to what is the "real" data (it could actually look like the figure in the bottom right – although this is not likely).

Let us consider the problem with

x	f
0	1
1	-0.2
2	0.05
3	0.01
4	0.01
5	0.01
6	0.005
7	0.005

In this case we have

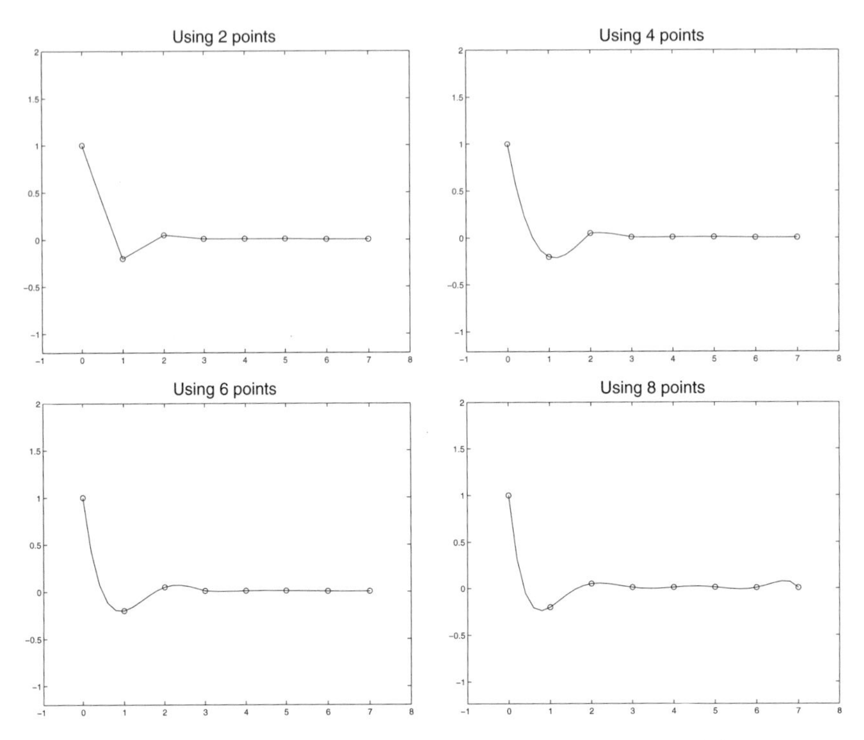

Here we note that the lower-order polynomials produce reasonable approximations; it is only when we try to fit an octic to the data we observe the problems at the right-hand end. We need to be aware the accuracy of our interpolated values is very dependent on the fidelity and resolution of the starting data.

5.8.1 Insufficient Data Points

Let us consider the example $f(x)$ approximated for an interval from -1 to 1, using N points and at each stage approximating the curve using all the points.

```
global x f
for n = 2:1:10
   x = linspace(-1,1,n);
   f = sin(2*x*pi);
   z = linspace(-1,1,100);
   if mod(n,2)==0
      ni = n;
   else
      ni = n-1;
   end
   for j = 1:100
       f_z(j) = poly_int(z(j),ni);
   end
   clf
   plot(x,f,'o','MarkerEdgeColor','k',...
                  'MarkerFaceColor',[.49 1 .63],...
                  'MarkerSize',12)
   hold on
   plot(z,f_z,z,sin(2*z*pi),'--')
   hold off
   yrange_max = max(max(f_z),1);
   yrange_min = min(min(f_z),-1);
   axis([-1.2 1.2 yrange_min-0.2 yrange_max+0.2])
   t = ['Interpolated using ' num2str(n) ' data points '];
   title(t)
   ti = ['int' num2str(n) '.ps'];
   h = gcf;
   print(h,'-dps2',ti)
   pause
end
```

This program waits for the user to press a key between each plot. It also uses `gcf` which returns a handle to the current figure. This allows a different form of the `print` command to be used which permits the filename to be passed as a string, namely `ti` which is changed depending on how many points are used. This code gives

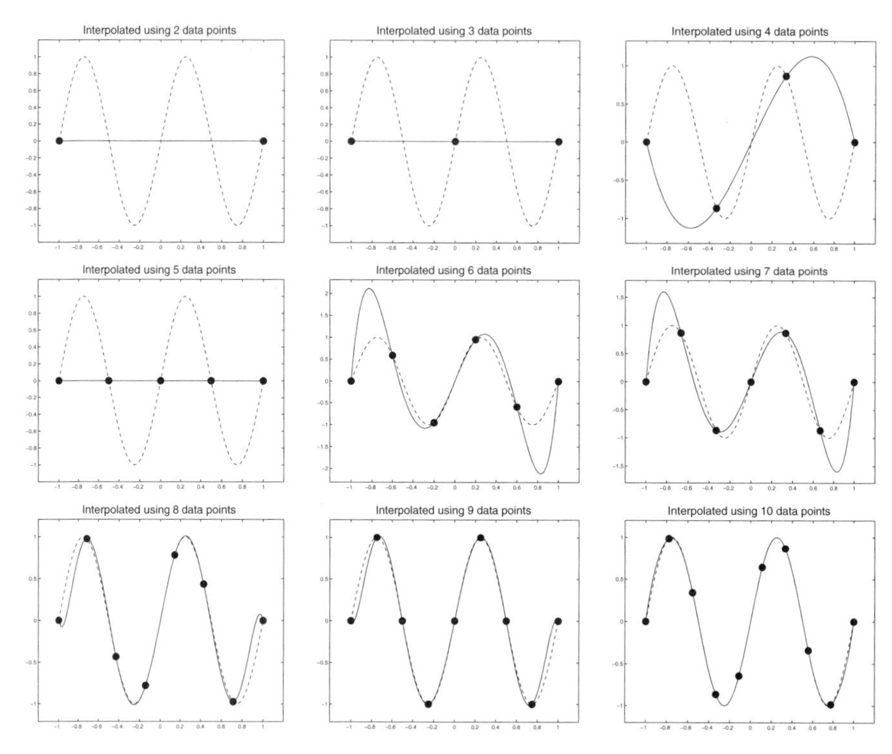

It should be noted that when we have very few grid points (denoted by the large circles) we cannot possibly hope to realise the exact form of the underlying data (shown as the dashed curve). To highlight this we have included the interpolated curve (shown as the solid line) using the highest-order polynomial possible. We note when only two points are used there is no hope of resolving any variation: similarly when three points are used the interpolation routine still thinks the data is equal to zero. Even when four points are used, the interpolation routine gives completely misleading results (these are far closer to being $-\sin \pi x$ rather than $\sin 2\pi x$). Unfortunately with five points these are coincident with the zeros of the function and again the interpolation routine believes it is looking at a set of zero data. It is only when ten points are used that the interpolation could be considered to be accurate. In fact five points per period are the minimum points required to resolve a sine wave, which in this case corresponds to nine points (due to the overlap).

It is also possible to use other functions than polynomials. We could have plotted the previous function using a sine expansion and got very accurate results, with effectively only one term, but that would require a knowledge of the form of the function. If the function was changed to $\sin(2.3\pi x)$ then trying to fit sine waves using $n \in \mathbb{Z}$ would be prone to severe errors. Some knowledge of where the underlying data was generated from and its likely form can help significantly. We can then consider the idea of parameter retrieval.

5.9 Minimisation of Functions and Parameter Retrieval

In the previous example we discussed the method of least squares: this is where the distance from the data to a function of a given form is minimised. We shall now discuss how we could minimise a function of several variables. This is a vast topic and rather than deriving the techniques from first principles we just exploit MATLAB from the outset.

We shall use the MATLAB command `fmins` (which is going to revert to `fminsearch` in future editions of MATLAB). Consider the simple function of two variables $f(x_1, x_2) = x_1^2 + x_2^2$, which has a global minimum at the origin. The minimum is such that both of the partial derivatives $\partial f/\partial x_1$ and $\partial f/\partial x_2$ are zero (together with other conditions to ensure that the point is a minimum). In this case the minimum is obvious and can be found using the code:

```
function [f] = func(x)
f = x(1)^2+x(2)^2;
```

and the command `[x] = fmins('func',[1 1])` (where the array `[1 1]` is the initial guess). This uses the default tolerances and returns a value of

```
x =

   1.0e-04 *

   -0.2102    0.2548
```

We shall now discuss how we could improve this value. The input options have values:

```
The parameters are:
    options(1)-Display parameter (Default:0). 1 displays some results.
    options(2)-Termination tolerance for X.(Default: 1e-4).
    options(3)-Termination tolerance on F.(Default: 1e-4).
    options(4)-Termination criterion on constraint violation.(Default: 1e-6)
    options(5)-Algorithm: Strategy:  Not always used.
    options(6)-Algorithm: Optimizer: Not always used.
    options(7)-Algorithm: Line Search Algorithm. (Default 0)
    options(8)-Function value. (Lambda in goal attainment. )
    options(9)-Set to 1 if you want to check user-supplied gradients
    options(10)-Number of Function and Constraint Evaluations.
    options(11)-Number of Function Gradient Evaluations.
```

```
options(12)-Number of Constraint Evaluations.
options(13)-Number of equality constraints.
options(14)-Maximum number of function evaluations.
           (Default is 100*number of variables)
options(15)-Used in goal attainment for special objectives.
options(16)-Minimum change in variables for finite difference gradients.
options(17)-Maximum change in variables for finite difference gradients.
options(18)-Step length. (Default 1 or less).
```

We shall now use the code

```
options = foptions;
options(2) = 1e-8;
options(3) = 1e-8;
[x] = fmins('func',[1 1],options);
```

which unsurprisingly gives a better value of `[0.2892e-8,-0.0997e-8]`. Here the `options(2)` and `options(3)` set the tolerances on the answers and the function respectively.

Example 5.6 *Determine the minimum value of the function*

$$f(x_1, x_2, x_3) = \sin\left(\frac{x_1}{x_2^2+1}\right) - \cos(x_3)$$

in the neighbourhood of the origin to within a tolerance of 10^{-6}.

Firstly we construct the code to evaluate the function:

```
function [f] = func2(x)
f = sin(x(1)/(x(2)^2+1))+cos(x(3));
```

and

```
options = foptions;
options(2) = 1e-6;
options(3) = 1e-6;
[x,opts] = fmins('func2',[0 0 0],options)
```

which gives the answer [-5.5897 1.5995 3.1416]. We can substitute this back into the function **func2** *to check the value associated with the minimum. This gives* -2 *so that both* cos *and* sin *take the value of* -1, *hence* $x_3 = -\pi$ *and* $x_1/(x_2^2+1) = -\pi/2$.

In the above example we have also returned details of the iteration which are stored in the array `opts`. We also note that using the option `option(1) = 1;` generates a vast amount of information concerning the iterative procedure.

This is a very powerful and useful function.

5.9.1 Parameter Retrieval

In this topic we will exploit the idea of minimising the distance between two functions. We have already met this where we can specify functions as linear combinations and then minimise the errors using the method of least squares. Let us presume we have some data which we know to be of the form $\sin ax$ on the range $x = 0$ to $x = \pi$, where a is unknown. Let us simply construct a function which conveys how close another function $\sin bx$ is to this.

```
a = 1.45; % Usually unknown
x = 0.0:pi/20:pi;
data = sin(a*x);
b = 0.0:0.02:3.0;
for j = 1:length(b)
    trial = sin(b(j)*x);
    f(j) = norm(trial-data);
end
```

The plot of `f` against `b` clearly shows a minimum near $b = 1.5$, which is unsurprising in this case.

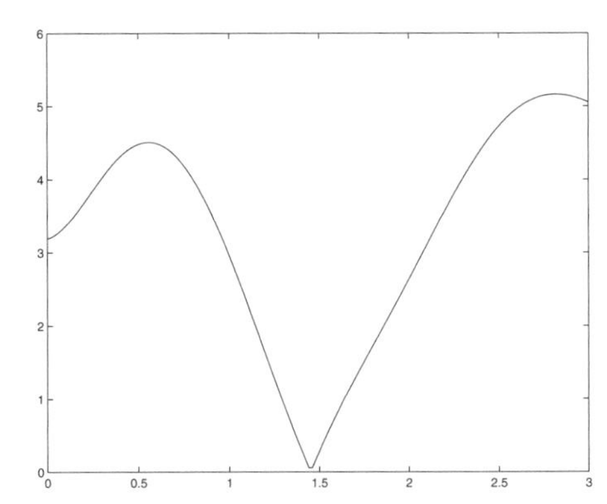

We now remark that in order to find the values of this minimum we could use search methods. But here we make the observation that one is seeking a minimum of the function, which is associated with a zero of the derivative of the function. We shall use the Newton–Raphson method modified for looking for zeros of $f'(x)$, which is

$$x_{n+1} = x_n - \frac{f'(x_n)}{f''(x_n)}.$$

Here we use

$$f'(x) \approx \frac{f(x+h) - f(x-h)}{2h} \text{ and } f''(x) \approx \frac{f(x+h) - 2f(x) + f(x-h)}{h^2}.$$

```
a = 1.45; % Usually unknown
x = 0.0:pi/20:pi;
data = sin(a*x);
b = 1.5;    % Initial guess
del = 0.1;
tol = 1e-6;
for its = 1:20
    f0 = norm(sin(b*x)-data);
    fp = norm(sin((b+del)*x)-data);
    fm = norm(sin((b-del)*x)-data);
    df = (fp-fm)/(2*del);
    if abs(df)<tol
       break
    end
    df2 = (fp-2*f0+fm)/(del^2);
    b = b - df/df2;
end
```

This program actually needs a good starting guess of b: the problem which is experienced is that the code will locate a local minimum but it cannot identify whether this is a global minimum.

5.9.2 Using `fmins` for Parameter Retrieval

We shall now discuss how we might use the aforementioned MATLAB function `fmins` to retrieve parameters, see page 161. Initially we shall consider the sample example again:

Example 5.7 *We shall try to determine the value of the constant* `a` *given that the data has been generated using* `sin(a*x)`*. In general this value will not be known.*

```
global x data
a = 1.45;
x = 0.0:pi/20:pi;
data = sin(a*x);
clear a
[a] = fmins('func',[1]);
```

used in conjunction with the program

```
function [f] = func(a)
global x data
trial = sin(a*x);
f = norm(trial-data);
```

This gives the value of `a` *very accurately (to within* 10^{-14}*).*

We shall now describe an example with more parameters:

Example 5.8 *We start with the data* $f(x) = 0.2\sin 3.2x + 0.45x$ *and using the fact that the data is generated using* $a\sin bx + c$ *we try to recover the above values.*

The main code is:

```
global x data
x = linspace(0,10);
data = 0.2*sin(3.2*x)+0.45*x;
[co] = fmins('func3',[0.1 3. 0.5]);
```

and the code `func3.m` *is*

```
function [f] = func3(co)
global x data

trial = co(1)*sin(co(2)*x)+co(3)*x;
f = norm(trial-data);
```

This recovers the starting values, since we have used a good starting guess. We note that using other starting locations will lead to other answers, which correspond to local minima. It is difficult to guarantee that we have obtained a global minimum.

5.10 Tasks

Task 5.1 *Calculate the quadratic which passes through the points* $(0, 0)$, $(2, -1)$ *and* $(5, 5)$ *(by hand).*

Task 5.2 *Generate the data* $x^2 + 3x + 2$ *on the points* $x = 0, 1, 2, \cdots, 10$ *using the code*

```
x = 0:10;
y = x.^2+3*x+2;
```

Calculate the error associated with using straight lines in the intervals at x *equals* $1/2$, $3/2$ *and* $5/2$. *This involves working out which intervals should be used to calculate the approximate values. The error is given by using the fact that the exact form of the data is actually known (i.e. as the above quadratic).*

Task 5.3 *Calculate the quadratic which passes through the points* $(0, 0)$, $(\pi/2, 1)$ *and* $(\pi, 0)$. *It would be a good idea to do this by hand and using MATLAB (using* `polyfit`*).*

Task 5.4 *Calculate the cubic which passes through the points* $(-\pi/2, -1)$, $(0, 0)$, $(\pi/2, 1)$ *and* $(\pi, 0)$.

Task 5.5 *Plot the spline which goes through the points* $(-\pi, 0)$, $(-\pi/2, -1)$, $(0, 0)$, $(\pi/2, 1)$ *and* $(\pi, 0)$ *at the points* $(-\pi, \pi)$ *in steps of* $\pi/20$. *You could use the MATLAB command* `spline` *for this purpose.*

Task 5.6 *Given that the data*

x	y
0.0	*3.16*
0.1	*3.01*
0.2	*2.73*
0.3	*2.47*
0.4	*2.13*
0.5	*1.82*
0.6	*1.52*
0.7	*1.21*
0.8	*0.76*
0.9	*0.43*
1.0	*0.03*

was generated using an expression of the form $a\sin x + b\cos x$ *determine the values of* a *and* b*.*

Task 5.7 (*) *Calculate the splines associated with the points* (x_j, y_j) $j = 1, \cdots, 5$*. Plot these splines for the points* $(-\pi, 0)$, $(-\pi/2, -1)$, $(0, 0)$, $(\pi/2, 1)$ *and* $(\pi, 0)$ *on a grid of points in steps of* $\pi/10$ *from* $-\pi$ *to* π*.*

Given that the initial data is given by $y = \sin x$*, determine the sum of the squares of the total errors over these points.*

Task 5.8 (D) *The following code should construct the function* $f(x) = x^3 + \sin x - 1$ *on the grid of integers from 2 to 11, and then works out the value of the function at* $x = 4.5$ *and* $x = 15$ *using linear interpolation and extrapolation.*

```
x = 1:11+1;
f = polyvalue([1 0 0 -1],x) + sin x;
% x = 4.5
r = 4:5;
c = polyfit(x(r),f(r),1);
yy = polyvalue(c,4.5)

% x = 15
r = 14:15;
c = polyfit(x(r),f(r),1);
yy = polyvalue(c,15);
```

6
Matrices

6.1 Introduction

As noted earlier MATLAB derives its name from the term MATrix LABoratory. It is therefore not surprising that it is in the area of matrix operations that the true power of MATLAB is brought to light. We now turn our attention to matrices within MATLAB but we provide a brief mathematical introduction to the subject of matrices in the Appendix (see page 322)[1]. We now describe how matrices are constructed and manipulated within MATLAB.

6.1.1 Initialising Matrices Within MATLAB

Before we proceed with a description of matrix operations it is important to first discuss how we set up and access matrices within MATLAB. The simplest method involves simply typing the elements:

```
>> A = [2 3 4; 5 4 3]

A =

     2     3     4
```

[1] Even if you are comfortable with the theory of matrices we still recommend you read this section as it provides the important details of the correct syntax for entering and manipulating matrices within MATLAB.

```
     5     4     3
```

Here the elements in the first row are separated by a space and the end of the first row is denoted by a semicolon. The second row is then typed, again with a space between successive elements. The whole matrix is contained within square brackets. The variable `A` is now initialised to be a two-by-three matrix; we can see this by using the command `size(A)`.

```
>> size(A)

ans =

     2     3
```

Here the command `size` returns two values; the number of rows followed by the number of columns. In the main we have just let the answer to our calculations appear on the screen (in fact they are also put into a variable called `ans`). We could have used the code:

```
>> A = [2 3 4; 5 4 3];
>> b = size(A);
>> disp(b)
>> [rows,columns] = size(A);
>> disp([rows columns])
```

The variable `b` is actually a one-by-two matrix (that is, a row vector containing two elements). The elements of this vector are `b(1)` and `b(2)` and these are used to store the number of rows and columns of `A`, respectively. Thus we can determine the number of rows of matrix **A**, in this case 2, by simply typing `b(1)` at the MATLAB prompt. We could also call the function `size(A)` and assign the values to a vector whose elements are called `rows` and `columns`. This is done in the fourth line of code. In this example we again encounter the MATLAB function `disp`, which literally displays its argument (which in this case is a row vector); we shall again return to the `disp` command a little later.

We now return to the question of initialising matrices (and vectors) using the method of just typing in the elements. We need to take some care with vectors, as MATLAB differentiates between column and row vectors (as should we). To enter a column vector we could use any of:

```
>> column_vector = [4
5
6];

>> column_vector = [4;5;6];
```

```
>> column_vector = [4 5 6]';
```

The first command is self-explanatory. In the second we enter the first row, terminated by a semicolon, and then the second row, terminated by a semicolon, etc. The third version first creates the row vector `[4 5 6]` and then takes its transpose (interchanges rows and columns) to produce the column vector. We can use any of these methods but since our ultimate aim will be to write computer code in MATLAB, rather than entering commands at the prompt, we will generally not use the first method. In general we shall, throughout the text, employ the second form.

We now turn our attention to how we access elements (or parts of a matrix). Let us start with the matrix

```
>> A = [11 12 13 14; 21 22 23 24; 31 32 33 34];
```

so that

$$\mathbf{A} = \begin{pmatrix} 11 & 12 & 13 & 14 \\ 21 & 22 & 23 & 24 \\ 31 & 32 & 33 & 34 \end{pmatrix}.$$

There is no significance to these numbers, other than that they provide an indication of their location within the matrix; we have purely used them so we will be able to check that we have got the correct elements. The simplest way of obtaining an element is using parentheses, so that `A(2,3)` returns the value in the second row, third column (in this case 23). Note also that it is necessary to use round brackets. This is exactly as we would expect, and follows the convention that $a_{i,j}$ refers to the element in the ith row and the jth column.

It is also possible to refer to a whole row (or column) of a matrix. For example

```
>> A(2,:)

ans =

    21    22    23    24
```

returns the second row of the matrix `A`. Here the colon indicates all the elements along a particular row. Alternatively to refer to a particular column we could use

```
>> A(:,4)

ans =

    14
    24
```

```
    34
```

to return the fourth column of matrix `A`. Note that in both cases the answers are the same "shape" as they would appear in the matrix, that is the second row has been returned as a row vector and the fourth column has been returned as a column vector.

With the colon operator we can create row vectors which can then be used as arguments in other commands. For instance

```
>> r = 1:3;
>> A(r,1)

ans =

    11
    21
    31
```

This yields the same result as `A(1:3,1)` or `A(:,1)`. We could also use `r = 1:2` to get the first two elements of the first column, or `r = 1:2:3` to get the first and last (noting that `r = 1:2:3` gives the row vector `r = [1 3]`). The possibilities are extensive; for example if we want to obtain the top left hand two-by-two corner of `A`, we could employ the following:

```
>> r = 1:2;
>> B = A(r,r)

B =

    11    12
    21    22
```

We now return to the problem of initialising matrices. Although it is not necessary in MATLAB, it is good programming practise to set up an empty matrix before accessing the elements. In many programming languages this is crucial and is referred to as dimensionalising variables and the syntax for how this is done varies considerably between languages. There are many special commands in MATLAB which aid in the initialisation of matrices and we start looking at these by using the colon (which refers to the entire row or column, depending on context) in some examples.

Example 6.1 *Enter the matrix in MATLAB*

$$\begin{pmatrix} 1 & 2 & 3 & 4 \\ 2 & 0 & 0 & 0 \\ 3 & 0 & 0 & 0 \\ 4 & 0 & 0 & 0 \end{pmatrix}$$

In order to achieve this we use the commands

```
>> A = zeros(4);
>> r = 1:4;
>> A(:,1) = r';          % First column
>> A(1,:) = r            % First row

A =

     1     2     3     4
     2     0     0     0
     3     0     0     0
     4     0     0     0
```

The first command here initialises the matrix `A` *to be a four-by-four matrix whose elements are all equal to zero. The matrix we want to enter has the first row and column having the same elements* $1, 2, 3, 4$. *We therefore set up a vector, in this case* r, *to contain these values. We now want to assign the first column of the matrix* `A` *which we can do with the third command. Notice here that we are now assigning to the matrix* `A` *to have the first column equal to the transpose of the vector* r; *we have used the transpose because we want a column vector. The final command assigns the first row of* `A` *to be equal to the vector* r *and to print out the resulting matrix* `A` *which, as we see, is precisely the matrix we required. We could have entered the matrix in many different ways. One way would be to simply type the matrix in element by element:*

```
>> A = [1 2 3 4;
   2 0 0 0;
   3 0 0 0;
   4 0 0 0];
```

This is fine for small matrices but not practical for when we need to set up larger matrices. Another variant which can be used to set up the matrix is

```
>> A = [1 2 3 4; 2 zeros(1,3); 3 zeros(1,3); 4 zeros(1,3)];
```

Here the command `zeros(1,3)` *sets up a row vector (one-by-three) full of zeros. This example serves to emphasise there is no unique way to go about setting up a particular matrix; some ways are more elegant (and sometimes less readable) than others.*

6.1.2 Matrix Operations

We shall now consider how matrix operations are performed within MATLAB, referring to the mathematical definitions given in the Appendix. We shall start with addition and subtraction, using simple examples:

Example 6.2 *Consider the addition* ($\mathbf{C} = \mathbf{A} + \mathbf{B}$) *and subtraction* ($\mathbf{D} = \mathbf{A} - \mathbf{B}$) *of the matrices*

$$\mathbf{A} = \begin{pmatrix} 1 & 2 \\ 3 & 4 \end{pmatrix} \text{ and } \mathbf{B} = \begin{pmatrix} 4 & 3 \\ 2 & 1 \end{pmatrix}.$$

We shall start by working through these by hand and then proceed to give the MATLAB code. Firstly the addition

$$\begin{aligned} \mathbf{C} &= \begin{pmatrix} 1 & 2 \\ 3 & 4 \end{pmatrix} + \begin{pmatrix} 4 & 3 \\ 2 & 1 \end{pmatrix}, \\ &= \begin{pmatrix} 1+4 & 2+3 \\ 3+2 & 4+1 \end{pmatrix} = \begin{pmatrix} 5 & 5 \\ 5 & 5 \end{pmatrix}; \end{aligned}$$

and now the subtraction

$$\begin{aligned} \mathbf{D} &= \begin{pmatrix} 1 & 2 \\ 3 & 4 \end{pmatrix} - \begin{pmatrix} 4 & 3 \\ 2 & 1 \end{pmatrix}, \\ &= \begin{pmatrix} 1-4 & 2-3 \\ 3-2 & 4-1 \end{pmatrix} = \begin{pmatrix} -3 & -1 \\ 1 & 3 \end{pmatrix}. \end{aligned}$$

The MATLAB code to achieve these operations is

```
>>A = [1 2; 3 4]; B = [4 3; 2 1];
>>C = A+B

C =

     5     5
     5     5
>>D = A-B

D =

    -3    -1
     1     3
```

We note that matrices need to be the same size to perform addition or subtraction since there needs to be a corresponding element in each matrix.

Example 6.3 *Consider the expression* $\mathbf{A} - 3\mathbf{B}^T$ *where*

$$\mathbf{A} = \begin{pmatrix} 3 & 8 & -1 \\ 5 & 2 & 0 \end{pmatrix} \text{ and } \mathbf{B} = \begin{pmatrix} -3 & 2 \\ 2 & 2 \\ -1 & 3 \end{pmatrix}.$$

The MATLAB code is:

```
A = [3 8 -1; 5  2 0];
B = [-3 2; 2 2; -1 3];
C = A-3*transpose(B);
```

Note that although the matrix $\mathbf{B}$ *is not the same size as* $\mathbf{A}$*, its transpose* $\mathbf{B}^T$ *is.*

Before we proceed we add this word of caution within the context of MATLAB. As we have already seen MATLAB will adapt its variable structure to accommodate solutions. However, it will object to calculations that are impossible (that is, not well defined). For instance if we try to add a three-by-three matrix to a two-by-two one (both full of 1's)[2]

```
>>A = ones(3); B = ones(2);
>>C = A+B


??? Error using ==> +
Matrix dimensions must agree.
```

This example demonstrates that MATLAB is not able to add a two-by-two matrix to a three-by-three one and it returns a sensible error message if we try to do so; in this case the dimension (or size) of the matrices are not the same.

The only operation of this kind which is possible within MATLAB involves

```
>>A = ones(1); B = ones(2);
>>C = A+B
```

[2] We will use the command `ones(n)` which sets up an n-by-n matrix full of ones. The other form of this command takes two arguments, `ones(m,n)`, which unsurprisingly gives a m-by-n matrix with all elements equal to one. This is similar to the command `zeros(n)` that we saw earlier. An associated command that generates an n-by-n matrix with entries made up of random numbers between zero and one is `rand(n)`. The precise structure of these commands can be found by using the `help` command.

```
C =

     2     2
     2     2
```

In this case although the variable `A` is a one-by-one matrix it is treated as a scalar. In this case the command is interpreted by MATLAB as

$$c_{i,j} = \lambda + b_{i,j}, \qquad i = 1, \cdots, m \quad \text{and } j = 1, \cdots, n;$$

where $\mathbf{C} = \lambda \mathbf{1} + \mathbf{B}$, with $\mathbf{1}$ a matrix full of ones (which can be obtained in MATLAB using `ones`). This operation is mathematically correct and it is also a well-defined operation within MATLAB. Similarly if a matrix is multiplied by a scalar then each element is multiplied by that scalar (that is $c_{i,j} = \lambda a_{i,j}$ where $\mathbf{C} = \lambda \mathbf{A}$ and again the calculation is viable in mathematical terms).

Before proceeding to matrix multiplication within MATLAB we note that this idea of applying and performing an operation on every element extends to functions. We can demonstrate this with a simple example:

```
>> A = [pi/4 pi/2; pi pi/3]

A =

    0.7854    1.5708
    3.1416    1.0472

>> B = sin(A)

B =

    0.7071    1.0000
    0.0000    0.8660
```

Here we have set up the matrix[3]

$$\mathbf{A} = \begin{pmatrix} \frac{\pi}{4} & \frac{\pi}{2} \\ \pi & \frac{\pi}{3} \end{pmatrix}$$

and it returns the matrix $\mathbf{B}$ as

$$\mathbf{B} = \begin{pmatrix} \sin\frac{\pi}{4} & \sin\frac{\pi}{2} \\ \sin\pi & \sin\frac{\pi}{3} \end{pmatrix} = \begin{pmatrix} \frac{1}{\sqrt{2}} & 1 \\ 0 & \frac{\sqrt{3}}{2} \end{pmatrix}.$$

[3] Notice that the variable `pi` is predefined in MATLAB and returns the value of $\pi = 3.1415926535897\cdots$.

We shall return to this method in due course, but we now turn our attention to matrix multiplication within MATLAB. The operation `C = A*B` is performed as one might expect and returns the "normal" result of matrix multiplication so that if `A` is n-by-m and `B` is m-by-p then `C` is n-by-p whose elements are defined on page 326. We need to remember that when taking the product of two matrices the number of columns of the left-hand matrix must match the number of rows in the right-hand matrix. If this is not so then MATLAB will give an error message: `Inner matrix dimensions must agree.` This is somewhere we need to be very careful and although most of the time our errors will be flagged by MATLAB, sometimes the calculations will proceed even though the results are not as we would expect: for instance

```
>> A = [1 2 3];
>> B = [4; 5; 6];
>> A*B

ans =

    32

>> B*A

ans =

     4     8    12
     5    10    15
     6    12    18
```

In each case MATLAB is prepared to make the calculation (as it should): however the exchange of the two matrices yields very different results. Because MATLAB is willing to proceed with calculations with scalars or matrices as arguments, it may well be some time before this error is realised.

Example 6.4 *Calculate* $3\mathbf{A} - \mathbf{B}$ *and* $\mathbf{AB}$ *where*

$$\mathbf{A} = \begin{pmatrix} 1 & 5 & 6 \\ 0 & 2 & 3 \\ -1 & 0 & 0 \end{pmatrix} \quad \textit{and} \quad \mathbf{B} = \begin{pmatrix} 0 & -3 & 2 \\ 1 & 0 & -2 \\ 1 & 0 & -4 \end{pmatrix}.$$

The code for this should be relatively self explanatory:

```
>> A = [1 5 6; 0 2 3; -1 0 0];
>> B = [0 -3 2; 1 0 -2; 1 0 -4];
```

```
>> disp(3*A-B)
>> disp(A*B)
```

Note that it is not necessary to use the command `disp` here.

Dot arithmetic can be readily extended to work with matrices. With matrices `A` and `B` the matrix `C = A.*B` is given by

$$c_{i,j} = a_{i,j} b_{i,j}, \qquad i = 1, \cdots, m \text{ and } j = 1, \cdots, n$$

and similarly those of `D = A./B` by

$$d_{i,j} = \frac{a_{i,j}}{b_{i,j}}, \qquad i = 1, \cdots, m \text{ and } j = 1, \cdots, n.$$

In addition to multiplication `.*` and division `./` we can also use dot arithmetic for exponentiation using `.^` as in

```
>> A = [1 2; 3 4];
>> B = [1 2; 3 4];
>> A.^B

ans =

     1     4
    27   256
```

Of course in using dot arithmetic in MATLAB we must ensure the matrices are of the same size for the operation to be defined. We note either of the arguments can be scalars:

```
>> A = [1 2; 3 4];
>> B = A.^2;          % square all entries
>> C = 2.^A;          % [2 4; 8 16]
```

We note these are effectively new binary operations and the addition of the dot should be thought of as changing the variables. It might help to think of them as separate operations: that is `C = A*B` performs the mathematical multiplication $\mathbf{A} \times \mathbf{B}$ whereas `C = A.*B` gives $c_{i,j} = a_{i,j} b_{i,j}$. With practice you will come to appreciate the difference between these two operations.

It is not necessary to have a `.+` or a `.-` command since using the definitions above these give the same results, since the elements of $\mathbf{C} = \mathbf{A} + \mathbf{B}$ are exactly $c_{i,j} = a_{i,j} + b_{i,j}$.

Example 6.5 *Given the matrices*

$$\mathbf{A} = \mathbf{I} \text{ and } \mathbf{B} = \begin{pmatrix} 1 & 2 \\ 3 & 4 \end{pmatrix},$$

compare the results of the MATLAB calculations `A*B`, `A.*B`, `A/B` *and* `A./B`.

For the multiplications we have the code:

```
>>A = eye(2); B = [1 2; 3 4];
>>C = A*B

C =

     1     2
     3     4

>> C = A.*B

C =

     1     0
     0     4
```

Here we have introduced the command `eye(n)` *which gives an n-by-n identity matrix*[4]*. For the division operations we have*

```
>> C = A/B

C =

   -2.0000    1.0000
    1.5000   -0.5000

>> C = A./B

C =

    1.0000         0
         0    0.2500
```

[4] This could also have been obtained using the command `diag`.

Hopefully this example makes it clear that the straightforward asterisk operation performs the mathematical operation of multiplication, with multiplication by the identity matrix **A** *leaving* **B** *unchanged, whereas multiplying using* `.*` *performs the operation element-wise, so that*

$$\mathbf{C} = \mathbf{A}.*\mathbf{B} = \begin{pmatrix} 1\times 1 & 0\times 2 \\ 0\times 3 & 1\times 4 \end{pmatrix} = \begin{pmatrix} 1 & 0 \\ 0 & 4 \end{pmatrix}.$$

We have previously skirted the issue of division of matrices, but now that we have introduced the element-wise operation this can be dealt with naturally and it works in the same way as multiplication. Above we have the results of the calculation

$$\texttt{A/B} = \begin{pmatrix} 1 & 0 \\ 0 & 1 \end{pmatrix} \text{'/'} \begin{pmatrix} 1 & 2 \\ 3 & 4 \end{pmatrix} = \begin{pmatrix} -2 & 1 \\ \frac{3}{2} & -\frac{1}{2} \end{pmatrix}$$

which is $\mathbf{AB}^{-1}$ *and the other calculation is*

$$\texttt{A./B} = \begin{pmatrix} 1 & 0 \\ 0 & 1 \end{pmatrix} \text{'./'} \begin{pmatrix} 1 & 2 \\ 3 & 4 \end{pmatrix} = \begin{pmatrix} \frac{1}{1} & \frac{0}{2} \\ \frac{0}{3} & \frac{1}{4} \end{pmatrix} = \begin{pmatrix} 1 & 0 \\ 0 & \frac{1}{4} \end{pmatrix}.$$

We note when using the division operator `/` *without the dot we are, in fact, constructing the inverse of the matrix* **B**.

In order to calculate the transpose of a matrix we can use the command `transpose` or we can use an apostrophe, `A'`. If the elements of the matrix are complex numbers, then this operation constructs the complex conjugate of the transpose of `A`. If the elements of the matrix are real this is then equal to the transpose of the matrix. To obtain the transpose without conjugation, in the case of a complex matrix, we use the construction `A.'`.

6.1.3 Operations on Elements of Matrices

In the previous section we have talked about dot arithmetic but we have also met other examples such as where we added a scalar quantity to each element of a matrix and where we took the sine of each element. We are now in a position to apply the functions we have discussed to all of the elements of a matrix (or pairs of elements for binary operations). We can use any unary operation on a matrix to return a matrix of the same size. We shall consider a few examples to illustrate this.

Example 6.6 *Investigate the behaviour of the unary operations* ***abs****,* ***sign****,* ***cos*** *,* ***sin****,* ***exp*** *and* ***sinh*** *on the matrix*

$$\mathbf{A} = \begin{pmatrix} 1 & -.5 & -3.2 \\ 0 & 1 & -\pi \\ \frac{\pi}{2} & -2 & 7.25 \end{pmatrix}.$$

This is done with the MATLAB code:

```
>> A = [1 -0.5 -3.2;
   0 1 -pi;
   pi/2 -2 7.25];
>> abs(A)
>> sign(A)
>> cos(A)
>> sin(A)
>> exp(A)
>> sinh(A)
```

We have not included the answers which MATLAB displays, but you should try entering these commands in order to understand how MATLAB interprets a function of a matrix.

We need to take some care when using the dot command as some commands may be allowable within MATLAB but may produce results which are not what we expected, or wanted. The following example provides a good illustration of this point.

Example 6.7 *Consider the matrix*

$$\mathbf{A} = \begin{pmatrix} \frac{\pi}{4} & \frac{\pi}{2} \\ \frac{\pi}{3} & \frac{\pi}{6} \end{pmatrix}.$$

We wish to determine the matrix whose elements take the values $\sin x/x$ *where* x *corresponds to the elements of the matrix* $\mathbf{A}$*. In other words we want to calculate the matrix*

$$\begin{pmatrix} \frac{4}{\pi}\sin\frac{\pi}{4} & \frac{2}{\pi}\sin\frac{\pi}{2} \\ \frac{3}{\pi}\sin\frac{\pi}{3} & \frac{6}{\pi}\sin\frac{\pi}{6} \end{pmatrix}.$$

This can be accomplished using the commands

```
>> A = [pi/4 pi/2; pi/3 pi/6];
>> sin(A)./A

ans =
```

```
    0.9003    0.6366
    0.8270    0.9549
```

However, if we had missed out the dot and simply typed `sin(A)/A` *we would still obtain an "answer":*

```
>> sin(A)/A

ans =

    0.5487    0.2637
    0.0569    0.7843
```

which is not what we had in mind. What this last calculation has done is to construct the matrix $\mathbf{BA}^{-1}$ *where the elements of* $\mathbf{B}$ *are sines of those of* $\mathbf{A}$.

6.1.4 More on Special Matrices

In the main, to this point in the text, we have constructed matrices by simply typing in their entries. However, there are commands in MATLAB which allow us to make use of *special matrices*. The simplest ones, and those that are most often used, are

`eye(n)` The identity matrix of size n-by-n, $\mathbf{I}_n$. This matrix has the property that for any square matrix $\mathbf{A}$ of size n-by-n $\mathbf{AI}_n = \mathbf{I}_n\mathbf{A} = \mathbf{A}$.

`ones(n,m)` A matrix filled with ones of size n-by-m. Note that `ones(n)` gives a square n-by-n matrix filled with ones and that `ones(n,1)` gives a column vector of ones and `ones(1,n)` a row vector.

`zeros(n,m)` A matrix filled with zeros of size n-by-m. This command operates in the same manner to `ones` when used as `zeros(n)`, `zeros(n,1)` and `zeros(1,n)`.

A matrix whose definition is slightly more involved is that given by the command `diag(x,n)`. This produces an appropriately sized matrix whose n^{th} diagonal is set to be the vector `x`. The commands `diag(x,0)` and `diag(x)` produce identical results, that is the vector `x` on the leading diagonal (that is the one running from the top left to the bottom right; the elements of the form $a_{i,i}$). The value of `n` being positive means that we are considering the nth super-diagonal and similarly negative values of `n` gives the nth sub-diagonal. It should be noted that the nth diagonal is shorter by $|n|$ than the leading diagonal, for instance if `x = ones(3,1)` then `diag(x)` produces a three-by-three matrix whereas `diag(x,1)` produces a four-by-four matrix as does `diag(x,-1)`.

Considering an N-by-N matrix: The nth super-diagonal consists of the elements $a_{i,i+n}$ where $i = 1, \cdots, N-n$ and the nth sub-diagonal consists of the elements $a_{i+n,i}$ where $i = 1, \cdots, N-n$. Notice it does not matter whether the vector is a row or column vector.

Example 6.8 *If* `x = ones(4,1)` *then we have:*

`diag(x,0)` *is given by*

$$\begin{pmatrix} 1 & 0 & 0 & 0 \\ 0 & 1 & 0 & 0 \\ 0 & 0 & 1 & 0 \\ 0 & 0 & 0 & 1 \end{pmatrix}$$

`diag(x(1:3),1)`

$$\begin{pmatrix} 0 & 1 & 0 & 0 \\ 0 & 0 & 1 & 0 \\ 0 & 0 & 0 & 1 \\ 0 & 0 & 0 & 0 \end{pmatrix}$$

and similarly

`diag(x(1:3),-1)`

$$\begin{pmatrix} 0 & 0 & 0 & 0 \\ 1 & 0 & 0 & 0 \\ 0 & 1 & 0 & 0 \\ 0 & 0 & 1 & 0 \end{pmatrix}.$$

We can use this command to produce combinations such as

`diag(x,0)+diag(x(1:3)*2,1)+diag(x(1:3)*(-2),-1)`

$$\begin{pmatrix} 1 & 2 & 0 & 0 \\ -2 & 1 & 2 & 0 \\ 0 & -2 & 1 & 2 \\ 0 & 0 & -2 & 1 \end{pmatrix}.$$

Notice we have been very careful to only try viable combinations. In the above we have used the fact that we can extract parts of a matrix or vector. The command `ones(4,1)` *sets up a column vector (four-by-one) full of ones and then* `x(1:3)` *gives the first three elements of that vector, that is* `[1 1 1]`.

Example 6.9 *Consider the matrices constructed using the commands:*

```
>> x = 2*ones(3,1);
>> y = -2*ones(4,1);

>> A = diag(y,0)+diag(x,1);
>> B = diag(y,1)+diag(x,-2);
```

The first command sets x to be a column vector (three-by-one) full of 2's and the second one sets y to be a column vector (four-by-one) full of −2's. It is worth typing these commands (and leaving the semicolon off to see what is happening). It is possible to see the contents of a variable by merely typing its name at the prompt and pressing return. Let's now dissect the command `A = diag(y,0)+diag(x,1);`*. The first component of the right-hand side puts the contents of the column vector y down the leading diagonal of a matrix (which will be a a four-by-four matrix since the vector y is four-by-one). The second command on the right-hand side constructs a matrix which has the components of the column vector x down the super-diagonal; notice the vector x is three-by-one which is one shorter than the leading diagonal. Finally, these two matrices are added together (to give A) which is a valid operation as both matrices are four-by-four. If you are having difficulty following this try typing* `diag(y,0)` *and* `diag(x,1)` *separately, and look at the resulting matrices.*

In the second command the matrix **B** *is constructed with a super-diagonal and a second sub-diagonal. These commands give*

```
>> A

A =

    -2     2     0     0
     0    -2     2     0
     0     0    -2     2
     0     0     0    -2

>> B

B =

     0    -2     0     0     0
     0     0    -2     0     0
     2     0     0    -2     0
     0     2     0     0    -2
     0     0     2     0     0
```

Notice the sizes of the matrices have been determined by the length of the diagonals. If we had attempted to use the command `diag(y,0)+diag(x,0)` *we would have been asking MATLAB to add two matrices of different sizes, and as such it would have produced an error message.*

There are several other special matrices which can be easily constructed using MATLAB. The command `magic(n)` gives a magic square of size n-by-n. This matrix has the property that the sum of all the elements along any row or column is identical (and along diagonals).

The commands `hilb(n)` and `invhilb(n)` produce what are known as a Hilbert matrix and its inverse. The elements of the Hilbert matrix `A` are given by $a_{i,j} = 1/(i+j)$; the inverse of these matrices are notoriously hard to calculate.

Other examples of special matrices exist within MATLAB; a full list of matrix functions can be found by using `lookfor matrix`. This lists all the functions which include the word `matrix` in their description.

6.1.5 Matrices Containing Strings

We briefly note we can form vectors containing strings. This is useful when we wish to construct arguments to produce prompts for the user. Let us consider a simple example

Example 6.10 *This code gets the user to input a value of* `a` *and simply print it out again.*

```
str = 'Please enter the value of a: ';
a = input(str);
sol = 'The value of a is equal to ';
soln = [sol num2str(a)];
disp(soln)
```

Here we have formed a row vector using the two strings `'The value of a is equal to '` *and* `num2str(a)`*; the latter of which is the conversion of the value of* `a` *to a string.*

We note the size of the row vector `sol` *is one-by-twenty seven and as such we could access parts of the vector* `sol(1,1:3)` *is the word* `The`*.*

There are numerous commands which can be used to operate on strings:

```
STR2MAT Form blank padded character matrix from strings.
STRCAT Concatenate strings.
STRCMP Compare strings.
STRCMPI Compare strings ignoring case.
STRINGS Character strings in MATLAB.
STRNCMP Compare first N characters of strings.
STRNCMPI Compare first N characters of strings ignoring case.
STRVCAT Vertically concatenate strings.
```

(this was generated using the command `lookfor strings`).

In this chapter we shall tackle the problems involved in the consideration of matrices and provide details of how they might be inverted. We shall mention a few concepts but shall not deal with them in depth, electing to leave those to subsequent books on linear algebra.

We will also discuss the concept of eigensolutions and also show how these occur within differential systems.

6.2 Properties of Matrices and Systems of Equations

We shall discuss the way certain characteristics of systems can be used to determine whether they have solutions or not. We start by considering a couple of simple systems:

$$x = 1 \tag{6.1a}$$

this system is in one variable (that is x) and has one solution, that is x equal to 1. The number of variables equals the number of unknowns. Now we consider the system

$$\left.\begin{array}{l} x = 1 \\ y = 2 \end{array}\right\} \tag{6.1b}$$

again this system has the same number of variables as equations, and hopefully we can work out the solution as x equals 1 and y equals 2. In both of these cases the solution exists and it is unique. These systems are ridiculously simple. Let us again consider a system with the same number of variables as equations

$$\left.\begin{array}{l} x - y = 1 \\ x + y = 2 \end{array}\right\} \tag{6.1c}$$

This has a unique solution $x = 3/2$ and $y = 1/2$: however the system

$$\left.\begin{array}{l} x + y = 1 \\ x + y = 2 \end{array}\right\} \tag{6.1d}$$

clearly does not have a solution, since $x + y$ cannot equal both one and two. Further considering the system

$$\left.\begin{array}{l} x + y = 1 \\ x + y = 1 \end{array}\right\} \qquad (6.1e)$$

has solutions: in fact $x = \alpha$ and $y = 1 - \alpha$ is a solution for any value of α. In this case we say there are an infinite number of solutions. The question we now pose is what is the difference between these sets of equations. We start with a couple of definitions:

Definition 6.1 (Linear Independence)

Consider the set of vectors $\mathbf{a}_j$, $j = 1, \cdots, n$ and the linear combination

$$\sum_{j=1}^{n} \mathbf{a}_j c_j.$$

This summation is zero if all the c_j's are zero, and the set of vectors $\mathbf{a}_j$ are said to be **linearly independent** if this is the only set of constants for which this is true.

Example 6.11 *Consider the vectors*

$$\begin{array}{rcl} \mathbf{a}_1 & = & (1 \quad 0 \quad 1 \quad 0), \\ \mathbf{a}_2 & = & (1 \quad 0 \quad -1 \quad 0), \\ \mathbf{a}_3 & = & (1 \quad 0 \quad 0 \quad 0), \\ \mathbf{a}_4 & = & (0 \quad 1 \quad 0 \quad 1). \end{array}$$

We note that $\mathbf{a}_1 + \mathbf{a}_2 - 2\mathbf{a}_3 = \mathbf{0}$, *that is* $c_1 = 1$, $c_2 = 1$, $c_3 = -2$ *and* $c_4 = 0$; *hence there are values of the* c_j*'s for which the linear combination of the* $\mathbf{a}_j$ *are zero, and hence these vectors are linearly dependent.*

We can now introduce a further term:

Definition 6.2 (Rank of a Matrix)

The **rank** of a matrix is the maximum number of linearly independent rows of a matrix.

This can be calculated using the MATLAB command `rank`. We can determine the rank of the system above using the commands

```
a = [1 0 1 0; ...
   1 0 -1 0; ...
   1 0 0 0; ...
   0 1 0 1];
rank(a)
```

This gives the answer 3, so this tells us that one of the rows depends on the others (since there are four rows): as noted above $\mathbf{a}_1 = 2\mathbf{a}_3 - \mathbf{a}_2$.

We now consider a system of linear equations:

$$\begin{aligned} a_{11}x_1 + a_{12}x_2 + \cdots + a_{1n}x_n &= b_1, \\ a_{21}x_1 + a_{22}x_2 + \cdots + a_{2n}x_n &= b_2, \\ \vdots \quad &\vdots \;\; \vdots \\ a_{m1}x_1 + a_{m2}x_2 + \cdots a_{mn}x_n &= b_m, \end{aligned}$$

which we can write as $\mathbf{Ax} = \mathbf{b}$, where $\mathbf{A}$ is an m-by-n matrix and the vectors $\mathbf{x}$ and $\mathbf{b}$ are column vectors of lengths n and m respectively. In simple terms this represents a system of m equations in n unknowns. We also construct the matrix $\mathbf{A}|\mathbf{b}$ which is the augmented matrix.
We now state the results that:

1. The system above has solutions if and only if $\text{rank}(\mathbf{A}) = \text{rank}(\mathbf{A}|\mathbf{b})$.
2. If this rank is equal to the number of variables then the system has a unique solution.
3. If the rank is less than the number of variables then there are an infinite number of solutions. In fact the difference between these numbers gives the number of degrees of freedom (which we can think of as the number of variables like α which are available to us).

We now use MATLAB and this idea to discuss the structure of the systems (6.1a–6.1e)

```
A1 = [1]; b1 = 1;
A2 = [1 0; 0 1]; b2 = [1; 2];
A3 = [1 -1; 1 1]; b3 = [1; 2];
A4 = [1 1; 1 1]; b4 = [1; 2];
A5 = [1 1; 1 1]; b5 = [1; 1];
```

and then use the function `solns.m`

```
function [val] = solns(A,b)
rankA = rank(A); rankAb = rank([A b]);
[m,n] = size(A);
disp(['There are ' int2str(m) ' equations'])
disp(['with ' int2str(n) ' variables'])
if rankA ~= rankAb
   disp('This system has no solutions')
elseif rankA == n
   disp('There is a unique solution')
elseif rankA < n
   dof = n-rankA;
   disp('There are an infinite number of solutions')
   disp(['with ' int2str(dof) ' degrees of freedom '])
end
```

This produces the results:

```
>> solns(A1,b1)
There are 1 equations
with 1 variables
There is a unique solution
>> solns(A2,b2)
There are 2 equations
with 2 variables
There is a unique solution
>> solns(A3,b3)
There are 2 equations
with 2 variables
There is a unique solution
>> solns(A4,b4)
There are 2 equations
with 2 variables
This system has no solutions
>> solns(A5,b5)
There are 2 equations
with 2 variables
There are an infinite number of solutions
with 1 degrees of freedom
```

This can be used for far more complex systems and is not limited to cases for which **b** is a single column, consider the example on page 199.

```
A = [2 3; 1 -1];
```

```
b = [7 -2;1 8];
solns(A,b)
```

This informs us that there is a unique solution.

6.2.1 Determinants of Matrices

We now mention the determinant of a matrix. We shall not define this quantity mathematically leaving that to a text on linear algebra. However we note the property that a square matrix $\mathbf{A}$ has an inverse if and only if $\det(\mathbf{A}) \neq 0$. In general this will be more useful to us than the previous section on rank, since we will not be able to deal with problems with an infinite number of solutions. We shall generally restrict our attention to systems which have a unique solution or identifying systems which have no solutions. This is particularly poignant when discussing eigensolutions, which will be done in due course.

We start with a general two-by-two matrix:

$$\mathbf{A} = \begin{pmatrix} a & b \\ c & d \end{pmatrix},$$

whose determinant is $ad - bc$. Notice if this quantity is zero then $a/b = c/d$ (presuming that $b, d \neq 0$). This means that the second row is merely a multiple of the first as we have in (6.1d) and (6.1e).

We construct the determinant of a three-by-three matrix by reducing it to three two-by-two determinants. We consider the matrix $\mathbf{A}$:

$$\mathbf{A} = \begin{pmatrix} a & b & c \\ d & e & f \\ g & h & i \end{pmatrix}.$$

We have

$$\det(\mathbf{A}) = +a \begin{vmatrix} e & f \\ h & i \end{vmatrix} - b \begin{vmatrix} d & f \\ g & i \end{vmatrix} + c \begin{vmatrix} d & e \\ g & h \end{vmatrix}.$$

We could now proceed to expand the two-by-two determinants. Here we have used the top row to expand the determinant, but we could have used any row or column, provided we premultiply by either plus or minus one, using

$$\begin{pmatrix} + & - & + \\ - & + & - \\ + & - & + \end{pmatrix}.$$

The sub-determinants are formed by removing the row and column containing the element we are using at the time.

We pick the row or column which is most convenient: for instance if one contains a couple of zeros this is an ideal choice.

Example 6.12 *Investigate the system*

$$\begin{pmatrix} 1 & 0 & 2 \\ 1 & -1 & 1 \\ r & 2 & 1 \end{pmatrix} \mathbf{x} = \mathbf{b}.$$

The determinant of this matrix is actually $1 + r$ *and in order for this system to have solutions we require* $r \neq -1$ *(we note that if* $r = -1$ *then the third row is equal to the first row minus two copies of the second row, so the rows are linearly dependent). Consequently we can invert the matrix and hence find that* $\mathbf{x} = \mathbf{A}^{-1}\mathbf{b}$. *We can work out the determinant by expanding using the top row as*

$$\begin{vmatrix} 1 & 0 & 2 \\ 1 & -1 & 1 \\ r & 2 & 1 \end{vmatrix} = 1 \begin{vmatrix} -1 & 1 \\ 2 & 1 \end{vmatrix} - 0 \begin{vmatrix} 1 & 1 \\ r & 1 \end{vmatrix} + 2 \begin{vmatrix} 1 & -1 \\ r & 2 \end{vmatrix}$$
$$= 1(-1-2) + 2(2+r) = 1 + r.$$

Note we could solve this example by evaluating the determinant for a variety of values of r and then using the command `polyfit` to the acquired data.

6.3 Elementary Row Operations

We shall briefly comment on how one might invert a matrix: we shall illustrate the methods and techniques using small matrices but the same methods will work for larger matrices.

Let us consider the general two-by-two matrix

$$\mathbf{A} = \begin{pmatrix} a & b \\ c & d \end{pmatrix}.$$

You may know that the inverse of this matrix is

$$\mathbf{A}^{-1} = \frac{1}{\Delta} \begin{pmatrix} d & -b \\ -c & a \end{pmatrix},$$

where $\Delta = ad - bc$ and this is the determinant of the matrix. You may just know this or remember it as:[5]

Exchange the two elements on the leading diagonal and multiply the other elements by minus one, and divide the whole matrix by the determinant.

[5] If you have any difficulty remembering which ones are exchanged and which ones are multiplied by minus one, simply remember it should leave the identity matrix unchanged.

The fact that we are dividing by the determinant means that this has to be non-zero for the inverse to exist. Let us consider the equations

$$ax + by = p, \tag{6.2a}$$
$$cx + dy = q. \tag{6.2b}$$

Hopefully it should be clear to you this is equivalent to the matrix equation

$$\begin{pmatrix} a & b \\ c & d \end{pmatrix} \begin{pmatrix} x \\ y \end{pmatrix} = \begin{pmatrix} p \\ q \end{pmatrix}.$$

The solution of which is

$$\begin{pmatrix} x \\ y \end{pmatrix} = \frac{1}{\Delta} \begin{pmatrix} d & -b \\ -c & a \end{pmatrix} \begin{pmatrix} p \\ q \end{pmatrix},$$

thus

$$x = \frac{dp - bq}{ad - bc} \quad \text{and} \quad y = \frac{-cp + aq}{ad - bc}.$$

We could also get this by solving the equations (6.2). Firstly we wish to eliminate x from the second equation (6.2b) so we subtract c/a times the first one (6.2a). Thus (6.2b) becomes

$$\begin{aligned} cx + dy - \frac{c}{a}(ax + by) &= q - \frac{c}{a}p, \\ \left(d - \frac{c}{a}b\right) y &= q - \frac{c}{a}p, \\ \hat{d}y &= \hat{q}, \end{aligned}$$

where for convenience we have introduced $\hat{d}$ and $\hat{q}$. The two equations are now

$$ax + by = p \tag{6.3a}$$
$$\hat{d}y = \hat{q}. \tag{6.3b}$$

We can now eliminate y from (6.3a) using (6.3b) (by subtracting $b/\hat{d}$ times (6.3b) from (6.3a)). This gives

$$\begin{aligned} ax + by - \frac{b}{\hat{d}}\left(\hat{d}y\right) &= p - \frac{b}{\hat{d}}(\hat{q}) \\ \hat{a}x &= \hat{p}. \end{aligned}$$

We can now get the solution as $x = \hat{p}/\hat{a}$ and $y = \hat{q}/\hat{d}$, which will agree with that attained using the inverse of the matrix.

We have been using *elementary row operations*, these are very useful in the manipulation of matrices. We are able to:

– exchange complete rows,

– multiply a row by a non-zero constant

– add a multiple of a row to any other row.

If we consider matrices it is not immediately obvious these operations do not change the system. However when considering systems of equations we are happy to perform exactly these operations. Consider the simple set of three equations in three unknowns:

$$a_{11}x_1 + a_{12}x_2 + a_{13}x_3 = b_1, \tag{6.4a}$$

$$a_{21}x_1 + a_{22}x_2 + a_{23}x_3 = b_2, \tag{6.4b}$$

$$a_{31}x_1 + a_{32}x_2 + a_{33}x_3 = b_3. \tag{6.4c}$$

Dealing with the operations in order:

Exchanging rows This operation is the same as swapping the equations around, so for instance we could write

$$\begin{aligned} a_{31}x_1 + a_{32}x_2 + a_{33}x_3 &= b_3, \\ a_{21}x_1 + a_{22}x_2 + a_{23}x_3 &= b_2, \\ a_{11}x_1 + a_{12}x_2 + a_{13}x_3 &= b_1. \end{aligned}$$

Here we have swapped the equations (6.4a) and (6.4c), and hopefully we can see that the solution remains the same; we have merely altered the order in which the equations are written down. This operation will be denoted by $R_1 \rightleftharpoons R_3$.

Multiply a row by a constant Let us multiply the equation (6.4b) by the constant λ, so that the equations become

$$\begin{aligned} a_{11}x_1 + a_{12}x_2 + a_{13}x_3 &= b_1, \\ \lambda(a_{21}x_1 + a_{22}x_2 + a_{23}x_3) &= \lambda b_2, \\ a_{31}x_1 + a_{32}x_2 + a_{33}x_3 &= b_3. \end{aligned}$$

Again hopefully we can see this does not alter the solution of the system. This operation would be denoted as $R_2 \to \lambda R_2$.

Combining rows We shall now perform the operation $R_2 \to R_2 - a_{21}R_1/a_{11}$, which gives

$$\begin{aligned} a_{11}x_1 + a_{12}x_2 + a_{13}x_3 &= b_1 \\ \left(a_{22} - \frac{a_{12}a_{21}}{a_{11}}\right)x_2 + \left(a_{23} - \frac{a_{13}a_{21}}{a_{11}}\right)x_3 &= b_2 - \frac{a_{21}b_1}{a_{11}} \\ a_{31}x_1 + a_{32}x_2 + a_{33}x_3 &= b_3. \end{aligned}$$

In this case Equations (6.4a) and (6.4c) remain unchanged and we have succeeded in eliminating x_1 from (6.4b). This is the step we would take en route to solving this system and naturally does not alter the solution of the system.

We now use these operations to construct the inverse of a two-by-two matrix.

Example 6.13 *Use elementary row operations to determine the inverse of the matrix*

$$\begin{pmatrix} a & b \\ c & d \end{pmatrix}.$$

Firstly we need to augment the matrix with the corresponding identity matrix and form $\mathbf{B} = \mathbf{A}|\mathbf{I}$. *We then use elementary row operations to reduce the part of* $\mathbf{B}$ *which corresponds to* $\mathbf{A}$ *to the identity matrix* $\mathbf{I}$. *We shall run through the steps. We start with the matrix*

$$\begin{pmatrix} a & b & 1 & 0 \\ c & d & 0 & 1 \end{pmatrix}.$$

To eliminate c *from the second row we use the operation* $R_2 \to R_2 - c/aR_1$ *so that we have*

$$\begin{pmatrix} a & b & 1 & 0 \\ 0 & d - \frac{cb}{a} & -\frac{c}{a} & 1 \end{pmatrix}.$$

Notice that the operation has been recorded in the third element along the second row. We now divide the second row through using the operation $R_2 \to R_2/(d - \frac{cb}{a})$, *which gives*

$$\begin{pmatrix} a & b & 1 & 0 \\ 0 & 1 & -\frac{c}{ad-bc} & \frac{a}{ad-bc} \end{pmatrix}.$$

Now, to eliminate the b *element we use* $R_1 \to R_1 - b \times R_2$ *so that we have*

$$\begin{pmatrix} a & 0 & 1 + \frac{bc}{ad-bc} & -\frac{ab}{ad-bc} \\ 0 & 1 & -\frac{c}{ad-bc} & \frac{a}{ad-bc} \end{pmatrix},$$

after slight simplification and dividing the first row by a, *using the operation* $R_1 \to R_1/a$ *we have*

$$\begin{pmatrix} 1 & 0 & \frac{d}{ad-bc} & -\frac{b}{ad-bc} \\ 0 & 1 & -\frac{c}{ad-bc} & \frac{a}{ad-bc} \end{pmatrix}.$$

This matrix is now $\mathbf{I}|\mathbf{A}^{-1}$. *This technique can be extended to larger systems. You can also compare this with the earlier solution of the system (6.2).*

There is a MATLAB command which will run through these operations, `rref` or in slow motion `rrefmovie`. This also includes pivoting, which is how this technique deals with zeros when it encounters them.

Example 6.14 *Construct the inverse of the matrix*

$$\begin{pmatrix} 1 & 2 & 1 \\ 3 & 2 & 1 \\ 1 & 1 & 1 \end{pmatrix}$$

using MATLAB.

This can be done by using the code:

```
A = [1 2 1; 3 2 1; 1 1 1];
B = [A eye(3)]
C = rref(B);
```

By way of illustration we show the actual operations using the command `rrefmovie`.

```
A = [1 2 1; 3 2 1; 1 1 1];
B = [A eye(3)]

B =

     1     2     1     1     0     0
     3     2     1     0     1     0
     1     1     1     0     0     1
```

This has produced an augmented matrix which we would write as $\mathbf{A}|\mathbf{I}$ *(not to be confused with the MATLAB 'or' command). Now try the command* `rrefmovie(B)`*, which yields*

```
  Original matrix

A =

      1          2          1          1          0          0
      3          2          1          0          1          0
      1          1          1          0          0          1

Press any key to continue. . .

  swap rows 1 and 2

A =

      3          2          1          0          1          0
```

```
      1          2          1          1          0          0
      1          1          1          0          0          1

Press any key to continue. . .

  pivot = A(1,1)

A =

      1        2/3        1/3          0        1/3          0
      1          2          1          1          0          0
      1          1          1          0          0          1

Press any key to continue. . .

  eliminate in column 1

A =

      1        2/3        1/3          0        1/3          0
      1          2          1          1          0          0
      1          1          1          0          0          1

Press any key to continue. . .

  eliminate in column 1

A =

      1        2/3        1/3          0        1/3          0
      0        4/3        2/3          1       -1/3          0
      0        1/3        2/3          0       -1/3          1

Press any key to continue. . .

  pivot = A(2,2)

A =

      1        2/3        1/3          0        1/3          0
      0          1        1/2        3/4       -1/4          0
      0        1/3        2/3          0       -1/3          1

Press any key to continue. . .

  eliminate in column 2

A =

      1        2/3        1/3          0        1/3          0
      0          1        1/2        3/4       -1/4          0
      0        1/3        2/3          0       -1/3          1
```

```
Press any key to continue. . .

  eliminate in column 2

A =

       1            0            0         -1/2          1/2            0
       0            1          1/2          3/4         -1/4            0
       0            0          1/2         -1/4         -1/4            1

Press any key to continue. . .

  pivot = A(3,3)

A =

       1            0            0         -1/2          1/2            0
       0            1          1/2          3/4         -1/4            0
       0            0            1         -1/2         -1/2            2

Press any key to continue. . .

  eliminate in column 3

A =

       1            0            0         -1/2          1/2            0
       0            1          1/2          3/4         -1/4            0
       0            0            1         -1/2         -1/2            2

Press any key to continue. . .

  eliminate in column 3

A =

       1            0            0         -1/2          1/2            0
       0            1            0            1            *           -1
       0            0            1         -1/2         -1/2            2

Press any key to continue. . .
```

This provides a movie of row echelon operations, giving a running commentary. This is used purely for illustration and MATLAB has used the `format rat` *command to present the numbers as rationals. Notice one number is given as a* `*`*: this is because MATLAB cannot adequately represent this as a rational. If we now use* `C = rref(B)`*, we find this element is in fact zero but rounding errors made it to be a very small number. Now extracting the last three columns* `X = C(:,4:6)` *where* $\mathbf{X}$ *is in fact the inverse of* $\mathbf{A}$*, which can be confirmed by typing* `X*A` *and* `A*X`*.*

The above technique involves other methods than those we used in the simple two-by-two case (for instance pivoting). It is not the intention of this text to provide a comprehensive discussion of these matrix operations, merely to make you aware of them. You may notice that the rôle of the augmented matrix is to keep track of the operations required to change `A` to the identity matrix. Have a look at the second matrix in which the command `swap rows 1 and 2` has been performed. This is to increase the size of the element which is going to be used for the elimination. If the original element is zero it will definitely need to be moved, as the algorithm is going to divide by this quantity. If the column is full of zeros then the inverse does not exist and the algorithm continues to work on the remaining rows. For instance

```
D=[0 1 3; 0 1 1; 0 1 2];
B=[D eye(3)];
rref(B)
```

yields

```
0    1    0    0    2   -1
0    0    1    0   -1    1
0    0    0    1    1   -2
```

This shows the original matrix does not have an inverse and in fact its rank is less than the number of rows. In this case we find $\text{rank}(\mathbf{D}) = 2$ and the determinant is zero. We note the third row of the matrix can be formed by the linear combination of the first and second ones, $(\mathbf{r}_1 + \mathbf{r}_2)/2$.

6.3.1 Solving Many Equations at Once

We remark it is possible to determine the solution of many systems simultaneously where only the right-hand sides of the equation changes. Consider the solution of the equation:

$$\mathbf{Ax} = \mathbf{b}.$$

We shall assume $\mathbf{A}$ is an n-by-n matrix and $\mathbf{x}$ is n-by-m and $\mathbf{b}$ is n-by-m. This allows m sets of equations to be solved simultaneously.

Example 6.15 *Consider the simple example*

$$\begin{aligned} 2x + 3y &= 7, \\ x - y &= 1, \end{aligned}$$

and

$$\begin{aligned} 2x + 3y &= -2, \\ x - y &= 8. \end{aligned}$$

These equations can be written as

$$\begin{pmatrix} 2 & 3 \\ 1 & -1 \end{pmatrix} \begin{pmatrix} x \\ y \end{pmatrix} = \begin{pmatrix} 7 & -2 \\ 1 & 8 \end{pmatrix}.$$

This system can be solved using the MATLAB code:

```
A = [ 2 3; 1 -1];
b = [7 -2; 1 8];
x = A\b;
```

This gives

```
x =

   2.0000    4.4000
   1.0000   -3.6000
```

so the solution to the first system is $(2, 1)^T$ *and the second is* $(22/5, -18/5)^T$.

6.4 Matrix Decomposition

As mentioned in the previous section we can think of the last three columns as recording the operations performed in order to reduce the matrix to row echelon form. Notice this process involves a forward sweep and a backward sweep. In general the forward sweep will produce a lower triangular matrix and an upper triangular matrix will result from the backward sweep[6]. If pivoting is used to determine this form it is necessary to also include a permutation matrix (which swaps rows). We use the MATLAB commands

```
A = [1 2 1; 3 2 1; 1 1 1];
[L,U,P] = lu(A)

L =
```

[6] The definition of upper and lower is based on the leading diagonal (which runs from the top left to the bottom right).

```
    1.0000         0         0
    0.3333    1.0000         0
    0.3333    0.2500    1.0000


U =

    3.0000    2.0000    1.0000
         0    1.3333    0.6667
         0         0    0.5000


P =

     0     1     0
     1     0     0
     0     0     1
```

Now we have the three matrices `L`, `U` and `P`, such that

$$\mathbf{PA} = \mathbf{LU}.$$

Notice it is possible to call this routine with only two outputs, in which case $\mathbf{L}$ is no longer lower triangular as it now contains pivoting information.

Let us now consider the inverse of the matrix $\mathbf{A}$. We need to recall that

$$(\mathbf{XY})^{-1} = \mathbf{Y}^{-1}\mathbf{X}^{-1},$$

to see this simply multiply both sides of the equation on the left by $(\mathbf{XY})$ (or on the right). In which case

$$\begin{aligned} (\mathbf{PA})^{-1} &= (\mathbf{LU})^{-1}, \\ \mathbf{A}^{-1}\mathbf{P}^{-1} &= \mathbf{U}^{-1}\mathbf{L}^{-1}, \\ \mathbf{A}^{-1} &= \mathbf{U}^{-1}\mathbf{L}^{-1}\mathbf{P}, \end{aligned} \tag{6.5}$$

multiplying both sides of the equation on the right by $\mathbf{P}$.

Consider the solution of the problem

$$\mathbf{Ax} = \mathbf{b}.$$

Multiplying both sides by $\mathbf{A}^{-1}$ on the left and substituting from (6.5), yields:

$$\begin{aligned} \mathbf{x} &= \mathbf{U}^{-1}\mathbf{L}^{-1}\mathbf{Pb} \\ \mathbf{x} &= \mathbf{U}^{-1}\mathbf{Eb}, \end{aligned}$$

where we have introduced $\mathbf{E} = \mathbf{L}^{-1}\mathbf{P}$. This actually occurs earlier in the series of operations since $\mathbf{EA} = \mathbf{U}$. The algorithm can be written as:

(i) Calculate $\mathbf{L}$, $\mathbf{U}$

(ii) Calculate $\mathbf{Ly} = \mathbf{b}$ $(\mathbf{y} = \mathbf{Eb})$

(iii) Solve $\mathbf{Ux} = \mathbf{y}$.

This method is efficient since it works with triangular matrices. If we need to solve the above system for different $\mathbf{b}$ then we only need to calculate $\mathbf{L}$ and $\mathbf{U}$ once. In fact this is the essence of how MATLAB solves matrix systems. Instead of actually constructing the inverse it works out the influence the inverse would have. This represents a dramatic saving in time. In order to test this we use the codes

```
%
% testmat.m
%
function [time1,time2] = testmat(n);
A = rand(n);
b = rand(n,1);
tnow = cputime;
x = A\b;
time1 = cputime-tnow;
tnow = cputime;
x = inv(A)*b;
time2 = cputime-tnow;
```

which is run using

```
%
% mtestmat.m
%
nn = 100:50:800;
n = length(nn);

for i = 1:n
    nn(i)
    [t1(i),t2(i)] = testmat(nn(i));
end

plot(nn,[t1; t2])
xlabel('Matrix size','FontSize',15)
ylabel('CPU time (secs)','FontSize',15)
text(600,10,'Decomposition','FontSize',14)
text(500,20,'Full inversion','FontSize',14)
print -dps2 testmat.ps
```

This gives the results

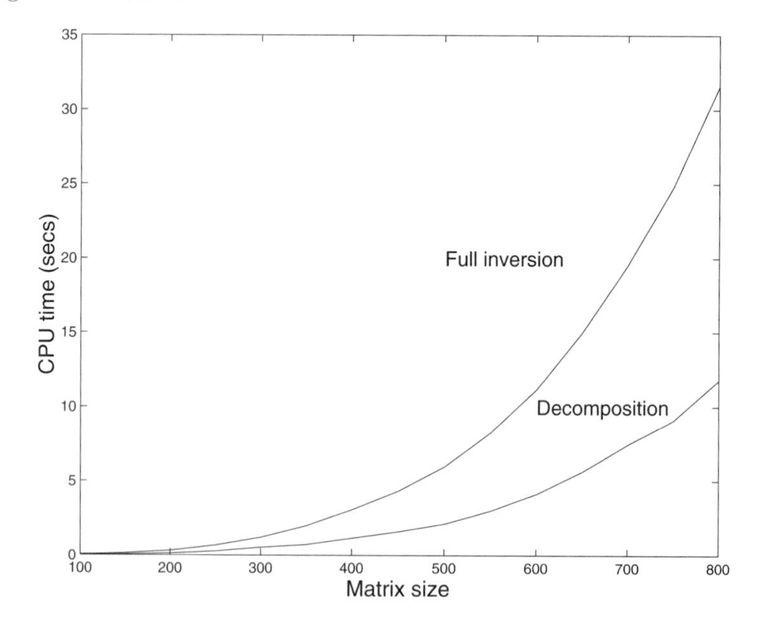

where the upper curve is the method using the construction of the inverse. In fact both of these curves grow like the size of the matrix to the power 3, but the lower one has a smaller constant of proportionality. Matrix inversion (and multiplication) is very expensive and any symmetry or special properties of a matrix should be exploited.

MATLAB has a few commands which can use the properties of these matrices: perhaps the most simple one is `sparse`. We will meet examples in due course but we shall simply show this example.

```
>> A = sparse(eye(4))

A =

   (1,1)        1
   (2,2)        1
   (3,3)        1
   (4,4)        1
>> B = sparse(diag(1,3))

B =

   (1,4)        1

>> A+B

ans =

   (1,1)        1
   (2,2)        1
   (3,3)        1
   (1,4)        1
   (4,4)        1
```

These matrices are

$$\mathbf{A} = \begin{pmatrix} 1 & 0 & 0 & 0 \\ 0 & 1 & 0 & 0 \\ 0 & 0 & 1 & 0 \\ 0 & 0 & 0 & 1 \end{pmatrix}, \ \mathbf{B} = \begin{pmatrix} 0 & 0 & 0 & 1 \\ 0 & 0 & 0 & 0 \\ 0 & 0 & 0 & 0 \\ 0 & 0 & 0 & 0 \end{pmatrix}$$

$$\text{and } \mathbf{A} + \mathbf{B} = \begin{pmatrix} 1 & 0 & 0 & 1 \\ 0 & 1 & 0 & 0 \\ 0 & 0 & 1 & 0 \\ 0 & 0 & 0 & 1 \end{pmatrix}.$$

Instead of storing the whole matrix it only stores the elements which are non-zero, and when operations are performed it returns a sparse matrix (or a full one if the result is full). A full matrix can be retrieved by using the `full` command.

6.5 Eigenvalues and Eigenvectors

Although this is slightly beyond the scope of this text we shall briefly discuss the ideas of eigenvalues and eigenvectors. These are very useful when solving matrix systems and understanding the influence of matrices. Consider the simple transformation

$$x \mapsto 3x, \qquad y \mapsto -2y,$$

which we could write as

$$x' = 3x \qquad y' = -2y,$$

or in matrix form

$$\begin{aligned} \begin{pmatrix} x' \\ y' \end{pmatrix} &= \begin{pmatrix} 3 & 0 \\ 0 & -2 \end{pmatrix} \begin{pmatrix} x \\ y \end{pmatrix}, \\ \mathbf{p}' &= \mathbf{A}\mathbf{p}. \end{aligned}$$

The action of this transformation is to multiply the x component by three and the y component by minus two. So the coordinate $(1, 1)$ is moved to $(3, -2)$.

Under this transformation which vectors' directions are left unchanged? Hopefully you should be able to see that anything on the x or y axes falls into this category. Along the x axis we have the general point $(\lambda, 0)$ which after the transformation becomes $(3\lambda, 0) = 3(\lambda, 0)$, so that $\mathbf{Ap} = 3\mathbf{p}$ (where $\mathbf{p} = (1, 0)^T$). Similarly along the y axis, we have the general point $(0, \mu)$ which transforms to $(0, -2\mu) = -2(0, \mu)$. The values 3 and -2 are referred to as the eigenvalues and $(1, 0)^T$ and $(0, 1)^T$ are the eigenvectors (we have taken λ and μ to both be unity).

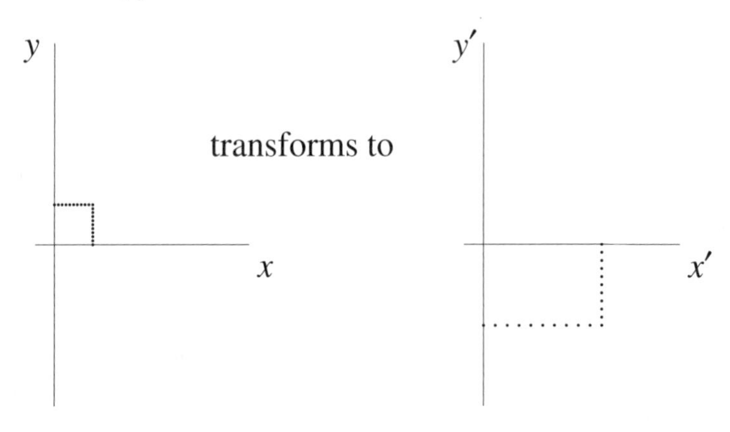

In this simple case it is easy to see how these quantities are determined but consider the more involved example:

Example 6.16 *Determine the eigenvalues and eigenvectors of the matrix*

$$\begin{pmatrix} 1 & 2 \\ 3 & 2 \end{pmatrix},$$

and comment on the effect of the matrix.

Let us briefly describe how this is done analytically. We assume that the eigenvalues are λ and as such for an eigenvector $\mathbf{x}$

$$\mathbf{A}\mathbf{x} = \lambda\mathbf{x},$$

which can be rearranged to give

$$(\mathbf{A} - \lambda\mathbf{I})\,\mathbf{x} = 0.$$

This equation only has trivial solutions for $\mathbf{x}$ unless the determinant of the matrix multiplying it is zero[7], that is

$$\det(\mathbf{A} - \lambda\mathbf{I}) = 0.$$

We have used the term 'det' to represent the determinant but we also use vertical lines, as below. In this case this leads to a quadratic:

$$\begin{aligned} \begin{vmatrix} 1-\lambda & 2 \\ 3 & 2-\lambda \end{vmatrix} &= (1-\lambda)(2-\lambda) - 6 \\ &= \lambda^2 - 3\lambda - 4 \\ &= (\lambda-4)(\lambda+1) = 0. \end{aligned}$$

So the eigenvalues are 4 and -1. Notice that we could have used a couple of MATLAB commands at this point.

```
>> p = poly(A)

p =

     1    -3    -4

>> roots(p)
```

[7] Notice that if the determinant is non-zero, then we could simply invert the matrix and obtain the solution $\mathbf{x} = \mathbf{0}$; that is the trivial solution.

```
ans =

     4
    -1

>>
```

The first command gives the coefficients of the characteristic polynomial (which is the name of the polynomial in λ, see page 214 for more details). The second one determines the roots of polynomial whose coefficients are stored in p.

In order to determine the eigenvectors we solve the equations

$$\begin{pmatrix} 1 & 2 \\ 3 & 2 \end{pmatrix}\begin{pmatrix} x \\ y \end{pmatrix} = \lambda \begin{pmatrix} x \\ y \end{pmatrix},$$

for each value of λ.

$\underline{\lambda = 4}$

$$\begin{pmatrix} 1 & 2 \\ 3 & 2 \end{pmatrix}\begin{pmatrix} x \\ y \end{pmatrix} = 4 \begin{pmatrix} x \\ y \end{pmatrix},$$

which gives

$$\begin{aligned} x + 2y &= 4x \qquad \Rightarrow 2y = 3x, \\ 3x + 2y &= 4y \qquad \Rightarrow 3x = 2y. \end{aligned}$$

Notice that both equations give the same result and in a sense there is a redundance (which is all linked to the fact that the determinant of the modified system is now zero). The eigenvector is defined by $x = 2y/3$ and hence choosing $y = 3$ gives $x = 2$:

$$\mathbf{e}_{\lambda=4} = \begin{pmatrix} 2 \\ 3 \end{pmatrix};$$

notice that we could have any scalar multiple of this. As such MATLAB will return the multiple of this such that

$$\hat{\mathbf{e}}_{\lambda=4} = \frac{\mathbf{e}_{\lambda=4}}{|\mathbf{e}_{\lambda=4}|},$$

which has unit modulus. The modulus $|\cdot|$ is defined as the square root of the sum of the squares of the elements.

$\underline{\lambda = -1}$

$$\begin{pmatrix} 1 & 2 \\ 3 & 2 \end{pmatrix}\begin{pmatrix} x \\ y \end{pmatrix} = (-1) \begin{pmatrix} x \\ y \end{pmatrix},$$

which gives

$$\begin{aligned} x + 2y &= -x \qquad \Rightarrow y = -x, \\ 3x + 2y &= -y \qquad \Rightarrow x = -y. \end{aligned}$$

Hence the eigenvector is

$$\mathbf{e}_{\lambda=-1} = \begin{pmatrix} 1 \\ -1 \end{pmatrix}.$$

We could have determined these using the command

```
[V,D] = eig(A)

V =

  -0.7071   -0.5547
   0.7071   -0.8321


D =

   -1     0
    0     4
```

This returns two matrices: one with the eigenvectors as columns and the other with the eigenvalues on the leading diagonal (this is discussed further on page 208). Notice that the eigenvectors are returned with the columns normalised such that $\mathbf{e}\mathbf{e}^T = 1$. In one case dividing through by $\sqrt{2}$ and in the other by $\sqrt{13} = \sqrt{2^2 + 3^2}$. It now remains for us to discuss the effect of this transformation: the eigenvectors are $(2,3)^T$ and $(1,-1)^T$ with corresponding eigenvalues 4 and -1. Hence any vector along the line $(2,3)$ is extended four times and along $(1,-1)$ is reflected in the origin.

We could consider a vector to be composed on components associated with each eigenvector, for instance consider the action on the point $(1,1)$.

Firstly solve the equation

$$\begin{pmatrix} 1 \\ 1 \end{pmatrix} = \alpha \mathbf{e}_{\lambda=4} + \beta \mathbf{e}_{\lambda=-1},$$

which gives the simultaneous equations

$$\begin{aligned} 1 &= 2\alpha + \beta \\ 1 &= 3\alpha - \beta, \end{aligned}$$

which are solved to give $\alpha = 2/5$ and $\beta = 1/5$. Hence

$$\begin{pmatrix} 1 \\ 1 \end{pmatrix} = \frac{2}{5}\mathbf{e}_{\lambda=4} + \frac{1}{5}\mathbf{e}_{\lambda=-1}.$$

Now we multiply both sides by $\mathbf{A}$ to give

$$\begin{aligned} \mathbf{A}\begin{pmatrix} 1 \\ 1 \end{pmatrix} &= \frac{2}{5}\mathbf{A}\mathbf{e}_{\lambda=4} + \frac{1}{5}\mathbf{A}\mathbf{e}_{\lambda=-1}, \\ &= \frac{2}{5}(4)\mathbf{e}_{\lambda=4} + \frac{1}{5}(-1)\mathbf{e}_{\lambda=-1}, \\ &= \frac{8}{5}\mathbf{e}_{\lambda=4} - \frac{1}{5}\mathbf{e}_{\lambda=-1}, \\ &= \frac{8}{5}\begin{pmatrix} 2 \\ 3 \end{pmatrix} - \frac{1}{5}\begin{pmatrix} 1 \\ -1 \end{pmatrix}, \\ &= \begin{pmatrix} 3 \\ 5 \end{pmatrix}. \end{aligned}$$

We could have used simple matrix multiplication but this kind of technique will allow us to perform far more complex calculations in which we may not know all the values (so in more general cases) and also in higher dimensions. This technique is especially useful when discussing manifolds in chaos (but that's material for another text book).

6.6 Specific MATLAB Commands

We remark that MATLAB has very powerful commands for determining eigenvalues (and eigenvectors) of these systems, `eig` and `eigs`. The former one has already been used on page 207.

We start with the command `eig`; in its simplest form it determines the eigenvalues of a square matrix, the syntax is

```
a = [1 3 2; 1 1 1; -1 0 0];
e = eig(a)
```

This gives the answer `e = [1.6180;-0.6180; 1.0000]` (the actual values are $(1 \pm \sqrt{5})/2$ and 1). A very simple change in syntax gives the eigenvectors as well `[V,D] = eig(a);`. In this case the eigenvectors are the columns of `V` and the eigenvalues are on the diagonal of the matrix `D`. We remark that

$$\mathbf{A} = \mathbf{P}\mathbf{D}\mathbf{P}^{-1},$$

where $\mathbf{P}$ is a matrix formed with the eigenvectors of the matrix $\mathbf{A}$ as columns (that is `V`) and $\mathbf{D}$ is a diagonal matrix with the eigenvalues on its diagonal. **Note**: the order of these must be the same.

This is verified for the above example using the code

```
>> [v,d] = eig(a)

v =

    0.7529    0.4004   -0.6396
    0.4653   -0.6479   -0.4264
   -0.4653    0.6479    0.6396


d =

    1.6180         0         0
         0   -0.6180         0
         0         0    1.0000

>> v*d*inv(v)

ans =

    1.0000    3.0000    2.0000
    1.0000    1.0000    1.0000
   -1.0000    0.0000         0
```

As a direct consequence of this we are able to work out $\mathbf{A}^n$ very cheaply. We note that

$$\begin{aligned}\mathbf{A}^2 = \mathbf{A}\mathbf{A} &= (\mathbf{P}\mathbf{D}\mathbf{P}^{-1})(\mathbf{P}\mathbf{D}\mathbf{P}^{-1}) \\ &= (\mathbf{P}\mathbf{D})(\mathbf{P}^{-1}\mathbf{P})(\mathbf{D}\mathbf{P}^{-1}) \\ &= (\mathbf{P}\mathbf{D})(\mathbf{I})(\mathbf{D}\mathbf{P}^{-1}) \\ &= (\mathbf{P}\mathbf{D}\mathbf{D}\mathbf{P}^{-1}) = \mathbf{P}\mathbf{D}^2\mathbf{P}^{-1}.\end{aligned}$$

It is quite simple to see that $\mathbf{A}^n = \mathbf{P}\mathbf{D}^n\mathbf{P}^{-1}$ (using proof induction for instance, see Task 6.22). We now note that $\mathbf{D}^n$ is very easy to work out, if $\mathbf{D}$ has the values $(\lambda_1, \cdots, \lambda_N)$ along the leading diagonal then $\mathbf{D}^n$ has $(\lambda_1^n, \cdots, \lambda_N^n)$ along that diagonal.

Example 6.17 *Calculate the matrix*

$$\mathbf{A} = \begin{pmatrix} 1 & 0 & -1 \\ 0 & 1 & 1 \\ 1 & 0 & 0 \end{pmatrix}$$

raised to the hundredth power.

This is accomplished using the code

```
a = [1 0 -1; 0 1 1; 1 0 0];
[v,d] = eig(a);
d = sparse(d);
d100 = d.^100;
a100 = v*d100*inv(v);
```

We also mention the command `eigs` which allows us to select only certain eigenvalues. This uses techniques akin to the power method mentioned earlier. We give an extract of the help page for `eigs`

```
[V,D,FLAG] = EIGS(A,B,K,SIGMA,OPTIONS)
[V,D,FLAG] = EIGS('Afun',N,B,K,SIGMA,OPTIONS)
```

```
where
```

```
B          A symmetric positive definite matrix the same size as A.
K          An integer, the number of eigenvalues desired.
SIGMA      A scalar shift or a two letter string.
OPTIONS    A structure containing additional parameters.
```

The viable values of `SIGMA` are

```
'LM'            Largest Magnitude  (the default)
'SM'            Smallest Magnitude (same as sigma = 0)
'LR'            Largest Real part
'SR'            Smallest Real part
'BE'            Both Ends.  Computes k/2 eigenvalues
                from each end of the spectrum (one more
                from the high end if k is odd.)
```

Example 6.18 *Determine the largest two eigenvalues of the matrix*

$$\begin{pmatrix} 1 & 0 & 0 & 1 & -1 \\ 0 & 2 & 3 & 5 & 0 \\ -1 & 0 & 0 & 0 & 1 \\ 6 & 8 & 1 & 2 & -2 \\ 1 & 1 & 1 & 1 & 1 \end{pmatrix}.$$

This is done using the code:

```
A = [1 0 0 1 -1; ...
   0 2 3 5 0; ...
   -1 0 0 0 1; ...
   6 8 1 2 -2; ...
   1 1 1 1 1];
[V,D] = eigs(A,2,'LM')
```

This iterates and gives:

```
iter =

     1


eigs =

    8.4127
   -4.8097


stopcrit =

   2.0072e-15

==========================

iter =

     2


eigs =
```

```
   8.4127
  -4.8097


stopcrit =

   3.2411e-15

==========================

V =

    0.0769    0.1400
    0.6050    0.5780
    0.0143    0.0270
    0.7674   -0.8034
    0.1974    0.0100


D =

    8.4127         0
         0   -4.8097
```

6.7 Characteristic Polynomials

We shall pause here to discuss the characteristic polynomial of a matrix, that is the polynomial $p(\lambda) = \det(\mathbf{A} - \lambda\mathbf{I})$. The zeros of this polynomial are the eigenvalues of the matrix. We quote the result

Theorem 6.19 (Cayleigh–Hamilton) *Every square matrix satisfies its own characteristic polynomial, so that* $p(\mathbf{A}) = \mathbf{0}$.

We also note that $p(0) = \det(\mathbf{A})$, which can be seen by setting λ equal to zero in the definition of $p(\lambda)$. The characteristic polynomial of an n-by-n matrix is of degree n. As a result of the Cayleigh–Hamilton theorem we have a further mechanism for determining the inverse of a matrix. Let us consider $p(\lambda)$ to have coefficients $a_0, \cdots, a_n$. Consequently we have

$$\sum_{i=0}^{n} a_i \mathbf{A}^i = \mathbf{0},$$

where we use the natural convention that $\mathbf{A}^0 = \mathbf{I}$. We can extract one term from the summation so that

$$a_0 \mathbf{I} + \sum_{i=1}^{n} a_i \mathbf{A}^i = \mathbf{0}.$$

Notice that $p(0) = a_0$ so that $a_0 = \det(\mathbf{A})$, which we assume is non-zero. Multiplying through by $\mathbf{A}^{-1}$ and dividing by $\det(\mathbf{A})$ we have:

$$\mathbf{A}^{-1} = -\frac{1}{\det(\mathbf{A})} \sum_{i=1}^{n} a_i \mathbf{A}^{i-1}.$$

Here we have written the inverse in terms of powers of $\mathbf{A}$ which we know can be written using the diagonalised form of the matrix. As such we have

$$\mathbf{A}^{-1} = -\frac{1}{\det(\mathbf{A})} \mathbf{P} \left\{ \sum_{i=0}^{n-1} a_{i+1} \mathbf{D}^i \right\} \mathbf{P}^{-1},$$

where $\mathbf{D}$ and $\mathbf{P}$ are defined on page 208.

We are aware that we could also construct the inverse using a variety of MATLAB commands, however it is important that one realises what is behind these commands.

We now note that we can determine the characteristic polynomial of a matrix by evaluating the polynomial $p(\lambda)$ at $n + 1$ points and then fitting a polynomial to these points:

```
function [co] = charpoly(A)
[m,n] = size(A);
if m ~= n
   disp('Matrix is not square')
   co = [];
   break
end

for i = 1:(n+1)
    x(i) = (i-1)*pi/n;
    y(i) = det(A-x(i)*eye(n));
end

co = polyfit(x,y,n);
```

We have elected to use the points $x_i = (i-1)\pi/n$ since these will not coincide with eigenvalues unless the elements of the matrix involve π. We can now consider the roots of this polynomial to determine the eigenvalues of the matrix (using `roots` for instance). There is also a MATLAB procedure to determine this polynomial, `poly(A)`.

6.8 Exponentials of Matrices

We shall motivate this section by discussing the simple differential equation:

$$\dot{x} = ax$$

which shall be solved subject to the initial condition that $x(t_0) = x_0$. This has the solution

$$x(t) = \mathrm{e}^{a(t-t_0)}x_0.$$

We now move on to discuss the matrix equivalent of this equation which is

$$\dot{\mathbf{x}} = \mathbf{A}\mathbf{x}$$

together with the initial condition $\mathbf{x}(t_0) = \mathbf{x}_0$ and this has the solution

$$\mathbf{x}(t) = \mathrm{e}^{\mathbf{A}(t-t_0)}\mathbf{x}_0.$$

This method can be used for solving constant coefficient second order systems.

Example 6.20 *Solve the equation*

$$\ddot{y} + y = 0$$

subject to the initial conditions $y(0) = 0$ and $\dot{y}(0) = 1$. We introduce the quantities $x_1(t) = y(t)$ and $x_2(t) = \dot{y}(t)$. As such this system can be rewritten as

$$\dot{\mathbf{x}} = \begin{pmatrix} 0 & 1 \\ -1 & 0 \end{pmatrix}\mathbf{x},$$

with $\mathbf{x}_0 = (0,1)^T$. This has the solution

$$\mathbf{x}(t) = \exp(\mathbf{A}t)\mathbf{x}_0.$$

We now have to define and construct $\exp(\mathbf{A})$. We do this using the summation

$$\exp(\mathbf{A}) = \sum_{i=0}^{\infty} \frac{1}{i!}\mathbf{A}^i.$$

Again we can use the diagonalised form of the matrix to yield

$$\exp(\mathbf{A}) = \mathbf{P}\left\{\sum_{i=0}^{\infty} \frac{1}{i!}\mathbf{D}^i\right\}\mathbf{P}^{-1},$$

but this gives

$$\exp(\mathbf{A}) = \mathbf{P}\exp(\mathbf{D})\mathbf{P}^{-1},$$

where $\exp(\mathbf{D})$ is the matrix formed with $(\mathrm{e}^{\lambda_1}, \mathrm{e}^{\lambda_2}, \cdots, \mathrm{e}^{\lambda_n})$ along its leading diagonal.

Example 6.21 *Construct the matrix* $\exp(\mathbf{A})$ *where*

$$\mathbf{A} = \begin{pmatrix} 1 & 2 \\ 0 & -1 \end{pmatrix}.$$

This matrix has eigenvalues of 1 *and* -1, *with corresponding eigenvectors* $(1,0)^T$ *and* $(-1,1)^T$. *Hence we have*

$$\mathbf{P} = \begin{pmatrix} 1 & -1 \\ 0 & 1 \end{pmatrix} \text{ so that } \mathbf{P}^{-1} = \begin{pmatrix} 1 & 1 \\ 0 & 1 \end{pmatrix}.$$

Hence

$$\exp(\mathbf{A}) = \begin{pmatrix} 1 & -1 \\ 0 & 1 \end{pmatrix}\begin{pmatrix} \mathrm{e}^1 & 0 \\ 0 & \mathrm{e}^{-1} \end{pmatrix}\begin{pmatrix} 1 & 1 \\ 0 & 1 \end{pmatrix} = \begin{pmatrix} \mathrm{e} & \mathrm{e} - 1/\mathrm{e} \\ 0 & \mathrm{e} \end{pmatrix}.$$

Notice that to construct $\exp(\mathbf{A}t)$ we simply note that $\exp(\mathbf{D}t)$ has diagonal elements of the form $\mathrm{e}^{\lambda_j t}$. In the previous example we would have

$$\exp(\mathbf{A}t) = \begin{pmatrix} \mathrm{e}^t & \mathrm{e}^t - \mathrm{e}^{-t} \\ 0 & \mathrm{e}^t \end{pmatrix}.$$

There is actually a command for this purpose in MATLAB called `expm` (together with `expm1`, `expm2` and `expm3` – which exploit different methods). Simply typing `exp(A)` will return a matrix full of the exponentials of the elements of the original elements.

There is another way of calculating the exponential of a matrix which occurs as a direct consequence of Theorem 6.19. We note that since an n-by-n matrix satisfies its own characteristic polynomial $\mathbf{A}^n$ can be written in terms of a polynomial of degree $n-1$. Consequently multiplying through by $\mathbf{A}$ and using the previous expression for $\mathbf{A}^n$ we find that $\mathbf{A}^{n+1}$ can also be expressed in terms of a similar polynomial. In fact $\mathbf{A}^n$ can be written as a polynomial of degree $n-1$ in $\mathbf{A}$. Since the exponential form of the matrix involves only powers of the matrix, we note that $\exp(\mathbf{A})$ can also be written as a polynomial of degree $n-1$ in $\mathbf{A}$. We now need to find the coefficients of this polynomial. We again rely on the Cayleigh–Hamilton theory (but in reverse), that is if $\mathbf{A}$ satisfies the equation then so do the eigenvalues.

Example 6.22 *Determine* $\exp(\mathbf{A}t)$ *where* $\mathbf{A}$ *is as given in Example 6.20. The eigenvalues of the matrix*

$$\begin{pmatrix} 0 & 1 \\ -1 & 0 \end{pmatrix}$$

are actually $\pm i = \pm\sqrt{-1}$. *We now suppose that*

$$\exp(\mathbf{A}t) = \alpha_0 \mathbf{I} + \alpha_1 \mathbf{A},$$

which is a first degree polynomial in $\mathbf{A}$. *Now substituting in* $\lambda = i$ *we find that*

$$\mathrm{e}^{it} = \alpha_0 + \alpha_1 i \tag{6.6a}$$

and for the second eigenvalue

$$\mathrm{e}^{-it} = \alpha_0 - \alpha_1 i. \tag{6.6b}$$

Solving Equations (6.6a) and (6.6b) we find that $\alpha_0 = (\mathrm{e}^{it} + \mathrm{e}^{-it})/2 = \cos t$ *and* $\alpha_1 = (\mathrm{e}^{it} - \mathrm{e}^{-it})/(2i) = \sin t$. *This gives*

$$\begin{aligned} \exp(\mathbf{A}t) &= \cos t\mathbf{I} + \sin t\mathbf{A}, \\ &= \begin{pmatrix} \cos t & \sin t \\ -\sin t & \cos t \end{pmatrix}. \end{aligned}$$

Now substituting this into the solution of Example 6.20 we have

$$\begin{pmatrix} y(t) \\ \dot{y}(t) \end{pmatrix} = \begin{pmatrix} \cos t & \sin t \\ -\sin t & \cos t \end{pmatrix} \begin{pmatrix} 0 \\ 1 \end{pmatrix} = \begin{pmatrix} \sin t \\ \cos t \end{pmatrix}.$$

This is exactly the solution one would expect (which is a good thing).

In general if a matrix has distinct eigenvalues λ_1 and λ_2 we can show that

$$\mathbf{x}(t) = \frac{1}{\lambda_2 - \lambda_1} \left\{ \left(\lambda_2 \mathrm{e}^{\lambda_1 t} - \lambda_1 \mathrm{e}^{\lambda_2 t}\right) \mathbf{I} + \left(\mathrm{e}^{\lambda_1 t} - \mathrm{e}^{\lambda_2 t}\right) \mathbf{A} \right\} \mathbf{x}_0.$$

We see that the eigenvalues of the matrix associated with a system are crucial in determining the fate of the solution. For instance if the real parts of the eigenvalues are real and positive the solution will expand as t increases. Similarly if both are real and negative the solution will contract. This analysis is very powerful and allows us to understand the structure of these problems without having to actually solve them in totality (something which gets harder and harder).

6.9 Tasks

Task 6.1 *Repeat the calculations presented in the example on page 328 using MATLAB.*

Task 6.2 *Using the vector* `r = 1:4` *construct the matrix*

$$\begin{pmatrix} 1 & 2 & 3 & 4 \\ 0 & 0 & 0 & 3 \\ 0 & 0 & 0 & 2 \\ 0 & 0 & 0 & 1 \end{pmatrix}$$

Note: *there are two commands in MATLAB,* `fliplr` *(flip left-right) and* `flipud` *(flip up-down), which will help with this task; you can use either (in order to see how they work use the* `help` *command).*

There are many commands we can use to set up larger matrices. However, in the majority of this text we shall limit our attention to relatively small matrices.

We can use the `diag` command to set up matrices which have a structure which lends itself to construction via vectors of diagonals. Such matrices are referred to as banded. Consider the following example:

```
>> r = 1:5;
>> A = diag(r)+diag(r(1:4),1)+diag(r(2:5),-1);
```

which gives the output

```
>> A

A =

     1     1     0     0     0
     2     2     2     0     0
```

```
0    3    3    3    0
0    0    4    4    4
0    0    0    5    5
```

You should recall the nth diagonal is shorter by $|n|$ than the leading diagonal. In order to understand this code it may help you to type `r(1:4)` (which gives the first four elements of `r`) and `r(2:5)` (which gives the last four elements of `r`).

Task 6.3 *Set up the ten-by-ten matrix which has +1's in the first super-diagonal (directly above the leading diagonal) and −1's in the sub-diagonal (directly below the leading diagonal).*

By this point you should be reasonably confident about how to set up a variety of matrices, either by directly typing in the elements or by exploiting structures within the matrix. We pause here to make sure that you are able to understand how to extract parts of matrices. Before proceeding make sure you understand what the commands `a(:,1)`, `a(2,:)`, `a(2:3,:)`, `a(:,:)`, `a(1:2:3,2)` and `a(1,3:-1:1)` give (you could try them for the matrix `a = [11 12 13; 21 22 23; 31 32 33]`, where this has elements which reflect the labels of the elements). To help you understand these commands type `1:2:3` which gives the result `[1 3]` and `3:-1:1` which gives `[3 2 1]`. You should now be able to see which elements these expressions correspond to in the matrix.

It is perfectly acceptable to combine matrices: however we need to make sure that the operation is well defined within MATLAB. This usually involves ensuring that the matrix (or vectors) involved have the correct dimensions for the operation to proceed.

Task 6.4 *Determine which of the following operations are viable (you can do this simply by typing them: however you should perhaps try to determine which ones are viable a priori).*

```
A = ones(3);
B = 2*ones(3,2);
C = 3*ones(2,3);

[A B]
[A B']
[A C B']
[A C']
[A; C]
[A; B']
```

At this stage you are now in a position to perform relatively simple operations on matrices; we shall start with a couple of examples of addition and scalar combination of matrices.

Task 6.5 *Perform these calculations both by hand and using MATLAB:*

$$\begin{pmatrix} 1 & 2 \\ 3 & 4 \end{pmatrix}\begin{pmatrix} 3 & 4 \\ -1 & 2 \end{pmatrix}$$

$$2\begin{pmatrix} 3 & 5 \\ 6 & -2 \end{pmatrix} - 4\begin{pmatrix} -1 & 0 \\ 2 & 1 \end{pmatrix}$$

$$\begin{pmatrix} 1 & 3 & 5 \end{pmatrix}\begin{pmatrix} 2 & -1 \\ -1 & 0 \\ 7 & -2 \end{pmatrix}$$

Task 6.6 *Calculate the product* **AB** *and* **BA** *for a variety of two-by-two matrices. Attempt to work out which matrices of the general form*

$$\mathbf{A} = \begin{pmatrix} a & b \\ c & d \end{pmatrix}$$

commute with matrices of the form

$$\mathbf{B} = \begin{pmatrix} \alpha & \beta \\ \beta & \alpha \end{pmatrix}$$

(that is, which matrices does **AB** = **BA** *hold for?). Verify that your answer is correct by trying some examples; the more obscure the numbers you use the more unlikely your result is not a fluke.*

We have also seen that it is possible to use "dot" arithmetic, which acts element by element on matrices and vectors.

Task 6.7 *Show that for all square matrices* $\mathbf{A}$, *the matrices* $\mathbf{B} = \mathbf{A} + \mathbf{A}^T$ *and* $\mathbf{C} = \mathbf{A} - \mathbf{A}^T$ *are symmetric and anti-symmetric respectively.*

Task 6.8 *Construct the matrix*

$$\mathbf{A} = \begin{pmatrix} \cos\theta & \sin\theta \\ -\sin\theta & \cos\theta \end{pmatrix}$$

for $\theta = 0$, $\theta = \pi/2$ *and* $\theta = \pi$ *(you can either do this directly or by using MATLAB). By considering the effect of multiplying by the matrix, infer how* $\mathbf{Ax}$ *compares to* $\mathbf{x}$ *for general* $\mathbf{x} \in \mathbb{R}^2$ *(that is points* (x, y) *where* $x, y \in \mathbb{R}$*). Construct the inverse of this matrix either using MATLAB, directly by hand (if you know how) or by using geometric arguments.*

Task 6.9 *Solve the linear system of equations*

$$\begin{aligned} 3x + 4y &= 2 \\ -x + 2y &= 0 \end{aligned}$$

using MATLAB.

Task 6.10 *Solve the linear system of equations*

$$\begin{aligned} x + y + 2z &= 1 \\ x - y - 3z &= 0 \\ -2x - 5y + z &= 4. \end{aligned}$$

If you are still not confident with matrix manipulation you should try the following tasks.

Task 6.11 *Given the matrices*

$$\mathbf{A} = \begin{pmatrix} 1 & -1 \\ 0 & 2 \\ 3 & 2 \end{pmatrix}, \quad \mathbf{B} = \begin{pmatrix} 2 & -1 \\ -1 & 0 \\ 3 & 2 \end{pmatrix} \text{ and } \mathbf{C} = \begin{pmatrix} -1 & 0 \\ 2 & 1 \end{pmatrix}.$$

Calculate where possible $\mathbf{A} + \mathbf{B}$, $\mathbf{AC}$, $\mathbf{CB}$, $(\mathbf{A} - \mathbf{B})\mathbf{C}$ *and* $\mathbf{AC} - \mathbf{BC}$.

Task 6.12 *Calculate the quantities*

$$\begin{pmatrix} 1 & -1 & 2 \\ 3 & 0 & 1 \end{pmatrix}\begin{pmatrix} 3 \\ 2 \\ 1 \end{pmatrix} \quad \textit{and} \quad \begin{pmatrix} 5 & -2 \\ -1 & 2 \end{pmatrix}\begin{pmatrix} 4 & 0 & 1 & -1 \\ 2 & 1 & -2 & -1 \end{pmatrix}.$$

Task 6.13 *Calculate, by hand, the quantities*

$$\begin{pmatrix} a & b \\ c & d \end{pmatrix}\begin{pmatrix} 1 & 0 \\ 0 & 1 \end{pmatrix} \quad \textit{and} \quad \begin{pmatrix} 1 & 0 \\ 0 & 1 \end{pmatrix}\begin{pmatrix} a & b \\ c & d \end{pmatrix}.$$

As we saw earlier a matrix with ones on the leading diagonal (running from top left to bottom right of a square matrix) is called the identity matrix; it is usually denoted by $\mathbf{I}$.

Task 6.14 *If*

$$\mathbf{A} = \begin{pmatrix} 3 & 2 & -1 \\ 0 & -1 & -2 \end{pmatrix}$$

calculate the quantities $\mathbf{A}\mathbf{A}^T$ *and* $\mathbf{A}^T\mathbf{A}$ *where a superscript* T *denotes the transpose and corresponds to a reflection of the matrix elements about the leading diagonal. If* $\mathbf{A}$ *has elements* $a_{i,j}$ *then* $\mathbf{A}^T$ *has elements* $a_{j,i}$, *and the number of rows of* $\mathbf{A}$ *equals the number of columns of* $\mathbf{A}^T$ *(and similarly for the number of columns of* $\mathbf{A}$ *and rows of* $\mathbf{A}^T$*).*

Task 6.15 *Show that the calculation* $\mathbf{x}\mathbf{x}^T$ *where* $\mathbf{x}$ *is a row vector with real entries always gives a positive scalar.*

Task 6.16 *Expand the matrix equation*

$$\begin{pmatrix} 1 & 4 \\ -2 & 3 \end{pmatrix}\begin{pmatrix} x \\ y \end{pmatrix} = \begin{pmatrix} 1 \\ -2 \end{pmatrix}$$

and write it as two simultaneous equations. Write the set of three simultaneous equations in matrix form:

$$\begin{aligned} x + y + z &= 0 \\ x - 2y - z &= 2 \\ -x + 3y - z &= -1. \end{aligned}$$

Task 6.17 *Determine whether the following systems have solutions (and if these are unique):*

$$\begin{aligned} 3x + 2y &= 7 \\ 3x - 2y &= 7; \end{aligned}$$

$$\begin{aligned} x + y + z &= 1 \\ x + y - z &= 0 \\ x + y &= 0; \end{aligned}$$

$$\begin{aligned} x + y + z + a + b + c &= 1 \\ x - y + z + a + b + c &= 1 \\ x + y - z + a + b + c &= 1 \\ x + y + z - a + b + c &= 1 \\ x + y + z + a - b + c &= 1 \\ x + y + z + a + b - c &= 1. \end{aligned}$$

Task 6.18 *Determine the solution of the systems*

$$\begin{aligned} x_1 - x_4 &= 0 \\ -x_1 + 2x_2 - x_3 &= 0 \\ -x_2 + 2x_3 - x_4 &= 0 \\ x_4 &= 1; \end{aligned}$$

and

$$\begin{aligned} x_1 - x_4 &= 1 \\ -x_1 + 2x_2 - x_3 &= 0 \\ -x_2 + 2x_3 - x_4 &= 0 \\ x_4 &= 0. \end{aligned}$$

Task 6.19 *By using the command* **`polyfit`** *determine the cubic polynomial which is the determinant of the matrix*

$$\begin{pmatrix} 0 & 1 & s \\ s & 0 & 1 \\ 1 & s & 0 \end{pmatrix}.$$

Consequently determine the values of s for which this matrix is singular.

Task 6.20 *Determine an analytical expression for $\mathbf{B}^n$ where*

$$\mathbf{B} = \begin{pmatrix} 0 & 1 \\ -1 & 0 \end{pmatrix}.$$

Task 6.21 *Determine the eigenvalues of the matrix*

$$\begin{pmatrix} 1 & 0 & 0 & -1 \\ 0 & 1 & 0 & 0 \\ 0 & 0 & 1 & 0 \\ -1 & 0 & 0 & 1 \end{pmatrix}.$$

Task 6.22 *Prove that* $\mathbf{A}^n = \mathbf{P}\mathbf{D}^n\mathbf{P}^{-1}$ *using induction where* $n \in \mathbb{N}$ *and* $\mathbf{P}$ *and* $\mathbf{D}$ *are comprised of the eigenvectors (as columns) and eigenvalues of* $\mathbf{A}$ *respectively.*

Task 6.23 *Determine the characteristic polynomial of the matrix in Task 6.21 using the code on page 214 and, by finding the roots of this polynomial, verify the answer to that task.*

Task 6.24 *Solve the differential system*

$$\dot{\mathbf{x}} = \begin{pmatrix} 2 & -1 \\ -1 & 1 \end{pmatrix} \mathbf{x}$$

subject to the initial condition $\mathbf{x}(0) = (1, -1)^T$.

7 Numerical Integration

7.1 Introduction

In this chapter we shall discuss techniques whereby functions can be integrated; these are quite classical. We will give the derivation of the approximations and will mention the likely failings of the techniques. Two versions of the code will be given for each of the main methods, so we can start to appreciate the power of MATLAB. It is of course possible to write the code as if it were Fortran or C, but that would waste the power of the package.

Before we start we note: if we consider two points we can fit a straight line through them; with three we can fit a quadratic and with four we can fit a cubic (this was discussed in more detail in Chapter 5).

We start by breaking down the integration régime into small intervals and approximating the area below the curve by slices. This method is tantamount to counting the squares and dealing with the parts of squares at the tops of the columns in sophisticated ways.

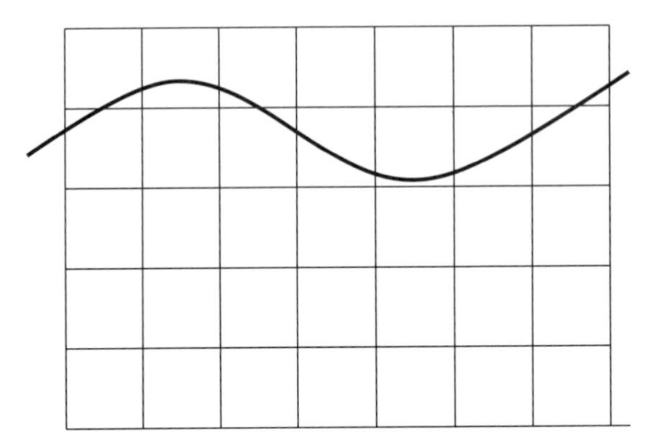

In this case the area under the curve is approximately

$$7 \times 4 \times \text{"the area of the boxes"}$$

(the fourth row up has a few part squares as does the fifth row). We obviously need a scheme which is slightly more robust.

7.2 Integration Using Straight Lines

In this chapter our objective is to calculate the value of the integral

$$I = \int_{x=a}^{b} f(x)\,\mathrm{d}x.$$

We shall assume we can calculate (easily) the value of $f(x)$ for all values of x between a and b. This means we are dealing with a function rather than a set of data values. We shall explore the latter case in due course but at this juncture we shall use a new routine `integrand.m`:

```
%
% integrand.m
% input a set of values (x)
% output function values f(x)
%
function [f] = integrand(x)
%  Here we use f = sin(x^2) as a
% sample function.
f = sin(x.^2);
```

We shall now take a while to derive the method we are going to use to integrate the function on the grid of points. We shall use N points and as such we use the code

```
step = (b-a)/(N-1);
x = a:step:b;
f = integrand(x);
```

In order to derive the form for the integration we introduce the nomenclature that the points we have defined above are (x_j, f_j), where j runs from 1 to N. Let us consider the consecutive points (x_j, f_j) and (x_{j+1}, f_{j+1}) and approximate the curve between them by a straight line. We use the formula for the line through the points (x_1, y_1) and (x_2, y_2) which is

$$\frac{y - y_1}{y_2 - y_1} = \frac{x - x_1}{x_2 - x_1} \text{ or } y = y_1 + \frac{y_2 - y_1}{x_2 - x_1}(x - x_1).$$

In our case this gives

$$f_L(x) = f_j + \frac{f_{j+1} - f_j}{h}(x - x_j),$$

where we have introduced $h = x_{j+1} - x_j$ and $f_L(x)$ the formula for the straight line (This is just a direct application of Newton's Forward Differences, (5.1).) Let us now perform the analytical integration of the function $f_L(x)$ between x_j and x_{j+1} to determine the area $A_{j,j+1}$:

$$A_{j,j+1} = \int_{x=x_j}^{x_{j+1}} f_L(x)\, dx = \int_{x=x_j}^{x_{j+1}} \frac{f_{j+1} - f_j}{h}(x - x_j) + f_j\, dx.$$

We introduce the linear transformation $X = x - x_j$, where $dX = dx$ and when $x = x_j$, $X = 0$ and $x = x_{j+1}$ correspond to $X = h$. The integral becomes

$$A_{j,j+1} = \int_{X=0}^{h} \left[\frac{f_{j+1} - f_j}{h} X + f_j\right] dX,$$

which can be integrated to yield

$$A_{j,j+1} = \frac{h}{2}(f_j + f_{j+1}).$$

This is the area of a trapezium with vertices at $(x_j, 0)$, (x_j, f_j), (x_{j+1}, f_{j+1}) and $(x_{j+1}, 0)$[1].

[1] The area of a trapezium is the mean of the length of the two parallel sides times the perpendicular distance between them; that is $(f_j + f_{j+1})/2$ times $(x_{j+1} - x_j)$.

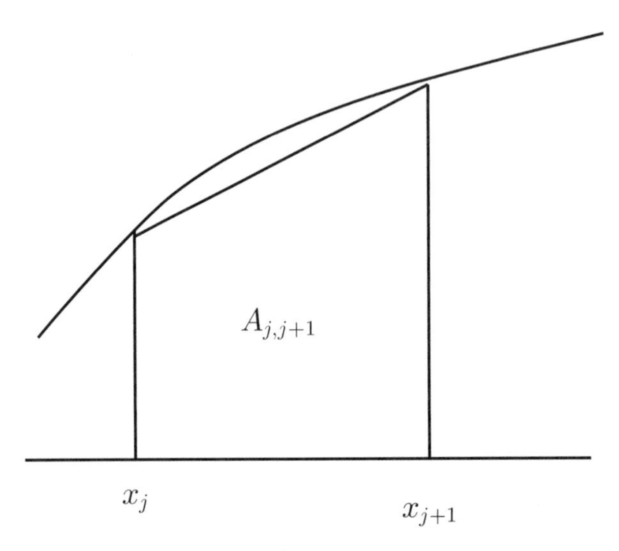

This method is unsurprisingly called the trapezium rule. In order to determine the total area from $x = a$ to $x = b$ we sum all the parts

$$\text{Area} = \sum_{j=1}^{j=N-1} A_{j,j+1} = \sum_{j=1}^{j=N-1} \frac{h}{2}(f_j + f_{j+1}).$$

In order to perform this summation it is instructive to write out the series

$$\begin{aligned}\text{Area} =& \frac{h}{2}(f_1 + f_2) + \frac{h}{2}(f_2 + f_3) + \frac{h}{2}(f_3 + f_4) + \cdots \\ &+ \frac{h}{2}(f_{N-2} + f_{N-1}) + \frac{h}{2}(f_{N-1} + f_N),\end{aligned}$$

so that

$$\text{Area} = \frac{h}{2}(f_1 + 2f_2 + 2f_3 + \cdots + 2f_{N-1} + f_N) \approx \int_{x=a}^{b} f(x)\,\mathrm{d}x.$$

Example 7.1 *We shall calculate the integral of the function* $f(x) = x^3 \sin x$ *between zero and one. We shall use* N *points*

```
N = 10;
x = linspace(0,1,N);
h = x(2)-x(1);
f = x.^3.*sin(x);
g = h*(sum(f)-f(1)/2-f(N)/2);
```

Here we have constructed a grid of points running from zero to one and then worked out the gap between successive points (that is `h`*). We now construct the function* `f` *and work out the expression for the trapezium rule, which is the sum of the values of* `f` *minus half of the end values. This gives the value of the integral as* `g`*.*

7.2.1 Errors in the Trapezium Method

The number of points required for a calculation depends on how exactly you need to know the answer (in general). The error in this scheme is encountered because we approximate the curve between the points (x_j, f_j) and (x_{j+1}, f_{j+1}) by a straight line. This can be reduced by using a quadratic instead over the interval spanned by the requisite three points.

We could have used the fundamental code which did not make use of the combinations of the terms:

```
integral = 0;
for i = 2:N
    integral = integral+(f(i)+f(i-1))/2*h;
end
```

The advantage of this form of the code is it is far easier to convert to one in which intermediate results are available. We note that the command `/2*h` divides by two and multiplies by `h` rather than dividing by `2h`: this would be accomplished via parentheses, that is `/(2*h)`.

Before we proceed we should consider the error involved in approximating the area by a set of trapezia.

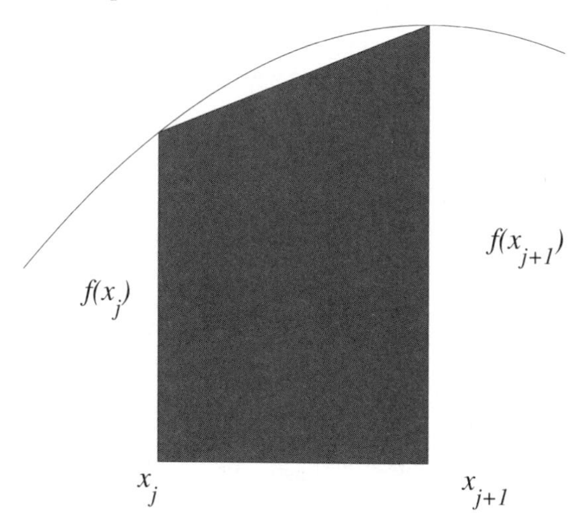

The error is the unshaded part below the curve. To estimate the extent of this we again use "Taylor Series" and note that

$$f(x) = f(x_j) + \frac{f(x_{j+1}) - f(x_j)}{h}(x - x_j) + \frac{(x - x_j)(x - x_{j+1})}{2} \left.\frac{\mathrm{d}^2 f}{\mathrm{d}x^2}\right|_{x=\xi},$$

where $\xi \in [x_j, x_{j+1}]$. This can be integrated to give

$$\int_{x=x_j}^{x_{j+1}} f(x)\,\mathrm{d}x = \frac{h}{2}\left(f(x_j) + f(x_{j+1})\right) + \frac{h^3}{6}\left.\frac{\mathrm{d}^2 f}{\mathrm{d}x^2}\right|_{x=\xi}.$$

Consequently the error in using just the first term is proportional to h^3 and $f''(\xi)$. Notice that if the second derivative is zero over the range the error is actually zero. Unsurprisingly this corresponds to $f(x)$ being a straight line.

7.3 Integration Using Quadratics

We need to construct a curve which passes through the points (x_{j-1}, f_{j-1}), (x_j, f_j) and (x_{j+1}, f_{j+1}). Let us consider a value of $j = 1$: this is purely to reduce the verbosity of our expressions. The quadratic passing through these three points can be written as

$$f_q(x) = f_0 + \Delta f_0 \frac{x - x_0}{h} + \Delta^2 f_0 \frac{(x - x_0)(x - x_1)}{2h^2},$$

using a truncation of Newton's forward difference formula, (5.1).

In passing we mention this series expansion is similar to a Taylor series expansion but instead of the terms being differentials they are differences. In order to derive the formula we now need to integrate the function $f_q(x)$ from $x = x_0$ to $x = x_2$. This is quite straightforward: however at this point we will exploit the symbolic capabilities of MATLAB

```
syms x f0 f1 f2 h x0
q = f0+(f1-f0)/h*(x-x0)+(f2-2*f1+f0)/(2*h^2)*(x-x0)*(x-(x0+h));
iq = int(q,x0,x0+2*h);
simplify(iq)


ans =


1/3*h*(f0+4*f1+f2)
```

We shall dissect this portion of code so you can appreciate what is happening.

- In the first line we assign the variables `x`, `f0`, `f1`, `f2`, `h` and `x0` to be symbolic. This means that MATLAB does not have to know the value of the variables and treats them as mathematical objects.
- In the second line we set up the function `q` (which is just the quadratic $f_q(x)$).
- This is integrated in the third line (between `x0` and `x2 = x0+2*h`. (This produces a verbose answer.)
- Finally we simplify our answer.

Hence we have the answer

$$\int_{x=x_0}^{x=x_2} f_q(x)\,\mathrm{d}x = \frac{h}{3}\left(f_0 + 4f_1 + f_2\right).$$

Although we know there is only one quadratic through a given set of points, it is instructive to re-derive this in another way. We shall start with the general quadratic

$$f_q(x) = a + b(x - x_0) + c(x - x_0)(x - x_1).$$

The requirement that the quadratic goes through the point (x_0, f_0) gives us that $a = f_0$. Using the point (x_1, f_1) gives us $b = (f_1 - f_0)(x_1 - x_0)$. Finally using (x_2, f_2) we have

$$f_2 = f_0 + \frac{f_1 - f_0}{x_1 - x_0}(x_2 - x_0) + c(x_2 - x_0)(x_2 - x_1),$$

which can be manipulated to give:

$$c = \frac{1}{x_2 - x_1}\left(\frac{f_2 - f_0}{x_2 - x_0} - \frac{f_1 - f_0}{x_1 - x_0}\right).$$

This gives us the general form of the quadratic through three points (which can be used for Task 7.10). We now return to the form for regularly spaced points so $\Delta x_j = h$ and integrating from $x = x_0$ to $x = x_2 = x_0 + 2h$:

$$\int_{x=x_0}^{x=x_0+2h} f_q(x)\,\mathrm{d}x = \int_{x=x_0}^{x=x_0+2h} f_0 + \frac{f_1 - f_0}{h}(x - x_0) + \frac{1}{h}\left(\frac{f_2 - f_0}{2h} - \frac{f_1 - f_0}{h}\right)(x - x_0)(x - x_1)\,\mathrm{d}x.$$

Now using the substitution $X = x - x_0$, so that $x = x_0$ corresponds to $X = 0$ and $x = x_2 = x_0 + 2h$ corresponds to $X = 2h$ the expression $x - x_1 =$

$X + x_0 - x_1 = X - h$ and $\mathrm{d}x = \mathrm{d}X$. Hence we have

$$
\begin{aligned}
&= \int_{X=0}^{2h} f_0 + \frac{f_1 - f_0}{h} X + \frac{1}{2h^2} (f_2 - f_0 - 2(f_1 - f_0)) X(X-h) \, \mathrm{d}X \\
&= \left[f_0 X + \frac{f_1 - f_0}{h} \frac{X^2}{2} + \frac{1}{2h^2} (f_0 - 2f_1 + f_2) \left(\frac{X^3}{3} - h \frac{X^2}{2} \right) \right]_0^{2h} \\
&= 2h f_0 + \frac{f_1 - f_0}{h} 2h^2 + \frac{1}{2h^2} (f_0 - 2f_1 + f_2) \left(\frac{8h^3}{3} - 2h^3 \right) \\
&= h \left(2f_0 + 2(f_1 - f_0) + \frac{1}{3} (f_0 - 2f_1 + f_2) \right) \\
&= \frac{h}{3} (f_0 + 4f_1 + f_2) .
\end{aligned}
$$

We now need to divide the range of integration into the appropriate number of subintervals. Notice the number of intervals needs to be even (and hence the number of points needs to be odd). The total integral is approximated by

$$
\begin{aligned}
\int_{x=a}^{b} f(x) \, \mathrm{d}x \approx & \frac{h}{3} (f_0 + 4f_1 + f_2) + \frac{h}{3} (f_2 + 4f_3 + f_4) + \frac{h}{3} (f_4 + 4f_5 + f_6) + \cdots \\
& + \frac{h}{3} (f_{N-4} + 4f_{N-3} + f_{N-2}) + \frac{h}{3} (f_{N-2} + 4f_{N-1} + f_N) .
\end{aligned}
$$

This can be simplified to give

$$
\int_{x=a}^{b} f(x) \, \mathrm{d}x \approx \frac{h}{3} (f_0 + 4f_1 + 2f_2 + 4f_3 + 2f_4 + \cdots + 2f_{N-2} + 4f_{N-1} + f_N) .
$$

This is called Simpson's $\frac{1}{3}$ rule.

We shall now construct a code to determine the integral. At this point we could use a code which used a conventional approach, but we shall try to use a version which exploits the power of MATLAB.

```
%
% Simpson's 1/3 rule.
%
x = 0.0:0.1:1.0;
h = x(2)-x(1);
N = length(x);
if mod(N,2) == 0
         disp('Routine needs an odd number of points')
         break     % Ensure the number of points is odd
end

rodd = 1:2:N;
reven = 2:2:(N-1);
weights(rodd) = 2; weights(1) = 1; weights(N) = 1;
weights(reven) = 4;

f = sin(x.^2);
integral = h/3*sum(weights.*f);
format long e
disp([integral])
```

This calculates the value of the expression

$$\int_0^1 \sin x^2 \, \mathrm{d}x.$$

The value which MATLAB comes out with is `3.102602344332209e-01`, where we have changed the way in which these numbers are displayed by using the command `format long e`. This answer can be checked using MATLAB's symbolic capabilities:

```
syms x f
f = sin(x^2);
f1 = int(f,0,1)


f1 =


1/2*FresnelS(2^(1/2)/pi^(1/2))*2^(1/2)*pi^(1/2)
```

This value can be compared with that attained using the symbolic toolbox, using the command `vpa` (variable precision arithmetic):

```
vpa(f1)

ans =


.310268301723381101808152423165
```

(Note we can change the number of digits using the command `digits(10)`). We further note that this is still a symbolic object: in order to obtain a value (which can be used for plotting, for example) we use the command `double`. This is a case of MATLAB being too clever: it has solved the integral and written it as a Fresnel integral (for further details see Abramowitz and Stegun – *Handbook of Mathematical Functions*). Our simple integration using ten points does quite well: in fact the error is proportional to h^4, which can be shown using the same technique as we used for the trapezium rule on page 230.

In the above code we have introduced the term `weights` which is applied to the terms before they are added together to obtain the integral. The MATLAB command `mod` allows the user to determine the remainder when the first argument is divided by the second. In this case when considering `mod` 2 we are checking for parity (that is whether N is even or odd).

In the code for Simpson's one third rule we have re-introduced another MATLAB command, namely `break`. This stops the code, or more exactly it exits the current level: for example if it is used in a nested loop it will terminate the current level and return to the previous level. This manual entry for `break` is instructive here

```
>> help break

 BREAK Terminate execution of WHILE or FOR loop.
    BREAK terminates the execution of FOR and WHILE loops.
    In nested loops, BREAK exits from the innermost loop only.
```

We now progress to consider cubic approximations to the function in the hope that this will provide even more accurate answers.

Again we can use a technique which involves summing the separate intervals, which you may prefer:

```
integral = 0;
for j = 1:2:N-2
    integral = integral + h*(f(j)+4*f(j+1)+f(j+2))/3;
end
```

7.4 Integration Using Cubic Polynomials

As in the previous sections we need to define an approximating curve and in order to do so we need four points (x_0, f_0), (x_1, f_1), (x_2, f_2) and (x_3, f_3). Using the same form as above we can write the cubic equation as

$$f_c(x) = f_0 + \Delta f_0 \frac{x - x_0}{h} + \Delta^2 f_0 \frac{(x - x_0)(x - x_1)}{2h^2} + \Delta^3 f_0 \frac{(x - x_0)(x - x_1)(x - x_2)}{6h^3}.$$

In addition to the terms defined earlier we have introduced the third-order forward difference. This can be defined recursively using Δ^2, so that

$$\begin{aligned} \Delta^3 f_0 = \Delta^2(\Delta f_0) &= \Delta^2 f_1 - \Delta^2 f_0 \\ &= (f_3 - 2f_2 + f_1) - (f_2 - 2f_1 + f_0) \\ &= f_3 - 3f_2 + 3f_1 - f_0, \end{aligned}$$

using the Newton forward difference for $\Delta^2 f_0$. We now need to integrate from $x = x_0$ to $x = x_3$, and we again exploit the symbolic tool box for this, using the code

```
syms x f0 f1 f2 f3 h x0
x1 = x0+h; x2 = x0+2*h; x3 = x0+3*h;
t1 = (f1-f0)/h*(x-x0);
t2 = (f2-2*f1+f0)/(2*h^2)*(x-x0)*(x-x1);
t3 = (f3-3*f2+3*f1-f0)/(6*h^3)*(x-x0)*(x-x1)*(x-x2);
q = f0+t1+t2+t3;
q1 = int(q,x,x0,x3);
simplify(q1)
```

This gives the answer `3/8*h*(f0+3*f1+3*f2+f3)`. Thus we have

$$\int_{x=x_0}^{x_3} f_c(x)\,\mathrm{d}x = \frac{3h}{8}\left(f_0 + 3f_1 + 3f_2 + f_3\right).$$

You should note it is possible to do all these integrals by hand, but we wish to demonstrate the power and the utility of this particular toolbox.

We now need to combine all the subintervals, so

$$\begin{aligned} \int_{x=a}^{b} f(x)\,\mathrm{d}x \;&\approx\; \frac{3h}{8}\left(f_0 + 3f_1 + 3f_2 + f_3\right) + \frac{3h}{8}\left(f_3 + 3f_4 + 3f_5 + f_6\right) + \cdots \\ &\quad + \frac{3h}{8}\left(f_{N-3} + 3f_{N-2} + 3f_{N-1} + f_N\right), \end{aligned}$$

which can be simplified to give

$$\int_{x=a}^{b} f(x)\,\mathrm{d}x \approx \frac{3h}{8}\left(f_0 + 3f_1 + 3f_2 + 2f_3 + 3f_4 + 3f_5 + \cdots \right.$$
$$\left. + 2f_{N-3} + 3f_{N-2} + 3f_{N-1} + f_N\right).$$

This is referred to as Simpson's $\frac{3}{8}$ rule.

We shall now give a MATLAB program based on the same structure as the previous one. Notice this time the number of points needs to be divisible by three.

```
%
% simpson's 3/8 rule.
%
N = 10;
x = linspace(0,1,N);
h = x(2)-x(1);
ms = 'Number of intervals should be divisible by three';

if mod(N-1,3) ~= 0
         disp(ms)
         break
end
m = (N-1)/3;
rdiff = 3*(1:(m-1))+1;
weights = 3*ones(1,N);
weights(1) = 1; weights(N) = 1;
weights(rdiff) = 2;
f = sin(x.^2);
integral = 3*h/8*sum(weights.*f);
disp([integral])
```

The answer given by this is `0.31024037588964`. In fact this is not quite as good as the previous method. The error is again proportional to h^4 but the constant of proportionality is larger. We could also write the integral in separate regions, that is without combination, so

```
integral = 0;
for j = 1:3:N-3
    integral = integral + h*(f(j)+3*f(j+1) ...
                          +3*f(j+2)+f(j+3))*3/8;
end
```

Each of these methods has a restriction on the numbers of points one can use. However these can be circumvented by using an amalgamation of the two schemes. There are other methods available for integrating functions and some of these will be met in due course. We could continue this process, especially with the symbolic toolbox at our disposal to perform the algebra. For instance the formula obtained by using a quartic over five points is

$$\int_{x=x_0}^{x_4} f(x)\,\mathrm{d}x \approx \frac{2h}{45}\left(7f_0 + 32f_1 + 12f_2 + 32f_3 + 7f_4\right).$$

Although you might think the higher the order the polynomial the more accurately you would know the answer, there are problems which are intrinsic to using high-order polynomials. In fact the optimum method is to use a combination of the two Simpson methods. If the number of points supplied is even we use the one third rule for the first $N-3$ points (which is necessarily an odd number of points and consequently an even number of intervals) and then we use the three-eighths rule on the remaining points.

7.5 Integrating Using MATLAB Commands

As with many examples in this text we can also use standard MATLAB commands, for instance `quad` and `quad8` (see `help quad`). These commands use similar techniques to those above but with the advantage of automation. For instance they exploit grids which can adapt. This means that in regions which are harder to integrate (perhaps with more variation in the function) the scheme adds extra points.

The syntax is:

```
tol = [1e-4 1e-5];
a = 0; b = pi;
trace = 1;
q = quad('sin',a,b,tol,trace)
```

These quantities are respectively the extent of the domain $x \in [a, b]$, the tolerances (relative and absolute) and whether the user wants to see a trace or not (setting trace as non-zero shows the evolution of the calculation).

Example 7.2 *We show the integration of the function $J_1(x)$ for $x = 0$ to $x = 10$. This is actually a Bessel function which occurs as one of the solutions* *to the differential equation*

$$x^2\frac{\mathrm{d}^2y}{\mathrm{d}x^2} + x\frac{\mathrm{d}y}{\mathrm{d}x} + (x^2-1)y = 0,$$

and is revisited in Chapter 8.

Fortunately MATLAB has a routine which evaluates Bessel functions but we need to write our own routine to make sure that it is available for **quad**:

```
function [value]=ourbess(x)
value = besselj(1,x);
```

and then we simply need the code:

```
q = quad('ourbess',0,10,[1e-5 1e-5],1);
```

This gives a value of 1.24593587184673. *The integral of the function* $J_1(x)$ *is* $-J_0(x)$ *and hence the value we seek is* $-J_0(10) + J_0(0) \approx 1.245935764$; *so as we can see the integration scheme does very well.*

7.6 Specific Examples of Integrals

We shall now describe how we can deal with other problems which arise when evaluating numerical integrals.

7.6.1 Infinite Integrals and Removable Singularities

Using the various methods we can evaluate integrals from a to b, but if one or more of these values is infinite we need a different treatment. This will vary depending on the form of the integral or more exactly the integrand.

Example 7.3 *For example let us consider the integral*

$$I = \int_0^\infty \mathrm{e}^{-x^2}\,\mathrm{d}x.$$

In this case the integrand e^{-x^2} *decays very quickly so we can adequately determine the value of the integral using*

$$I_X = \int_0^X \mathrm{e}^{-x^2}\,\mathrm{d}x,$$

for a suitable value of X *(in this case* $X = 10$ *is more than sufficient). We can evaluate the integral for a few values of* X *until* I_X *is constant (to within a defined tolerance). We could develop an algorithm to decide what value of* X *to use, but in general we will use the above method. We can also exploit the symmetry of problems, for instance*

$$\int_{-\infty}^{\infty} \mathrm{e}^{-x^2}\,\mathrm{d}x = 2\int_0^{\infty} \mathrm{e}^{-x^2}\,\mathrm{d}x \approx 2\int_0^X \mathrm{e}^{-x^2}\,\mathrm{d}x,$$

provided $X \gg 1$*. We can also use transformations to rescale the regions of integration.*

In fact in this example we could have used the MATLAB command `erf` which gives the error function

$$E(q) = \frac{2}{\sqrt{\pi}} \int_{q=0}^{x} \mathrm{e}^{-q^2}\,\mathrm{d}q.$$

We can use `erf(Inf)` which gives us the value unity.

We now consider how we might perform an integral for which the integrand is singular at an end-point of the range. For instance:

$$\int_{q=0}^{1} \frac{1}{q^{1/2}}\,\mathrm{d}q = \left[2q^{1/2}\right]_{q=0}^{1} = 2.$$

Example 7.4 *We consider the integral of the function* $f(x) = \mathrm{e}^{-x}/\sqrt{x}$ *between* $x = 0$ *and* $x = 1$*. Firstly we separate the range into the two disjoint ranges* $[0,\epsilon)\cup[\epsilon,1]$ *where* $\epsilon \ll 1$ *(that is it is very small). We now consider the integral:*

$$\int_0^{\epsilon} \frac{\mathrm{e}^{-x}}{\sqrt{x}}\,\mathrm{d}x$$

and note that over this range $\mathrm{e}^{-x} \approx 1 - x + x^2/2 + \cdots$. *Thus*

$$\int_0^{\epsilon} \frac{\mathrm{e}^{-x}}{\sqrt{x}}\,\mathrm{d}x \approx \int_0^{\epsilon} \frac{1}{\sqrt{x}}\left(1 - x + \frac{x^2}{2} + \cdots\right)\mathrm{d}x = \left[2x^{1/2} - \frac{2}{3}x^{3/2} - \frac{1}{5}x^{5/2} + \cdots\right]_0^{\epsilon}$$

$$= \left(2\epsilon^{1/2} - \frac{2}{3}\epsilon^{3/2} - \frac{1}{5}\epsilon^{5/2}\right).$$

The integral can now be evaluated using:

```
epsil = 0.01;
val = 2*sqrt(epsil)-2/3*epsil^1.5-1/5*epsil^2.5;
x = linspace(epsil,1,100);
f = exp(-x)./sqrt(x);
h = x(2)-x(1);
N = length(x);
int = val;
for j = 1:N-1
    int = int + h/2*(f(j+1)+f(j));
end
```

This gives a value `1.49764481658781` *(the value given by MATLAB is* `erf(1)*sqrt(pi)` *which is* $\mathit{erf}(1)\sqrt{\pi} \approx 1.493648266$*) where*

$$\mathit{erf}(x) = \frac{2}{\sqrt{\pi}} \int_{t=0}^{x} \mathrm{e}^{-t^2}\,\mathrm{d}t.$$

7.6.2 Indefinite Integrals

So far we have only been interested in definite integrals, but we shall now consider how one might determine

$$I(x) = \int_{q=a}^{x} f(q)\,\mathrm{d}q.$$

Instead of using a scalar variable to store the cumulative total, we exploit a vector to store intermediate results, so that

```
integral = zeros(size(x));
step = x(2)-x(1);
N = length(integral);
for j = 2:N
    integral(j) = integral(j-1)+step*(f(j)+f(j-1))/2;
end
disp(['Value of total integral ' num2str(integral(N))]);
```

Example 7.5 *We now plot the function*

$$g(x) = \int_{q=0}^{x} \mathrm{e}^{-q^2}\,\mathrm{d}q.$$

We determine $g(x)$ for $x = 0$ to $x = 10$ using the code

```
xv = linspace(0,10);
f = exp(-xv.^2);
N = length(xv);
step = xv(2)-xv(1);
g(1) = 0.0;
for j = 2:N
    g(j) = g(j-1)+step*(f(j)+f(j-1))/2;
end
plot(xv,g)
text(5,max(g)/2,'Integral of e^{-x^2}','FontSize',20)
```

which gives

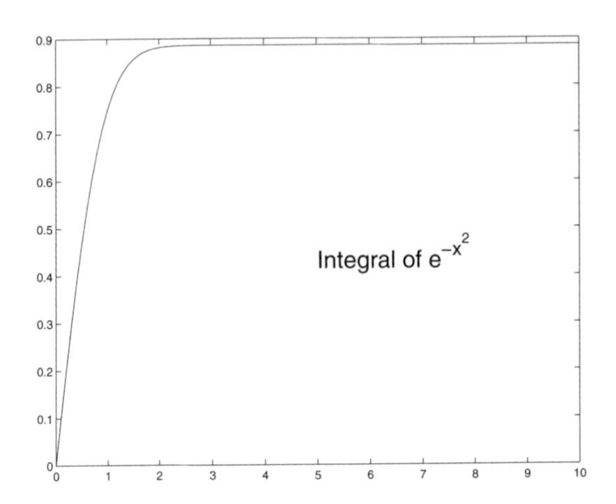

7.7 Tasks

Task 7.1 *Construct the sequence*

$$f(i) = \begin{cases} 1 & \text{if} \quad \bmod(i,3) = 0 \\ 2 & \text{if} \quad \bmod(i,3) = 1 \\ 3 & \text{if} \quad \bmod(i,3) = 2 \end{cases}$$

up to $N = 12$.

Task 7.2 *In Chapter 7 for Simpson's one third rule we used*

```
rodd = 1:2:N;
reven = 2:2:(N-1);
weights(rodd) = 2; weights(1) = 1; weights(N) = 1;
weights(reven) = 4;
```

and for Simpson's three eighths rule

```
m = (N-1)/3;
rdiff = 3*(1:(m-1))+1;
weights = 3*ones(1,N);
weights(1) = 1; weights(N) = 1;
weights(rdiff) = 2;
```

Write out these coefficients for $N = 9$ for the one third rule and $N = 10$ for the three eighths rule. You should construct each of the vectors by hand.

By now you should be able to read the codes `trap.m`, `simp13.m` *and* `simp38.m` *and understand what they do. You should also be able write your own program which returns the value of a function evaluated at a given point. Try this task.*

Task 7.3 *Write a routine which takes an input x and returns the value of $f(x) = \ln\left(x + \sqrt{x^2+1}\right)$.*

Task 7.4 *Using the trapezium rule calculate the integral of the quadratic $x^2 - 3x + 2$ between $x = 1$ and $x = 3$ (check your answer with the exact answer).*

Task 7.5 *Using Simpson's one third rule integrate the cubic $x^3 - x + 1$ between the limits $x = 0$ and $x = 1$ (check your answer against the exact answer).*

Task 7.6 *Using Simpson's one third rule integrate the function $f(x) = \sin x$ between the limits $x = 0$ and $x = \pi$ (check your answer against the exact answer). You might want to change the number of points you use and see what happens to the error.*

Task 7.7 *Calculate the value of the integral*

$$\int_0^\infty \frac{1}{\sqrt{x^2+1}}\,\mathrm{d}x.$$

You will need to truncate the domain, and you should investigate the effect of this truncation as well as the number of points required to accurately calculate the integral.

Task 7.8 *The length along a curve $y = y(x)$ from $x = a$ to $x = b$ is given by the expression*

$$S = \int_{x=a}^{b} \sqrt{1 + \left(\frac{\mathrm{d}y}{\mathrm{d}x}\right)^2}\,\mathrm{d}x.$$

In many cases this expression can be determined analytically: however there are some very simple cases for which it can't be. Consider the problem of a sine curve truncated over the range $[\theta, \pi - \theta]$:

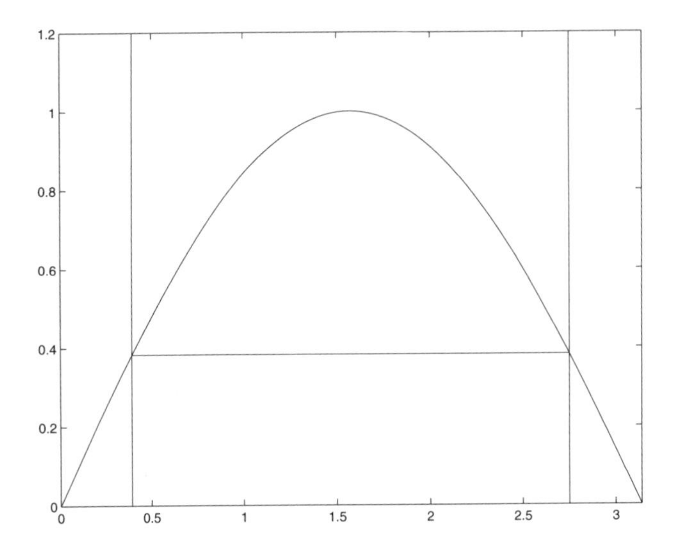

The expression for the length of this curve is

$$S = \int_{\phi=\theta}^{\pi-\theta} \sqrt{1+\cos^2 x}\,\mathrm{d}x.$$

Unfortunately this integral is intractable using analytic means, but determine the value numerically using the trapezium rule for a variety of values of θ.

Task 7.9 (*) *Determine the integral*

$$\int_0^{10} \frac{\cos x}{x^{1/2}}\,\mathrm{d}x$$

by splitting the integral into two ranges $[0,\epsilon]$ *and* $[\epsilon,10]$ *where* ϵ *is taken to be small. In this first range the function* $\cos x$ *can be approximated by its Taylor series and this form can be used to work out the contribution from the "singular part" of the integral.*

Task 7.10 (*) *Derive an expression for the integral of the quadratic passing through the points* (x_i, f_i) $i = 0, 1$ *and 2 from* $x = x_0$ *to* $x = x_2$ *(use the general quadratic on page 231). This gives you Simpson's scheme for variably spaced points this can also be extended to four points with a cubic.*

If you are feeling very brave you could also work out the errors associated with these approximations.

Task 7.11 *Integrate the function* $x \ln x$ *between the limits* 1 *and* 2 *using the MATLAB function* `quad`.

8
Solving Differential Equations

8.1 Introduction

In this book we have introduced you to a variety of techniques and hopefully given you a taster for how one might approach the solution of certain types of problem. In this chapter we shall tackle the problem of solving differential equations using numerical methods. This is perhaps the most prolific use of these techniques and it can involve all of the previous chapters and also those following.

Initially we will limit our attention to problems which we can solve analytically. This is purely so we can compare the solutions we obtain with those derived directly. It is very rare to be able to do this, although as your knowledge of the structure of differential equations increases you will realise more problems are tractable than one might initially think, but these still represent quite idealised situations. Understanding how methods work, or more specifically how they fail, is absolutely crucial when using them. This chapter will introduce quite a few terms.

8.2 Euler's Method and Crank–Nicolson

We shall start with first-order differential equations and we will use a formula which we used when we discussed the modification of the Newton–Raphson method (page 117). Consider the differential equation

$$\frac{\mathrm{d}y}{\mathrm{d}t} = f(y, t). \tag{8.1}$$

Using the formal definition we can write the derivative as

$$\frac{\mathrm{d}y}{\mathrm{d}t} = \lim_{\Delta t \to 0} \frac{y(t + \Delta t) - y(t)}{\Delta t}.$$

As $\Delta t \to 0$ this tends to the actual derivative (provided certain conditions are met), but in reality we will need to use a finite value for Δt. We shall denote the time level t by subscript n and the time level $t + \Delta t$ by the subscript $n+1$, in which case

$$\left.\frac{\mathrm{d}y}{\mathrm{d}t}\right|_{t=t_n} \approx \frac{y_{n+1} - y_n}{\Delta t},$$

which can be seen using the Taylor series expansion for $y(t + \Delta t)$. We can now evaluate the differential equation (8.1) at the time $t = t_n$, to give

$$\frac{y_{n+1} - y_n}{\Delta t} = f(y_n, t_n).$$

This is easily rearranged to give

$$y_{n+1} = y_n + \Delta t f(y_n, t_n).$$

This is called Euler's method and is an example of an explicit method. Let us use it to solve a couple of problems:

Example 8.1 *Consider the differential equation*

$$\frac{\mathrm{d}y}{\mathrm{d}t} = t$$

subject to the initial condition that $y(0) = 0$ *from 0 to 2. We can integrate this equation directly to give* $y(t) = t^2/2$. *Or using the explicit Euler method we have*

$$\frac{y_{n+1} - y_n}{\Delta t} = t_n,$$

which can be rearranged to give

$$y_{n+1} = y_n + (n-1)(\Delta t)^2$$

where we have used $t_n = (n-1)\Delta t$ *(so that* $t_1 = 0$*).*

Writing out the first few equations:

$$
\begin{aligned}
y_2 &= y_1, \\
y_3 &= y_2 + (\Delta t)^2 = y_1 + (\Delta t)^2, \\
y_4 &= y_3 + 2(\Delta t)^2 = y_1 + 3(\Delta t)^2, \\
y_5 &= y_4 + 3(\Delta t)^2 = y_1 + 6(\Delta t)^2, \\
y_6 &= y_5 + 4(\Delta t)^2 = y_1 + 10(\Delta t)^2 \ldots
\end{aligned}
$$

In fact we have the result that

$$y_n = y_1 + \frac{(n-1)(n-2)}{2}(\Delta t)^2$$

and this should be $y_1 + \frac{(n-1)^2}{2}(\Delta t)^2$ *for the exact solution: so we can see that this scheme does a reasonable job.*

Example 8.2 *Solve the differential equation*

$$\frac{\mathrm{d}y}{\mathrm{d}t} = (1-t)y$$

subject to the initial condition $y(0) = 1$ *from* $t = 0$ *to* $t = 5$*. Firstly, let us determine the analytical solution. This can be found by dividing through by* y

$$\frac{1}{y}\frac{\mathrm{d}y}{\mathrm{d}t} = 1 - t$$

and then integrating with respect to t

$$
\begin{aligned}
\int \frac{1}{y}\frac{\mathrm{d}y}{\mathrm{d}t}\,\mathrm{d}t &= \int 1 - t\,\mathrm{d}t \\
\int \frac{\mathrm{d}y}{y} &= t - \frac{t^2}{2} + C \\
\ln y &= t - \frac{t^2}{2} + C \\
y &= A\mathrm{e}^{t-t^2/2}.
\end{aligned}
$$

Now applying the initial condition we find that $A = 1$ *so the actual solution to the problem is:*

$$y = \mathrm{e}^{t-t^2/2}.$$

Now let us determine the numerical solution. We shall firstly set up the vectors for the time grid and the solution: the actual coding of the algorithm is very simple. The function $f(y,t) = (1-t)y$*, so*

```
% Euler's method
% euler.m
dt = 0.1;
t = 0.0:dt:5.0;
y = zeros(size(t));
y(1) = 1;
for ii = 1:(length(t)-1)
      y(ii+1) = y(ii) + dt * (1-t(ii))*y(ii);
end
exact = exp(t-t.^2/2);
plot(t,y,t,exact,'--')
```

This solution compares well with the exact solution shown as the dashed curve.

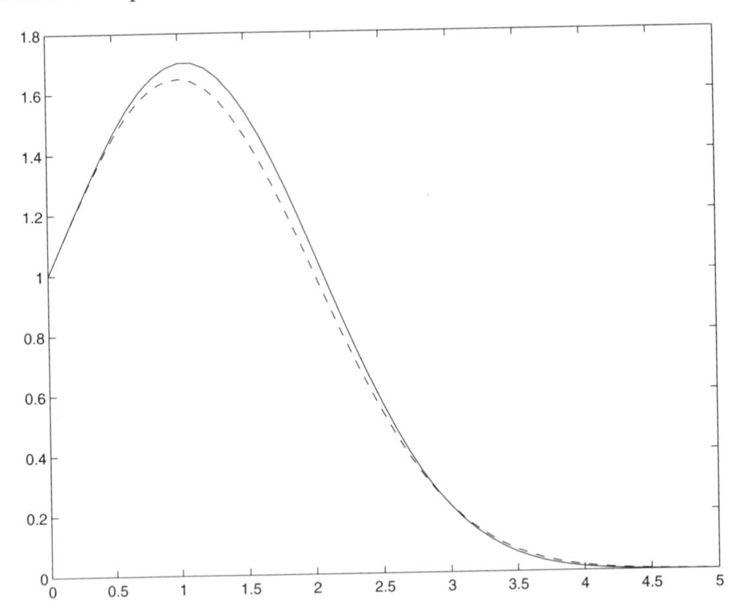

We have used fifty-one points to perform this integration and if we used more then we could have improved the accuracy of this solution.

Consider the example:

Example 8.3 *Solve the differential equation*

$$\frac{\mathrm{d}y}{\mathrm{d}t} = y$$

subject to the initial condition $y(0) = 1$*. The solution of this is* $y(t) = \mathrm{e}^t$ *and in this case* $f(y) = y$*, and we reuse the above codes changing the appropriate line in the loop.*

This method is dubbed **explicit**. One way of thinking of this is the value of y_{n+1} is available explicitly from the equation. We have taken the equation to be evaluated at $t = t_n$ but equally it could have been evaluated at $t = t_{n+1}$. Instead of using the formal definition to give the derivative at t we use it at $t + \Delta t$ (which is equally valid – representing a backward difference rather than a forward difference). The formula is now

$$y_{n+1} = y_n + \Delta t f(y_{n+1}, t_{n+1}).$$

Depending on the form of f this can be coded easily or it may involve subtle manipulation or use of the root finding techniques detailed in Chapter 4. These techniques are called **implicit** since the value (that is y_{n+1}) is contained within the equation; that is it is implicit to it. As an illustration let us consider the same example as above, that is

$$\frac{\mathrm{d}y}{\mathrm{d}t} = y, \qquad \text{subject to} \qquad y(0) = 1.$$

Using this new scheme this equation would be discretised as

$$\frac{y_{n+1} - y_n}{\Delta t} = y_{n+1},$$

so that

$$y_{n+1} = \frac{y_n}{1 - \Delta t}.$$

This type of solution generally improves matters. In fact in this case the schemes are very similar (especially for small values of Δt, since $1/(1 - \Delta t) \sim 1 + \Delta t$). We could improve things further by using a technique called Crank–Nicolson. In the previous two cases we have taken the formal definition of the derivative to represent the time derivative at either the previous (forward difference) or the next point (backward difference). We could also take it to be valid at the mid-point (central difference). The errors in the former two cases are of order Δt (that is first-order accurate) whereas the latter has errors of order Δt^2 (that is second-order accurate). The code for this purpose is:

```
% crank.m
function [error] = crank(dt)
t = 0.0:dt:5.0;
y = zeros(size(t));
y(1) = 1;
for ii = 1:(length(t)-1)
      y(ii+1) = y(ii)*(1+dt/2)/(1-dt/2);
end
exact = exp(t);
errors = abs(exact-y);
error = max(errors);
```

which is derived by using:

$$\frac{y_{n+1} - y_n}{\Delta t} = \frac{y_{n+1} + y_n}{2},$$

which can be rewritten as:

$$y_{n+1} = y_n \frac{1 + \Delta t/2}{1 - \Delta t/2}.$$

In general this is derived from

$$\left.\frac{\mathrm{d}y}{\mathrm{d}t}\right|_{t_{n+1/2}} = \frac{1}{2}\left(f(y_n, t_n) + f(y_{n+1}, t_{n+1})\right).$$

Notice if the inhomogeneity (or the coefficients) involves functions of t they may be evaluated at $t_{n+1/2}$.

By using the MATLAB command `polyfit` we can confirm that this scheme behaves as we predicted; that is the errors are of order Δt^2. We can improve on these methods and they are the basis for schemes which can solve partial differential equations. At this point we have dealt with systems for which we can easily "invert" the function $f(y, t)$. However if this is not so we will need to guess the new solution (using a predictor step) and then iterate until a solution is attained at each step (using a corrector step). Sometimes it is sufficient to use fixed-point iteration for this process, in which case the predictor and corrector steps are synonymous. However with some systems it is necessary to use more of their history.

8.2.1 Analytical Comparisons

Let us pause and compare the solutions for the equation $y' = y$ subject to the initial conditions $y(0) = 1$. The three schemes are:

$$\text{Explicit Euler}: \quad y_{n+1} = y_n(1 + \Delta t), \tag{8.2a}$$

$$\text{Implicit Euler}: \quad y_{n+1} = \frac{y_n}{1 - \Delta t}, \tag{8.2b}$$

$$\text{Crank–Nicolson}: \quad y_{n+1} = y_n \frac{1 + \Delta t/2}{1 - \Delta t/2}. \tag{8.2c}$$

All of which are of the form $y_{n+1} = \alpha y_n$ which have the solution $y_n = \alpha^n y_0$. We know that the exact solution is $y(t) = \mathrm{e}^t$ so that if $t = t_n = n\Delta t$, this gives

$$\begin{aligned} y_n =& \mathrm{e}^{n\Delta t} = \sum_{j=0}^{\infty} \frac{(n\Delta t)^j}{j!}, \\ y_n =& 1 + n\Delta t + \frac{(n\Delta t)^2}{2!} + O((n\Delta t)^3). \end{aligned}$$

We can now compare the solutions from each of the schemes:

Explicit Euler The solution of Equation (8.2a) gives

$$y_n = (1 + \Delta t)^n = 1 + n\Delta t + \frac{n(n-1)}{2}\Delta t^2 + O((n\Delta t)^3)$$

and as we can see this differs from the exact solution in the second term.

Implicit Euler The solution of Equation (8.2b) gives

$$y_n = (1 - \Delta t)^{-n} = 1 + n\Delta t + \frac{n(n+1)}{2}\Delta t^2 + O((n\Delta t)^3).$$

Again this solution differs in the second term.

Crank–Nicolson Finally the solution of Equation (8.2c) gives

$$y_n = \left(\frac{1 + \Delta t/2}{1 - \Delta t/2}\right)^n,$$

which can be manipulated to give

$$y_n = 1 + n\Delta t + \frac{(n\Delta t)^2}{2} + \left(\frac{1}{6}n^3 + \frac{n}{12}\right)(\Delta t)^3 + O((n\Delta t)^4).$$

In this case the solution differs in the third term and the first two are accurate.

So far we have only considered an explicit method: now let us consider an example using the implicit method. Solve the differential equation

$$\frac{\mathrm{d}y}{\mathrm{d}t} = t + y,$$

subject to $y(0) = 0$ from $t = 0$ to $t = 1$ in steps of $\Delta t = 1/4$. Firstly we discretise the equation

$$\frac{y_{n+1} - y_n}{\Delta t} = t_{n+1} + y_{n+1},$$

which can be rearranged to give:

$$y_{n+1} = \frac{1}{1 - \Delta t} \left\{ y_n + \Delta t \, t_{n+1} \right\}.$$

Now construct the solution where $t_i = i\Delta t$, so with $n = 0$ we have

$$y_1 = \frac{1}{1 - \frac{1}{4}} \left(y_0 + \frac{1}{4}\frac{1}{4} \right),$$

but the initial value $y_0 = y(0) = 0$ so that

$$y_1 = \frac{1}{12}.$$

And at the next point

$$y_2 = \frac{4}{3}\left(\frac{1}{12} + \frac{1}{4}\frac{2}{4} \right) = \frac{5}{18},$$

and for $n = 2$

$$y_3 = \frac{4}{3}\left(\frac{5}{18} + \frac{1}{4}\frac{3}{4} \right) = \frac{67}{108}$$

and finally with $n = 3$ to give y_4 (or $y(1)$)

$$y_4 = \frac{4}{3}\left(\frac{67}{108} + \frac{1}{4}\frac{4}{4} \right) = \frac{94}{81}.$$

Now we should solve the original problem, which we do using an integrating factor. Multiply through by e^{-t} to give

$$\mathrm{e}^{-t}\frac{\mathrm{d}y}{\mathrm{d}t} - \mathrm{e}^{-t}y = \mathrm{e}^{-t}t,$$

the left hand side of which can be written as an exact differential so that

$$\frac{\mathrm{d}}{\mathrm{d}t}\left(\mathrm{e}^{-t}y\right) = \mathrm{e}^{-t}t,$$

which can be integrated and manipulated to give:

$$y = C\mathrm{e}^t + \mathrm{e}^t \int_t \mathrm{e}^{-t} t \,\mathrm{d}t.$$

The integral can be determined using integration by parts to give

$$y(t) = C\mathrm{e}^t - t - 1.$$

The constant is found to be unity and hence $y(1) = \mathrm{e} - 2 \approx 0.718$. We can see that the above solution is quite bad. We note it is possible to perform this analysis in a far more rigorous manner. We should also note that the stability of these schemes is crucial.

We now consider an inhomogeneous example.

Example 8.4 *Solve the equation*

$$\frac{\mathrm{d}y}{\mathrm{d}t} + y = \sin t$$

subject to the initial condition $y(0) = 1$ *for the range* $[0, 1]$ *in steps of 0.1 using the Crank–Nicolson technique.*

We discretise this equation as

$$\frac{y_{n+1} - y_n}{\Delta t} + \frac{y_{n+1} + y_n}{2} = \sin t_{n+1/2}.$$

It is tempting to approximate the right-hand side of this expression by $(\sin t_n + \sin t_{n+1})/2$. *However the method we have chosen above actually is the value at the mid-point, whereas this other treatment is merely an approximation. The difference is only of order* Δt^2, *but we aim to reduce errors where possible (this error can be shown by using Taylor series). Rearranging the above gives:*

$$y_{n+1} = \frac{1}{\frac{1}{\Delta t} + \frac{1}{2}} \left(y_n \left\{ \frac{1}{\Delta t} - \frac{1}{2} \right\} + \sin t_{n+1/2} \right).$$

This can be coded using:

```
n = 11;
t = linspace(0,1,n);
dt = t(2)-t(1);
y = zeros(size(t));
y(1) = 1;
f1 = 1/dt+1/2;
f2 = 1/dt-1/2;
for j = 1:(n-1)
    y(j+1) = (y(j)*f2+ ...
     sin((t(j)+t(j+1))/2)/f1;
end
```

The exact solution is

$$y(t) = \frac{1}{2}(\sin t - \cos t) + \frac{3}{2}\mathrm{e}^{-t}.$$

Plotting both solutions on the same graph, there does not appear to be any difference.

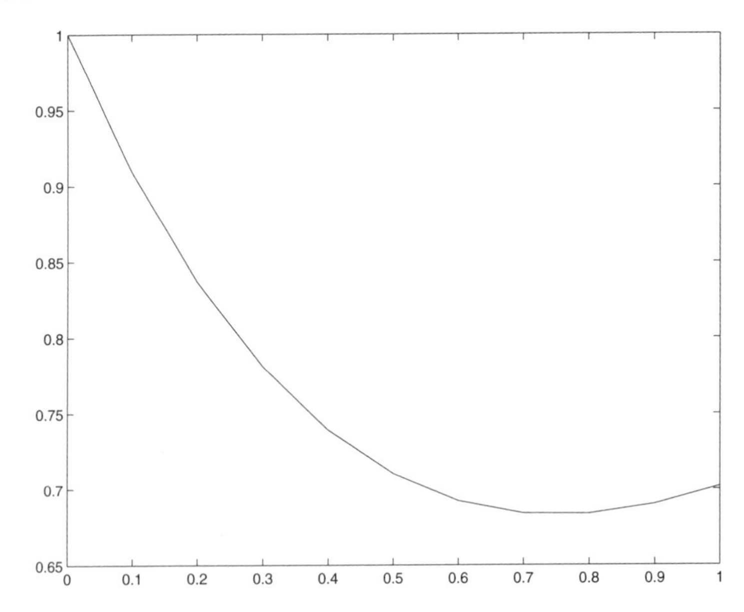

However plotting their difference shows that the errors are of order 10^{-4}.

The explicit Euler method is not limited to linear equations. For instance we could solve the equation

$$\frac{\mathrm{d}y}{\mathrm{d}t} = t^2 + y^2.$$

This has the solution

$$y(t) = -t\frac{CY_{-3/4}\left(\frac{t^2}{2}\right) + J_{-3/4}\left(\frac{t^2}{2}\right)}{CY_{1/4}\left(\frac{t^2}{2}\right) + J_{1/4}\left(\frac{t^2}{2}\right)},$$

which is in terms of Bessel functions. These are solutions to the second-order differential equation

$$t^2\frac{\mathrm{d}^2y}{\mathrm{d}t^2} + t\frac{\mathrm{d}y}{\mathrm{d}t} + \left(t^2 - \nu^2\right)y = 0.$$

In this case the parameter ν takes the value $1/4$ or $-3/4$. You will meet the functions in due course but at the moment let us presume that the solution to the above problem is unknown in which case numerical methods are our only way forward. In fact in some cases even where the solution is known its form may be so complicated that it is better to use the numerical solution straight away (however if an analytical solution is available it is worth checking that we have the correct value).

In order to solve the above problem (subject to the boundary condition that $y(0) = 0$) the code `euler.m` needs to be modified to (`mod_eul.m`):

```
dt = 0.1;
t = 0.0:dt:1.0;
y = zeros(size(t));
y(1) = 0;
for ii=1:(length(t)-1)
      y(ii+1) = y(ii) + dt * (t(ii)^2+y(ii)^2);
end

figure(1)
plot(t,y,'--')
```

This gives

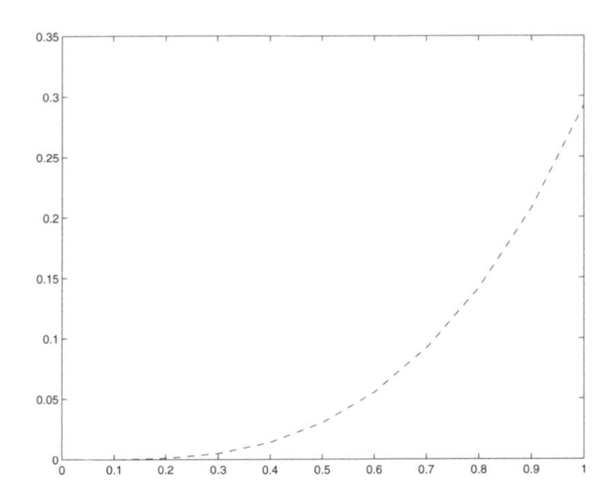

The "exact" solution can be plotted and is given by

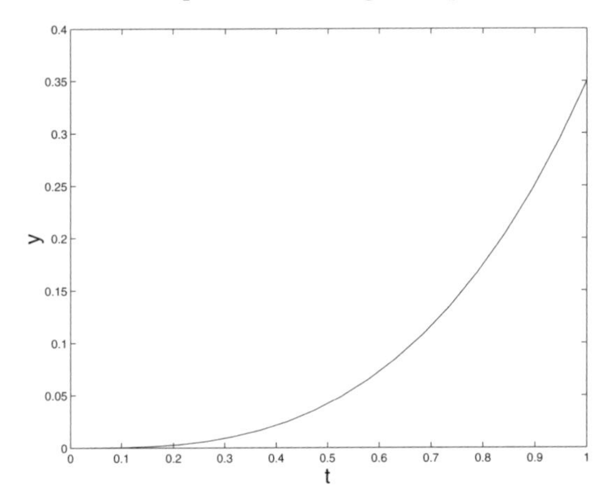

This was obtained using the symbolic toolbox within MATLAB with the code

```
syms t y
y = dsolve('Dy = t^2+y^2','y(0)=0','t');
N = 20;
tv = linspace(eps,1,N);
for i = 1:N
   f(i) = double(subs(y,t,tv(i)));
end
```

Here we have used the commands

`dsolve` Which solves the differential equation defined as its first argument, subject to the initial condition `y(0)=0` and finally we use the third argument to tell MATLAB that `t` is the independent variable.

`subs` This substitutes values from the array `tv` in the solution of the differential equation.

`double` This returns a double precision value (that is it converts a symbolic value to a double precision value).

We shall now discuss a class of methods which can be derived using Taylor series.

8.3 Banded Matrices

In many problems matrices will not only be sparse but they will also have very well-defined structure. Let us consider a tri-diagonal matrix, so that the only non-zero entries are in the super- and sub-diagonal. We consider the set of equations

$$\begin{aligned} a_1 y_2 + b_1 y_1 &= r_1, \\ a_2 y_3 + b_2 y_2 + c_2 y_1 &= r_2, \\ \vdots &= \vdots \\ a_{n-1} y_n + b_{n-1} y_{n-1} + c_{n-1} y_{n-2} &= r_{n-1}, \\ b_n y_n + c_n y_{n-1} &= r_n. \end{aligned}$$

Notice that we can solve the first equation to give y_1 in terms of y_2, so that

$$y_1 = \frac{r_1}{b_1} - \frac{a_1}{b_1} y_2$$

which can then be substituted into the next equation to eliminate y_1. This gives

$$\begin{aligned} a_2 y_3 + b_2 y_2 + c_2 \left(\frac{r_1}{b_1} - \frac{a_1}{b_1} y_2 \right) &= r_2, \\ a_2 y_3 + \left(b_2 - c_2 \frac{a_1}{b_1} \right) y_2 &= r_2 - c_2 \frac{r_1}{b_1}. \end{aligned}$$

This amounts to a redefinition of b_2 and r_2. This equation can then be used to express y_2 in terms of y_3 and then substitute it into the next equation, until we reach the final equation at which point we have an equation solely in y_n which is easily solved. The system at this point has only got a super-diagonal (we have eliminated the sub-diagonal) and hence the values of all the preceding y's can be found by back substitution. This is written in code as:

```
% Set up system
x = 0.0:pi/6:pi;
N = length(x);
h = x(2)-x(1);
a = 1/h^2*ones(size(x));
b = -2/h^2*ones(size(x));
c = 1/h^2*ones(size(x));
r = x.*sin(x);
a(1) = 0; b(1) = 1; r(1) = 0;
c(N) = 0; b(N) = 1; r(N) = 0;
% Forward sweep

for j = 2:N
     b(j) = b(j) - c(j)*a(j-1)/b(j-1);
     r(j) = r(j) - c(j)*r(j-1)/b(j-1);
end
% Final equation
y(N) = r(N)/b(N);
for j = (N-1):-1:1
     y(j) = r(j)/b(j)-a(j)*y(j+1)/b(j);
end

xf = 0:pi/50:pi;
sol = 2*(1-cos(xf))-xf.*sin(xf)-4*xf/pi;
clf
plot(x,y,'.','MarkerSize',24)
hold on
plot(xf,sol,'b')
xlabel('x','FontSize',15)
ylabel('f','FontSize',15)
text(0.5,-0.2,'Finite difference solution','FontSize',12)
text(0.7,-0.3,'shown as blobs','FontSize',12)
```

This is actually solving the following problem

$$\frac{\mathrm{d}^2y}{\mathrm{d}x^2} = x\sin x \qquad y(0) = y(\pi) = 0.$$

We can solve this by hand by integrating twice and applying the boundary conditions to obtain

$$y(x) = -2\cos x - x\sin x - \frac{4x}{\pi} + 2.$$

As you can see even using 7 points determines the solution quite well

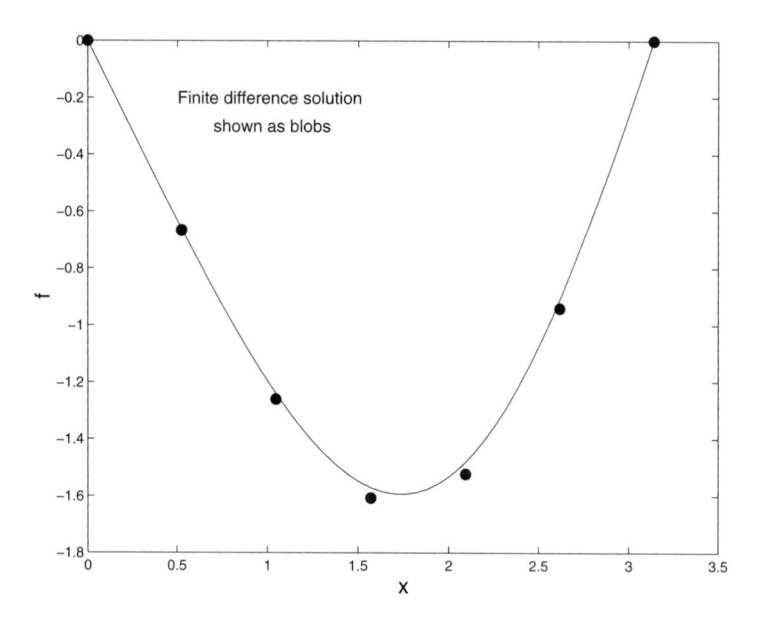

This method is called a **Thomas algorithm**.

Example 8.5 *In order to solve the system*

$$\frac{\mathrm{d}^2 y}{\mathrm{d}x^2} + 4\frac{\mathrm{d}y}{\mathrm{d}x} - y = \cos x$$

subject to the boundary conditions $y(0) = 0$ *and* $y(\pi) = 1$, *we use the code*

```
% Set up system
x = 0.0:pi/60:pi;
N = length(x);
h = x(2)-x(1);
a = (1/h^2+4/(2*h))*ones(size(x));
b = (-2/h^2-1)*ones(size(x));
c = (1/h^2-4/(2*h))*ones(size(x));
r = cos(x);
a(1) = 0; b(1) = 1; r(1) = 0;
c(N) = 0; b(N) = 1; r(N) = 1;
% Forward sweep

for j = 2:N
     b(j) = b(j) - c(j)*a(j-1)/b(j-1);
     r(j) = r(j) - c(j)*r(j-1)/b(j-1);
end
% Final equation
y(N) = r(N)/b(N);
for j = (N-1):-1:1
     y(j) = r(j)/b(j)-a(j)*y(j+1)/b(j);
end

clf
plot(x,y)
xlabel('x','FontSize',15)
ylabel('y','FontSize',15)
```

which gives

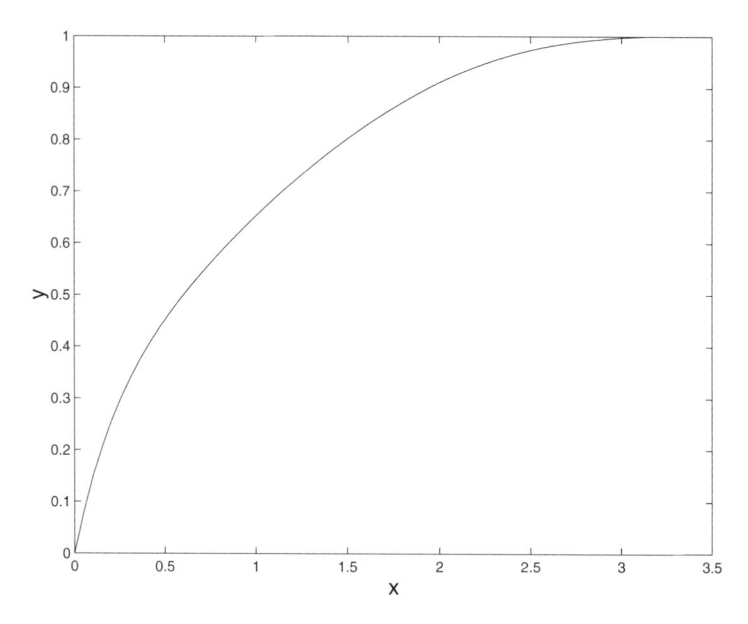

We have used the discretisation

$$\frac{y_{n+1} - 2y_n + y_{n-1}}{h^2} + 4\frac{y_{n+1} - y_{n-1}}{2h} - y_n = \cos x_n.$$

This can be extended to penta-diagonal systems, which can again be done using a Thomas algorithm but this becomes more and more complicated. However, we can also exploit the `sparse` command. We can also solve periodic problems which give rise to sparse problems, but with elements in the top right and bottom left corners in addition to the diagonals.

8.4 Runge–Kutta Methods

We shall not derive these methods in fullness but shall give a flavour of how this may be done. In order to solve the system

$$\frac{\mathrm{d}y}{\mathrm{d}t} = f(t, y),$$

we use Taylor series and as such we write the scheme as

$$y_{n+1} = y_n + ak_1 + bk_2 \tag{8.3}$$

where

$$\begin{aligned} k_1 &= \Delta t\, f(t_n, y_n), \\ k_2 &= \Delta t\, f(t_n + \alpha\Delta t, y_n + \beta k_1). \end{aligned}$$

In order to derive an equation of this form we start with the Taylor series for $y_{n+1} = y(t + \Delta t)$

$$y_{n+1} = y_n + \Delta t f(t_n, y_n) + \frac{\Delta t^2}{2} f'(t_n, y_n) + \cdots,$$

where the prime denotes a derivative with respect to t. Using the chain rule we can write

$$\frac{\mathrm{d}f}{\mathrm{d}t} = \frac{\partial f}{\partial t} + \frac{\partial f}{\partial y}\frac{\mathrm{d}y}{\mathrm{d}t},$$

but using the original equation ($\dot{y} = f$)

$$\frac{\mathrm{d}f}{\mathrm{d}t} = \frac{\partial f}{\partial t} + \frac{\partial f}{\partial y} f,$$

so that

$$y_{n+1} = y_n + \Delta t f(t_n, y_n) + \frac{\Delta t^2}{2}\left(\frac{\partial f}{\partial t} + \frac{\partial f}{\partial y} f\right) \cdots.$$

Let us now substitute the forms for k_1 and k_2 into Equation (8.3), so that

$$y_{n+1} = y_n + a\Delta t f(t_n, y_n) + b\Delta t f\left(t_n + \alpha\Delta t, y_n + \beta\Delta t f(t_n, y_n)\right).$$

We now need to expand the last term in this equation, so that

$$f\left(t_n + \alpha\Delta t, y_n + \beta\Delta t f(t_n, y_n)\right) \approx f(t_n, y_n) + \alpha\Delta t \left.\frac{\partial f}{\partial t}\right|_{t_n, y_n} + \beta\Delta t f(t_n, y_n) \left.\frac{\partial f}{\partial y}\right|_{t_n, y_n}.$$

By substitution and comparison with the other Taylor series we find that

$$\begin{aligned} &\text{in } \Delta t f && a + b = 1, \\ &\text{in } \Delta t^2 \frac{\partial f}{\partial t} && \alpha b = \frac{1}{2}, \\ &\text{in } \Delta t^2 f \frac{\partial f}{\partial y} && \beta b = \frac{1}{2}. \end{aligned}$$

We have only three equations but with four unknowns, so we are free to choose one value. For instance $a = 1/2$ implies $b = 1/2$ with in turn gives $\alpha = \beta = 1$. The scheme is then

$$y_{n+1} = y_n + \frac{\Delta t}{2}\left(f(t_n, y_n) + f\left(t_n + \Delta t, y_n + \Delta t f(t_n, y_n)\right)\right).$$

We can of course choose other values of a (or any of the other variables). In order to code this algorithm we use:

```
%
% runge_kutta.m
%
t = 0.0:0.1:5.0;
del_t = 0.1;
a = 0.5; b = 1-a;
alpha = 0.5/b; beta = 0.5/b;
n = length(t);
y = zeros(size(t));
y(1) = 0;
for ii = 1:n-1
    time = t(ii);
    k_1 = del_t*func(t(ii),y(ii));
    k_2 = del_t*func(t(ii)+alpha*del_t,y(ii)+beta*k_1);
    y(ii+1) = y(ii) + a * k_1 + b * k_2;
end
exact = exp(t-t.^2/2);
```

using

```
function [value] = func1(t,y)

value = y*(1-t);
```

Using this number of points there is no visible difference between the numerical and the exact solution. We can investigate the effect of altering the value of a. This method can be extended to higher orders: for instance the fourth-order scheme is given by

$$y_{n+1} = y_n + \frac{1}{6}\left(k_1 + 2k_2 + 2k_3 + k_4\right)$$

where

$$\begin{aligned}
k_1 &= \Delta t f(t_n, y_n) \\
k_2 &= \Delta t f\left(t_n + \frac{\Delta t}{2}, y_n + \frac{k_1}{2}\right) \\
k_3 &= \Delta t f\left(t_n + \frac{\Delta t}{2}, y_n + \frac{k_2}{2}\right) \\
k_4 &= \Delta t f\left(t_n + \Delta t, y_n + k_3\right).
\end{aligned}$$

MATLAB offers two commands `ode23` and `ode45` which perform these integrations. It uses a more advanced numerical scheme which iterates to correct the solution as it progresses. The above code could be replaced by

```
y0 = 1; tspan = [0 5];
[tt,yy] = ode45('func1',tspan,y0);
```

This uses an adaptive step which is based on the local gradients. Interestingly this chooses to use 57 points to attain the default accuracy (which is defined in terms of relative and absolute errors, see the manual pages `help ode45` for details).

8.5 Higher-Order Systems

8.5.1 Second-Order Systems

At the moment we shall only briefly mention these systems and shall not dwell on their solution using the so-called finite difference techniques. However we shall discuss how they might be solved using Runge–Kutta methods (specifically using the intrinsic MATLAB commands). Let us consider a simple example:

Consider the solution of the system

$$\frac{\mathrm{d}^2 y}{\mathrm{d}t^2} + y = 0,$$

subject to the conditions that $y(0) = 1$ and $y'(0) = 0$ from $t = 0$ to $t = \pi$.

Firstly we introduce a vector $\mathbf{z}$ such that

$$\mathbf{z} = \begin{pmatrix} z_1 \\ z_2 \end{pmatrix} = \begin{pmatrix} y \\ y' \end{pmatrix},$$

so that

$$\mathbf{z}' = \begin{pmatrix} y' \\ y'' \end{pmatrix} = \begin{pmatrix} y' \\ -y \end{pmatrix} = \begin{pmatrix} z_2 \\ -z_1 \end{pmatrix}.$$

We can now construct the corresponding MATLAB code (`func2.m`):

```
function [out] = func2(t,in)
out=zeros(2,1);
out(1) = in(2);
out(2) = -in(1);
```

```
y0=[1; 0];
ts=[0 pi];
[t,y]=ode45('func2',ts,y0);
plot(t,y)
```

which gives

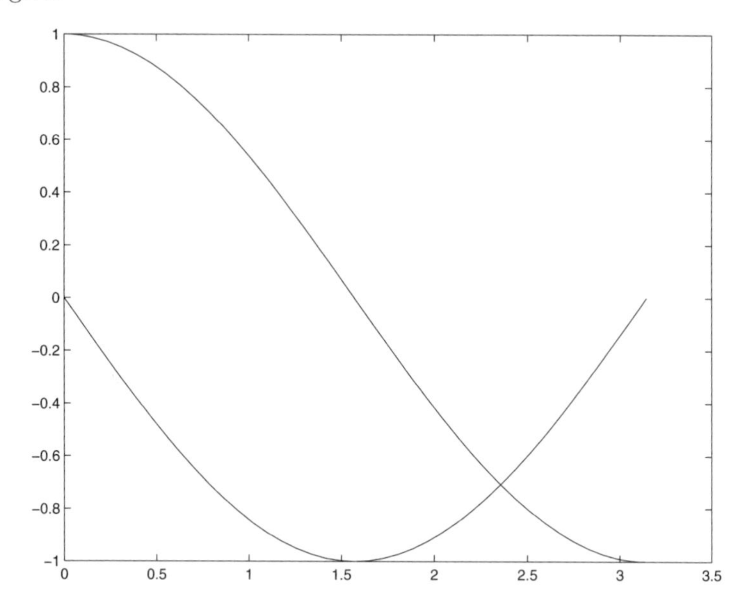

Notice that `y` not only contains the solution $z_1(t)$ but also its derivative $(z_2(t))$. The exact solution here is $\cos t$ (and consequently its derivative is $-\sin t$). This technique can be used equally for non-autonomous systems (that is involving t explicitly). At this stage we will not consider all the options which are available to use when using these methods: however we merely mention that using the `ode` routines one can set the absolute and relative errors.

Before moving on we remark that it is possible to consider higher-order systems in a variety of ways. For instance in order to solve

$$\frac{\mathrm{d}^2 y}{\mathrm{d}t^2} + 2\frac{\mathrm{d}y}{\mathrm{d}t} + y = \sin t,$$

(which is a second-order linear ordinary differential equation), we can either use finite difference representations of the derivatives, for instance

$$\left.\frac{\mathrm{d}^2 y}{\mathrm{d}t^2}\right|_{t=t_n} \approx \frac{y_{n+1} - 2y_n + y_{n-1}}{\Delta t^2},$$

or we can rewrite the system as coupled first-order systems (which is more conducive to Runge–Kutta techniques). We introduce $z_1(t) = y$ and $z_2(t) = \mathrm{d}y/\mathrm{d}t$. Hence the above system is rewritten as

$$\begin{aligned} \frac{\mathrm{d}z_1}{\mathrm{d}t} &= z_2 \\ \frac{\mathrm{d}z_2}{\mathrm{d}t} &= -2z_2 - z_1 + \sin t. \end{aligned}$$

Example 8.6 *We write the system*

$$\ddot{x} + x^2 = \cos t$$

as a system of first-order equations, defining z_1 *as* x *and* z_2 *and* $\dot{x}$ *so that*

$$\dot{z}_1 = z_2 \qquad \dot{z}_2 = -z_1^2 + \cos t.$$

We shall now discuss the solution of second-order systems and start with an example of a constant coefficient problem.

Example 8.7 *Solve the differential equation*

$$\frac{\mathrm{d}^2 y}{\mathrm{d}t^2} + 3\frac{\mathrm{d}y}{\mathrm{d}t} + 2y = 0$$

subject to the initial conditions that $y(0) = 1$ *and* $y'(0) = 0$*, from 0 to 1.*

We discretise the above equation to yield

$$\frac{y_{n+1} - 2y_n + y_{n-1}}{\Delta t^2} + 3\frac{y_{n+1} - y_{n-1}}{2\Delta t} + 2y_n = 0,$$

which is rearranged to give

$$\left(\frac{1}{\Delta t^2} + \frac{3}{2\Delta t}\right) y_{n+1} = -\frac{-2y_n + y_{n-1}}{\Delta t^2} + \frac{3}{2\Delta t} y_{n-1} - 2y_n.$$

The initial conditions tell us that $y_1 = 1$ *and* $(y_2 - y_1)/\Delta t = 0$ *(so that* $y_2 = 1$ *also). The equation has the exact solution* $y(t) = \mathrm{e}^{-t} - \mathrm{e}^{-2t}$*, which is used for comparison.*

```
n = 10;
t = linspace(0,1,n);
dt = t(2)-t(1);
y = zeros(size(t));
y(1) = 1; y(2) = 1;

fact = (1/dt^2+3/(2*dt));

for j = 2:(n-1)
    y(j+1) = (-(-2*y(j)+y(j-1))/dt^2 ...
             +3/(2*dt)*y(j-1) ...
             -2*y(j))/fact;
end

true = 2*exp(-t)-exp(-2*t);
clf
plot(t,true,t,y,'.','MarkerSize',15)
legend('Exact','Approx',-1)
```

This produces the results:

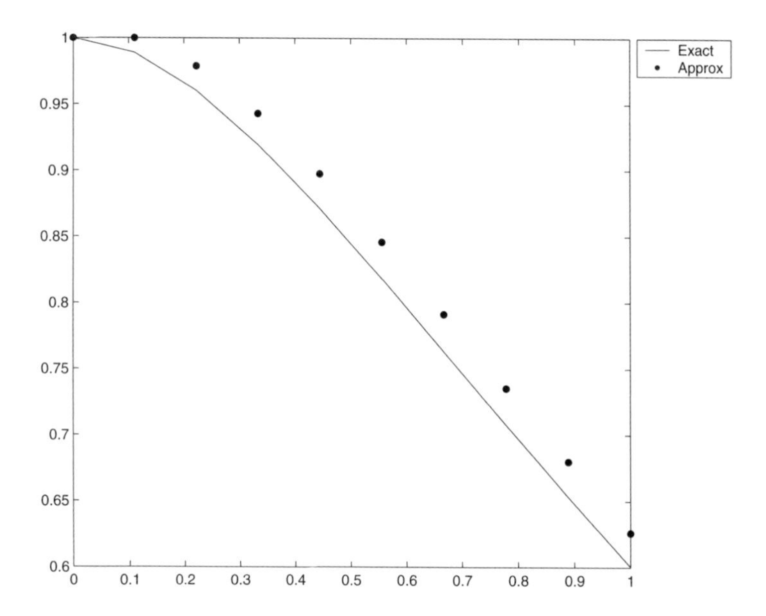

Notice: only using ten points we do not get a particularly accurate answer. This can be rectified by using more points.

We shall now discuss the solution of two problems, which are quite similar. Both of the examples can actually be solved using analytical methods: however at this stage the solutions are not accessible to us, so we shall treat them as problems for which we do not know the answers[1]. The first problem will be solved using finite difference methods and the second one will be tackled using the second-order Runge–Kutta scheme described on page 263.

8.5.2 Bessel's equation

We shall now discuss the solution of the mathematical problem, whereby we seek to integrate the differential equation

$$t^2\frac{\mathrm{d}^2y}{\mathrm{d}t^2}+t\frac{\mathrm{d}y}{\mathrm{d}t}+\left(t^2-\frac{1}{4}\right)y=0,$$

from $t=0$ to $t=10$ subject to the initial conditions $y(0)=0$ and $y'(0)=1$.

We shall solve this problem using finite-difference techniques, in order to do this we need to exploit the discretised form for the second derivative given on page 267. We introduce the grid such that $t_n=(n-1)\Delta t$, so that t_1 corresponds to the initial time, $t=0$. There is also the corresponding values of the function y_n. We shall presume that there are N points in the grid. At time $t=t_n$ we can write the above differential equation as:

$$t_n^2\frac{y_{n+1}-2y_n+y_{n-1}}{\Delta t^2}+t_n\frac{y_{n+1}-y_{n-1}}{2\Delta t}+\left(t_n^2-\frac{1}{4}\right)y_n=0$$

for $n=2$ upto $n=N-1$. In this expression we have used the central difference formula for both the first and second derivatives, which are both second-order accurate expressions. We now consider the initial condition $y(0)=0$ implies that $y_1=0$ and $y'(0)=1$ means that $(y_2-y_1)/\Delta t=1$ so that $y_2=\Delta t$. Here we have used the forward difference formula, which is necessarily biased. By using the above discretised equation with $n=2$ we obtain an expression for y_3 in terms of the known quantities y_2 and y_1. Similarly by using the equation with $n=3$ we can obtain y_4 and so on. To this end we rewrite the equation as:

$$\left(\frac{t_n^2}{\Delta t^2}+\frac{t_n}{2\Delta t}\right)y_{n+1}=-t_n^2\frac{-2y_n+y_{n-1}}{\Delta t^2}+\frac{t_ny_{n-1}}{2\Delta t}-\left(t_n^2-\frac{1}{4}\right)y_n.$$

We now use MATLAB code to solve the problem as posed above.

[1] The first one is called the Bessel equation and has solutions $J_\nu(t)$ and $Y_\nu(t)$ and in this case the parameter $\nu=1/2$. These functions are discussed extensively in a book by G. N. Watson *'A Treatise on Bessel Functions'*. The second equation has solutions which are Airy functions, affectionately called the Airy and Bairy functions.

```
%
% Code to determine Bessel functions of order 1/2.
%
y_zero = 0;
yp_zero = 1;
delta_t = 0.1;
t=0:delta_t:10;
N = length(t);
y=zeros(size(t));
y(1) = y_zero;
y(2) = y(1) + delta_t * yp_zero;
for j = 2:N-1
    factor = t(j)^2/delta_t^2+t(j)/(2*delta_t);
    y(j+1) = (-t(j)^2*(-2*y(j)+y(j-1))/delta_t^2 ...
              +t(j)*y(j-1)/(2*delta_t)...
              -(t(j)^2-1/4)*y(j))/factor;
end
```

The solution is found to be:

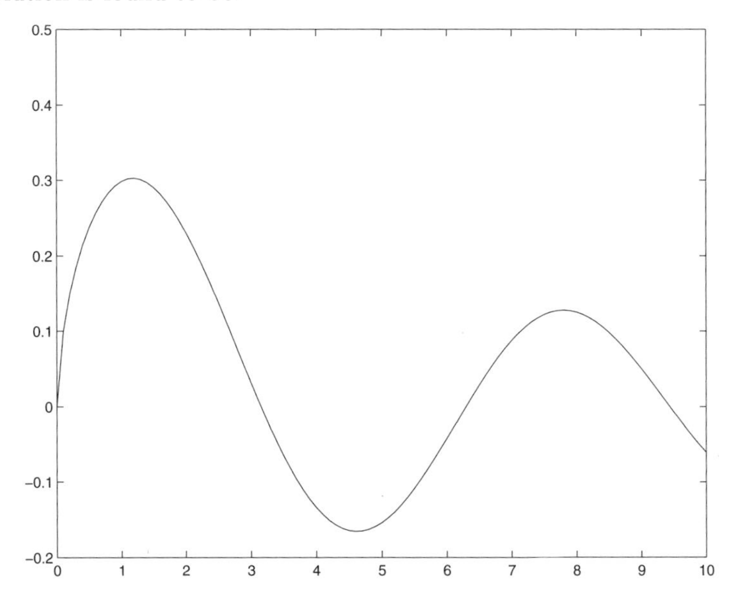

In this case we have elected to use the higher-order derivatives directly: however we could have used the idea of introducing a second function so that the system is reduced to two first-order differential equations. For this purpose we define

$$z^{(1)}(t) = y(t) \text{ and } z^{(2)}(t) = \frac{\mathrm{d}y}{\mathrm{d}t}.$$

Using these functions the differential equation becomes:

$$t^2 \frac{dz^{(2)}}{dt} + tz^{(2)} + \left(t^2 - \frac{1}{4}\right) z^{(1)} = 0$$

subject to the initial conditions $z^{(1)}(0) = 0$ and $z^{(2)}(0) = 1$. Again we discretise the equation so that

$$t_n^2 \frac{z_{n+1}^{(2)} - z_n^{(2)}}{\Delta t} + t_n z_n^{(2)} + \left(t_n^2 - \frac{1}{4}\right) z_n^{(1)} = 0.$$

We also need to discretise the relation between $z^{(1)}$ and $z^{(2)}$ so that

$$z_n^{(2)} = \frac{z_{n+1}^{(1)} - z_n^{(1)}}{\Delta t}.$$

We are now ready to use MATLAB code.

```
%
% Solve Bessel's equation using two first-order systems.
%
y_zero = 0;
yp_zero = 1;
delta_t = 0.1;
t=0:delta_t:10;
nt = length(t);
z=zeros(nt,2);
z(1,1) = y_zero;
z(1,2) = yp_zero;

for j = 2:N
    time = (t(j)+t(j-1))/2;
    factor = time^2/delta_t;
    z(j,1) = delta_t*z(j-1,2)+z(j-1,1);
    z(j,2) = (time^2*z(j-1,2)/delta_t-time*z(j-1,2)...
             - (time^2-1/4)*z(j-1,1))/factor;
end
```

Notice that we have elected to evaluate the time at the midpoint of the interval. If we had just used it at the start of the range, then the singularity (obtained by dividing by t^2) would cause our scheme to fail. In fact this method does not do very well and this sensitivity to initial conditions is partly the reason for this. Also we have actually used first-order accurate discretisations. We will see that this reduction to first-order systems is necessary for Runge–Kutta methods in the next section; fortunately the failings of this approach in finite differences are not reproduced.

8.5.3 Airy's Equation

We shall now discuss the solution of the problem

$$\frac{\mathrm{d}^2 y}{\mathrm{d}x^2} - xy = 0$$

subject to the initial conditions $y(0) = 1$ and $y'(0) = 0$. Again we use the reduction to two first-order systems defining

$$z^{(1)}(x) = y(x) \text{ and } z^{(2)}(x) = \frac{\mathrm{d}y}{\mathrm{d}x}.$$

We also introduce the column vector $\mathbf{z}$ with these functions as elements, so that

$$\mathbf{z}' = \mathbf{f}(x, \mathbf{z}) = \begin{pmatrix} z^{(2)} \\ xz^{(1)} \end{pmatrix};$$

derived using the fact that $(z^{(1)})' = z^{(2)}$ by definition, and $(z^{(2)})' = xz^{(1)}$ directly from the differential equation. The Runge–Kutta method derived on page 263 can equally be applied to vectors. We recall that we have

$$\mathbf{z}_{n+1} = \mathbf{z}_n + \frac{\delta x}{2} \left(\mathbf{f}(x_n, \mathbf{z}_n) + \mathbf{f}(x_n + \delta x, \mathbf{z}_n + \delta x \mathbf{f}(x_n, \mathbf{z}_n)) \right).$$

We can use a similar code to the previous example, but we shall use a slightly different format. Let us define a `function` to work out the function $\mathbf{f}$:

```
%
function [out] = func(x,in)

out(1) = in(2);
out(2) = x*in(1);
```

and the main routine

```
x=0.0:0.1:5.0;
delta_x = 0.1;
N = length(x);
z=zeros(N,2);
z(1,1) = 1; z(1,2) = 0;

for j=1:(N-1)
    k1 = delta_x*func(x(j),z(j,:));
    k2 = delta_x*func(x(j)+delta_x,z(j,:)+k1);
    z(j+1,:) = z(j,:) + 1/2*(k1+k2);
end
```

In these two examples we have solved problems which have two boundary (or initial) conditions at one location. Let us now discuss how we might solve problems which have conditions at two different locations.

8.5.4 Shooting Methods

Here we solve problems by starting at a point where we need to guess certain quantities and these guesses can subsequently be improved upon. We shall illustrate the technique with a couple of examples.

At the outset we note that for linear problems there are other ways of doing what we are about to discuss (see for instance the example in the section on banded matrices on page 259).

Example 8.8 (A Linear Boundary-Value Problem)
We shall use the Runge–Kutta scheme detailed above to solve the problem

$$\frac{\mathrm{d}^2 y}{\mathrm{d}x^2} + y = 0$$

subject to the boundary conditions $y(0) = 1/2$ *and* $y(\pi/3) = 1/2$. *We can solve this problem analytically to give*

$$y(x) = \frac{1}{2}\cos x + \frac{1}{2\sqrt{3}}\sin x.$$

Let us solve the equation subject to $y(0) = 1/2$ *and* $y'(0) = \lambda$, *where we are free to choose* λ *(at the moment). We then integrate from 0 to* $\pi/3$: *if* $y(\pi/3)$ *is not equal to 1/2 then we can adjust the value of* λ. *In order to do this we use the Newton–Raphson scheme. The main routine is:*

```
%
%
global x z
lambda = 1; % Initial guess
delta = 1e-2;
for its = 1:40
   f_lambda = int_eqn(lambda);
   if abs(f_lambda)<1e-8
      break
   end
   f_lambda_del = int_eqn(lambda+delta);
   lambda = lambda ...
        - f_lambda*delta/(f_lambda_del-f_lambda);
end
exact = 1/2*cos(x)+1/(2*sqrt(3))*sin(x);
```

```
%
% int_eqn.m
function [err] = int_eqn(lambda)
global x z
delta_x = (pi/3)/100;
x = 0.0:delta_x:pi/3;
N = length(x);
z = zeros(N,2);
z(1,1) = 1/2;
z(1,2) = lambda;
for j=1:(N-1)
    k1 = delta_x*func1(x(j),z(j,:));
    k2 = delta_x*func1(x(j)+delta_x,z(j,:)+k1);
    z(j+1,:) = z(j,:) + 1/2*(k1+k2);
end
err = z(N,1) - 1/2;
```

```
function [out] = func1(x,in)
out(1) = in(2);
out(2) = -in(1);
```

This method is very successful and the exact solution is retrieved (albeit approximately). Let us now discuss the solution of a problem for which an analytical solution is not available, and in fact this is how it would actually be solved. The other technique hinted at earlier is not available as this is a nonlinear problem.

Example 8.9 (A Nonlinear Boundary-Value Problem)
This problem comes from fluid dynamics and its solution provides a description of the flow profile within a boundary layer on a flat plate. We shall not dwell on its derivation, since it is far beyond the scope of this text, but we are required to solve

$$\frac{\mathrm{d}^3 f}{\mathrm{d}\eta^3} + f\frac{\mathrm{d}^2 f}{\mathrm{d}\eta^2} = 0,$$

subject to the boundary conditions $f(0) = f'(0) = 0$ *and*

$$\lim_{\eta\to\infty} f'(\eta) = 1.$$

In this case we have a third-order equation and as such we shall introduce

$$z^{(1)} = f(\eta),\ z^{(2)} = \frac{\mathrm{d}f}{\mathrm{d}\eta} \text{ and } z^{(3)} = \frac{\mathrm{d}^2 f}{\mathrm{d}\eta^2}.$$

As in the previous example we are short of a boundary condition at one end, so we solve the problem using the initial conditions

$$z^{(1)}(0) = 0,\ z^{(2)}(0) = 0 \text{ and } z^{(3)}(0) = \lambda.$$

We then integrate towards infinity (in fact in this case a value of 10 is fine for infinity, even if we integrated further nothing would change). The discrepancy between the value of f' *at this point and unity is used to iterate on the value of* λ. *The codes we shall use for this purpose are*

```
%
%
global x z
lambda = 0.5; % Initial guess
delta = 1e-2;
for its = 1:40
   f_lambda = int_blas(lambda);
   if abs(f_lambda)<1e-8
      break
   end
   f_lambda_del = int_blas(lambda+delta);
   lambda = lambda ...
         - f_lambda*delta/(f_lambda_del-f_lambda);
end
plot(z(:,2),x)
xlabel('Flow velocity')
ylabel('Distance from the wall')
axis([-0.5 1.5 0 10])
```

```
%
% int_blas.m

function [err] = int_blas(lambda)
global x z
delta_x = 0.1;
x = 0.0:delta_x:10;
N = length(x);
z = zeros(N,3);
z(1,1) = 0;
z(1,2) = 0;
z(1,3) = lambda;


for j=1:(N-1)
    k1 = delta_x*funcb(x(j),z(j,:));
    k2 = delta_x*funcb(x(j)+delta_x,z(j,:)+k1);
    z(j+1,:) = z(j,:) + 1/2*(k1+k2);
end
err = z(N,2) - 1;
```

```
function [out] = funcb(x,in)
out(1) = in(2);
out(2) = in(3);
out(3) = -in(1)*in(3);
```

The actual value of λ is approximately 0.4689. The flow profile looks like this:

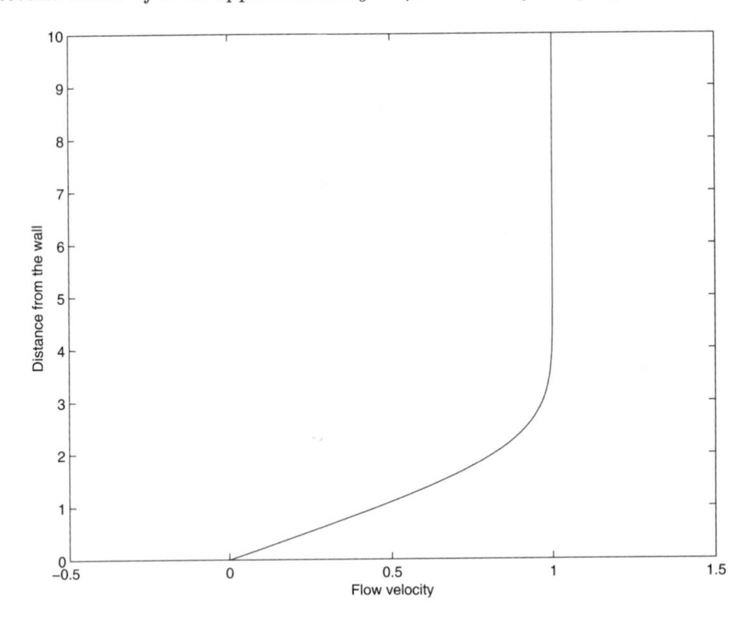

8.6 Boundary-Value Problems

We shall now consider more general boundary-value problems. These can be solved by shooting methods (as presented above): however if the problems are linear they can generally be solved directly. Rather than starting at one end of a domain we simply set up a set of equations for the entire problem and solve these as a matrix problem.

We shall demonstrate this via a few simple examples:

Example 8.10 *Solve the differential equation*

$$\frac{\mathrm{d}^2 y}{\mathrm{d}x^2} - y = 0$$

subject to the conditions that $y(0) = 0$ *and* $y(1) = 1$. *(The exact solution of this can be found as* $y(x) = \sinh x / \sinh 1$.*)*

Firstly we consider a grid of points running from 0 to 1 in steps of h: for the sake of argument we shall take h to be 0.1 (so that there are 11 points). This grid can be set up using

```
x = linspace(0,1,11);
y = zeros(size(x));
h = x(2)-x(1);
```

Here we have used the command `linspace` *to set up the grid and also set aside a vector* `y` *to contain the answer. The step size,* `h`*, has also been defined here.*

Let us consider the above equation at the point $x = x_j$. *Then we can approximate the quantities to give*

$$\frac{y_{j+1} - 2y_j + y_{j-1}}{h^2} - y_j = 0$$

or rearranging

$$y_{j+1} + y_j(-2 - h^2) + y_{j-1} = 0.$$

This equation holds for all the interior points, that is $j = 2$ *through to* $j = 10$. *At the end points we need to impose the boundary conditions, so at* $x = 0$ *(which corresponds to* $x = x_1$*) we have* $y_1 = 0$ *and at the other end, that is* $x = x_{11} = 1$, *we have* $y_{11} = 1$. *Combining these we have eleven equations in eleven unknowns which can be put together to form the matrix*

$$\begin{pmatrix} 1 & 0 & 0 & 0 & 0 & 0 & 0 & 0 & 0 & 0 & 0 \\ 1 & \tau & 1 & 0 & 0 & 0 & 0 & 0 & 0 & 0 & 0 \\ 0 & 1 & \tau & 1 & 0 & 0 & 0 & 0 & 0 & 0 & 0 \\ 0 & 0 & 1 & \tau & 1 & 0 & 0 & 0 & 0 & 0 & 0 \\ 0 & 0 & 0 & 1 & \tau & 1 & 0 & 0 & 0 & 0 & 0 \\ 0 & 0 & 0 & 0 & 1 & \tau & 1 & 0 & 0 & 0 & 0 \\ 0 & 0 & 0 & 0 & 0 & 1 & \tau & 1 & 0 & 0 & 0 \\ 0 & 0 & 0 & 0 & 0 & 0 & 1 & \tau & 1 & 0 & 0 \\ 0 & 0 & 0 & 0 & 0 & 0 & 0 & 1 & \tau & 1 & 0 \\ 0 & 0 & 0 & 0 & 0 & 0 & 0 & 0 & 1 & \tau & 1 \\ 0 & 0 & 0 & 0 & 0 & 0 & 0 & 0 & 0 & 0 & 1 \end{pmatrix} \begin{pmatrix} y_1 \\ y_2 \\ y_3 \\ y_4 \\ y_5 \\ y_6 \\ y_7 \\ y_8 \\ y_9 \\ y_{10} \\ y_{11} \end{pmatrix} = \begin{pmatrix} 0 \\ 0 \\ 0 \\ 0 \\ 0 \\ 0 \\ 0 \\ 0 \\ 0 \\ 0 \\ 1 \end{pmatrix},$$

where $\tau = -2 + h^2$. *This system can be set up using the MATLAB commands:*

```
A = diag(ones(11,1)*(-2-h^2),0)+diag(ones(10,1),1) ...
   +diag(ones(10,1),-1);
A(1,:) = 0; A(1,1) = 1;
A(11,:) = 0; A(11,11) = 1;
b = zeros(11,1);
b(11) = 1;
y = A\b;
```

This gives us the solution to the above problem to within approximately 10^{-4}, *which can be seen by comparing it with the exact solution. We have used the expression* `A\b` *to return* $\mathbf{A}^{-1}\mathbf{b}$.

In this case both of the boundary conditions were imposed on the function, whereas it is plausible to consider problems in which conditions are imposed on the derivative (or even mixed conditions).

Example 8.11 *Solve the differential equation*

$$\frac{\mathrm{d}^2 y}{\mathrm{d}x^2} - y = 0$$

subject to the conditions that $y'(0) = 0$ *and* $y(1) = 1$. *In this case the exact solution is* $y(x) = \cosh x / \cosh 1$. *We have used the same equation as the last example and hence we only need to change the lines which impose the boundary conditions. At the lower end of the domain we need to impose* $y'(0) = 0$. *The derivative at this end is given by*

$$\left.\frac{\mathrm{d}y}{\mathrm{d}x}\right|_{x=x_1} \approx \frac{y_1 - y_0}{h}.$$

This condition translates to $y_0 = y_1$. *Hence the first line of the matrix is set using*

```
A(1,:) = 0;
A(1,1) = 1; A(1,2) = -1;
```

This solution has errors of approximately 10^{-2} *which are still acceptable but are nevertheless larger than in the previous example.*

At this point it is worth pausing to consider the source of these errors. We can choose to use more points to reduce the errors but it would be better to understand how they arise, since increasing the number of points is not always an option. In the first example we used the approximation

$$\left.\frac{\mathrm{d}^2 y}{\mathrm{d}x^2}\right|_{x=x_j} \approx \frac{y_{j+1} - 2y_j + y_{j-1}}{h^2}. \tag{8.4}$$

We now wish to predict the probable error associated with this formula. To do this we exploit Taylor series. We recall that

$$y(a+h) = y(a) + h\left.\frac{\mathrm{d}y}{\mathrm{d}x}\right|_{x=a} + \frac{h^2}{2!}\left.\frac{\mathrm{d}^2 y}{\mathrm{d}x^2}\right|_{x=a} + \frac{h^3}{3!}\left.\frac{\mathrm{d}^3 y}{\mathrm{d}x^3}\right|_{x=a} + \cdots .$$

The key here is how many terms to take in the expansion. We note that the entire series can be written as

$$y(a+h) = y(a) + \sum_{n=1}^{\infty} \frac{h^n}{n!}\left.\frac{\mathrm{d}^n y}{\mathrm{d}x^n}\right|_{x=a}$$

or using the intermediate value theorem

$$y(a+h) = y(a) + \sum_{n=1}^{N} \frac{h^n}{n!}\left.\frac{\mathrm{d}^n y}{\mathrm{d}x^n}\right|_{x=a} + \frac{h^{N+1}}{(N+1)!}\left.\frac{\mathrm{d}^{N+1} y}{\mathrm{d}x^{N+1}}\right|_{x=\xi}$$

where $\xi \in [a, a+h]$. Different values of ξ in the interval obviously give differing values of the errors and we usually talk about the maximum error and the bound on this value.

Consider the simple example of approximating a curve by a straight line: the source of error is going to come from the fact that the curve is not necessarily a straight line. A straight line has a zero second derivative so the errors will be proportional to this quantity, that is if the curve is a straight line then the approximation will be exact.

Now we shall substitute the Taylor series into the finite difference formula (8.4) which gives:

$$\begin{aligned}
y_{j+1} &= y_j + h y_j' + \frac{h^2}{2!} y_j'' + \frac{h^3}{3!} y_j^{(3)} + \frac{h^4}{4!} y_j^{(4)} + \cdots , \\
y_j &= y_j, \\
y_{j-1} &= y_j - h y_j' + \frac{h^2}{2!} y_j'' - \frac{h^3}{3!} y_j^{(3)} + \frac{h^4}{4!} y_j^{(4)} + \cdots .
\end{aligned}$$

Thus we have

$$\frac{y_{j+1} - 2y_j + y_{j-1}}{h^2} = y_j'' + \frac{2h^2}{4!} y_j^{(4)} + \cdots .$$

We can eliminate the dots and write this as

$$= y_j'' + \frac{2h^2}{4!} y^{(4)}(\xi),$$

where $\xi \in [x_{j-1}, x_{j+1}]$. We now have a useful estimate of the error, namely that it is proportional to h^2 and the maximum value of the fourth derivative in the interval.

This shows us that we can reduce the error by increasing the number of points (which you might think anyway). Obviously the fourth derivative is beyond our control.

In the second example we had to use what is called a biased stencil at the edge to evaluate the derivative there. Again we can use the Taylor series to find out the form of the error

$$\left.\frac{\mathrm{d}y}{\mathrm{d}x}\right|_{x=x_1} \approx \frac{y_2 - y_1}{h}$$
$$= y_1' + \frac{h}{2} y''(\xi)$$

where $\xi \in [y_1, y_2]$. In this case we note that the error is proportional to h (rather than h^2) and as such the errors will be larger (as indeed they are).

It is possible to reduce these errors by using wider stencils (that is using the values of $y_{j\pm 2}$ for example). However, there is a computational cost to pay. In fact it is possible to use all the points to work out the derivative at a point and these methods are very accurate and require very few points (due to their enhanced accuracy). This class of techniques is called **spectral methods**, but their structure is beyond the scope of this text.

8.7 Population Dynamics

We now consider the solution of the equations which arise within the description of population models. Let us consider the solution of the problem of a population which depends on the environment in which it lives, namely the logistic model

$$\frac{\mathrm{d}N}{\mathrm{d}t} = rN - \frac{rN^2}{K}, \tag{8.5}$$

where r is assumed to be positive and the constant K represents the capacity of the environment to supply the population with resources (for instance food and water). The solution of this system is

$$N(t) = \frac{N_0 K}{N_0 + (K - N_0)\mathrm{e}^{-rt}},$$

where N_0 is the initial population. It is also observed that as $t \to \infty$ $N(t) \to K$ (the *equilibrium population*). Let us now presume that we do not know this solution and determine the solution of the equation numerically. Let us start with a simple Euler integration scheme.

```
r = 0.5;
K = 2;
N0 = 1;
t = 0:0.1:10.0;
delta_t = t(2)-t(1);
nt = length(t);
N = zeros(size(t));
N(1) = N0;
for j = 1:(nt-1)
    N(j) = N(j-1) + delta_t*(r*N(j)-r*N(j)^2/K);
end
exact = N0*K./(N0+(K-N0)*exp(-r*t));
```

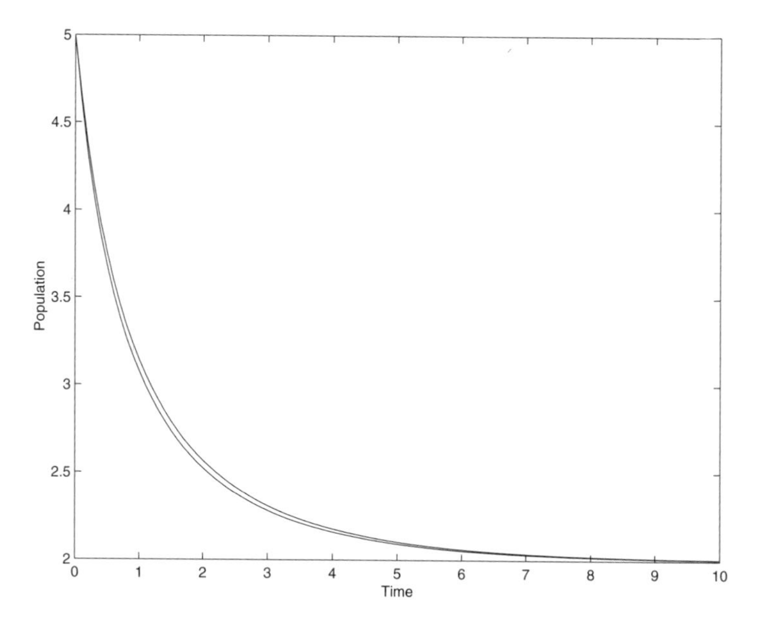

As you can see this numerical solution matches that acquired analytically very well.

We now move on to discuss the numerical solution of multi-dimensional systems, for instance:

$$\frac{\mathrm{d}N}{\mathrm{d}t} = N(a - bP)$$
$$\frac{\mathrm{d}P}{\mathrm{d}t} = P(cN - d),$$

where P reflects the number of predators and N the number of prey. The values of the constants a, b, c and d reflect the reproductive and predatory habits of the two species. For instance a larger value of a corresponds to an increased rate of reproduction of the prey. We again use Eulerian integration to give the numerical solution.

```
a = 1; b = 2;
c = 1; d = 2;
N0 = 4;
P0 = 0.2;
t = 0:0.1:10.0;
delta_t = t(2)-t(1);
nt = length(t);
N = zeros(size(t));
P = zeros(size(t));
N(1) = N0;
P(1) = P0;
for j = 1:(nt-1)
    N(j+1) = N(j)+delta_t*(N(j)*(a-b*P(j)));
    P(j+1) = P(j)+delta_t*(P(j)*(c*N(j)-d));
end
```

The solution here is oscillatory and looks like this

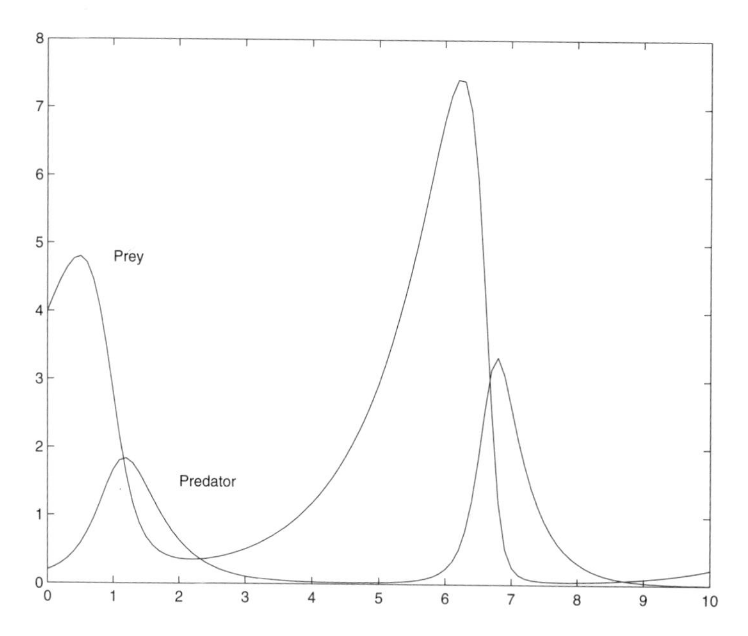

Here we observe that the population of the prey grows and then the numbers of predators are able to increase until there are too many predators and the supply of prey is virtually destroyed. Enough survive to rebuild the population and the cycle repeats.

8.8 Eigenvalues of Differential Systems

We shall now discuss the solution of equations like

$$\frac{\mathrm{d}^2 y}{\mathrm{d}x^2} + \lambda y = 0$$

subject to the conditions that $y(0) = y(\pi) = 0$. This is an example of a simple differential eigenvalue problem. It can be solved analytically and we find that there are only solutions when $\lambda = n^2$ for $n \in \mathbb{Z}$. There are many problems where an analytical solution is not accessible and we need to use finite difference techniques. To this end we rewrite the equation as

$$-\frac{\mathrm{d}^2 y}{\mathrm{d}x^2} = \lambda y.$$

We shall now discretise the equation at a set of points $\{y_j\}$ $j = 1, \cdots, N$, so that

$$-\frac{y_{j+1} - 2y_j + y_{j-1}}{h^2} = \lambda y_j$$

for $j = 2, \cdots, N-1$ and $y_j = (j-1)h$ with $h = \pi/(N-1)$. We also need the conditions that $y_1 = y_N = 0$, which reflect the conditions that $y(0) = y(\pi) = 0$. This gives a system of the form $\mathbf{Ay} = \lambda\mathbf{y}$. Consequently we expect the values of λ to be the eigenvalues of the matrix $\mathbf{A}$.

```
n = 200;
h = pi/(n-1);
A = diag(2*ones(n-2,1)/h^2,0) ...
   +diag(-ones(n-3,1)/h^2,1) ...
   +diag(-ones(n-3,1)/h^2,-1);
[V,D] = eigs(A,3,'SM');
y = linspace(0,pi,n);
for i = 1:3
    eigf = [0 V(:,i)' 0];
    plot(y,eigf)
    pause
end
```

This code gives the first three eigenvalues and eigenfunctions, which should be 1, 4 and 9 and correspondingly $\sin x$, $\sin 2x$ and $\sin 3x$.

8.9 Tasks

Task 8.1 *In this chapter we have seen an example which uses Euler's method to solve a first-order equation, namely*

$$\frac{\mathrm{d}y}{\mathrm{d}t} = (1-t)y,$$

subject to $y(0) = 1$. This was solved using the code `euler.m`*:*

```
%
% Euler's method
% euler.m
%
dt = 0.1;
t = 0.0:dt:5.0;
y = zeros(size(t));
y(1) = 1;
for ii=1:(length(t)-1)
        y(ii+1) = y(ii) + dt * (1-t(ii))*y(ii);
end
exact = exp(t-t.^2/2);
figure(1)
plot(t,y,t,exact,'--')
```

Modify the above code to solve the equation

$$\frac{\mathrm{d}y}{\mathrm{d}t} = -y\sqrt{t},$$

again subject to the boundary condition $y(0) = 1$ *from* $t = 0$ *to* $t = 1$. *(You should also work out the exact solution and compare your answers.)*

In the notes we then go on to discuss the errors intrinsic to this kind of calculation. In order to work out the error it is crucial to know the exact solution. There are two main ways of expressing the error, which are defined earlier (see page 52).

Task 8.2 *Calculate both the above errors involved in the solution of the equation*

$$\frac{\mathrm{d}y}{\mathrm{d}t} = ty,$$

subject to the boundary condition $y(0) = 1$ *between* $t = 0$ *and* $t = 1$, *using* $\Delta t = 1/4$. *(You should do these calculations by hand either using a calculator or typing the numbers into MATLAB). Compare the errors for the same calculation applied to the differential equation*

$$\frac{\mathrm{d}y}{\mathrm{d}t} = -ty,$$

everything else remaining the same.

Task 8.3 *Further modify the above code (`mod_eul.m`) to solve the differential equation*

$$\frac{\mathrm{d}y}{\mathrm{d}t} = \sin t + \sin y,$$

subject to the condition that $y(0) = 0$ from $t = 0$ to $t = 10\pi$. You can use a variety of step lengths: however you must ensure that they are less than unity in magnitude.

Task 8.4 *Solve the differential equation*

$$\frac{\mathrm{d}y}{\mathrm{d}t} = -\frac{t}{2} - \frac{y}{3},$$

using the implicit Euler scheme with $\Delta t = 1/3$, from $t = 0$ to $t = 1$ using the boundary condition $y(0) = 0$. Compare this solution to the exact solution and one obtained using a MATLAB code using 1001 points in this interval. (You will need to solve the above equation for y_{n+1} and then modify one of the previous codes. It might be best to go back to `euler.m`.)

We can also use MATLAB's intrinsic functions `ode23` or `ode45`, which exploit Runge–Kutta techniques. These routines are more accurate than the Euler schemes we have used above. In order to use them it is necessary to write a code which evaluates the right-hand side of the equation

$$\frac{\mathrm{d}y}{\mathrm{d}t} = f(t, y).$$

One such example is given on page 265. Let us give a further example. In order to solve the problem posed in Task 8.3 we should use the code (`odef.m`)

```
function [value] = odef(t,y)
value = sin(t)+sin(y);
```

together with the code

```
y0 = 0;
tspan = [0 10*pi];
[t,y] = ode45('odef',tspan,y0);
```

Task 8.5 *Modify the above codes to solve the problem*

$$\frac{\mathrm{d}y}{\mathrm{d}t} = t^2 - y^2,$$

from $t = 0$ to $t = 2$ subject to the boundary condition that $y(0) = 0$, using the MATLAB routine `ode45`.

It should be noted that even if the system does not explicitly vary with t (that is a dependence of the right-hand side on t) the first line of the corresponding function should include t in its arguments. For example in solving the system

$$\frac{\mathrm{d}y}{\mathrm{d}t} = y,$$

the corresponding function would be

```
function [value] = func(t,y)
value  = y
```

(Note that these equations are referred to as autonomous if there is no explicit variation with t.)

Task 8.6 *Compare the numerical and exact solution of the equation*

$$\frac{\mathrm{d}y}{\mathrm{d}t} = -y + t^2$$

subject to the condition that $y(0) = 1$ integrated from $t = 0$ to $t = 2$.

Task 8.7 *Modify the code on page 260 to solve the problems:*

1. $y'' = x \cos x$ *subject to the conditions $y(0) = 1$ and $y(1) = 0$.*
2. $y'' = x \cos x$ *subject to the conditions $y'(0) = 0$ and $y(1) = 0$.*
3. $y'' + 2y' = 0$ *subject to the conditions $y(0) = 0$ and $y(1) = 1$.*

Task 8.8 *Solve the differential equation*

$$\frac{\mathrm{d}^2y}{\mathrm{d}t^2} + 3y = t$$

subject to the initial conditions $y(0) = y'(0) = 0$ both numerically and analytically over the range $t \in [0, 1]$.

Task 8.9 *Solve the equation*

$$\frac{\mathrm{d}^2y}{\mathrm{d}t^2} + ty = \sin t,$$

for the initial conditions $y(0) = 1$ *and* $y'(0) = 0$ *over the range* $t \in [0, 2]$. *Either using a shooting method or by direct solution solve the same equation subject to the boundary conditions* $y(0) = y(2) = 0$.

Task 8.10 *Solve the first-order differential equation*

$$(t^2 + 1)y' + 2ty = 0,$$

subject to the initial condition that $y(0) = 1$ *(again both analytically and numerically) over the range* $t \in [0, 5]$.

Task 8.11 (*) *Solve the third-order differential equation*

$$y''' - 2y'' - y' + 2y = 0,$$

subject to the conditions that $y(0) = y'(0) = 0$ *and* $y''(0) = 1$, *by converting it to a system of three first-order equations.*

Task 8.12 (*) *Using the methods of finite differences determine the viable values of the eigenvalue* λ *in the differential equation*

$$y'' + (\sin x - \lambda)\, y = 0,$$

subject to the homogeneous boundary conditions $y(0) = y(\pi) = 0$. *Consider the three eigenfunctions which correspond to the eigenvalues with the smallest magnitude.*

9
Simulations and Random Numbers

9.1 Introduction

We shall use MATLAB to generate random numbers so that we can discuss its statistical capabilities and also explore some simulations. We also show how it may be used to analyse sets of data.

We start with a definition of some statistical quantities.

9.2 Statistical quantities

In this section we consider a set of data which we shall refer to as x_i where $i = 1$ to N. At this point we shall not make any further assumptions about the data both in terms of its source or its form.

9.2.1 Averages

Perhaps the most popular choice of average is the mean (which is usually the arithmetic mean): this is given by adding up all the data and dividing by the number of objects in the set. This is written as:

$$\bar{x} = \frac{1}{N} \sum_{i=1}^{N} x_i,$$

and can also be denoted as $E(X)$, the expected value of the random variable X. If we presume that our data is stored in a vector, the MATLAB code to determine the mean is simply:

```
xbar = sum(x)/length(x);
```

In fact, as you might expect, there is a MATLAB command which does exactly this, namely `mean(x)`.

In fact there are two further averages, the median and the mode. The latter is the data value which occurs the most often and we shall not discuss this further. However the median is the value in the "middle" when the data has been put in order. If there are an odd number of pieces of data this is well defined: however if it is even then the mean of the data at each end of the central interval is given, for example the median of $(1, 2, 3, 4)$ is $(2+3)/2 = 2\frac{1}{2}$ whereas the median of $(1, 5, 7)$ is 5. We note that the median can be equal to the mean (as it is in the former case) but it is not necessarily so. The median can be determined using the code:

```
xs = sort(x);
lx = length(x);
if mod(lx,2)==0
   median = (xs(lx/2)+xs(lx/2+1))/2;
else
   median = xs((lx+1)/2);
end
```

Alternatively we can use the command `median(x)`.

Example 9.1 *Determine the mean and median of the following data:*

$$3, 4, 5, 3, 2, 1, 3, 5, 6, 3, 2, 5, 6, 8, 2, 0.$$

Although it would be a simple matter to do this by hand we use the code:

```
x = [3 4 5 3 2 1 3 5 6 3 2 5 6 8 2 0];
mean(x)
median(x)
```

The mean is 3.625 and the median is 3.

9.2.2 Other Statistical Measures

In the previous section we have discussed the various averages of a set of data but there are more measures available to us, for instance these two distributions have the same mean but are obviously quite different in form:

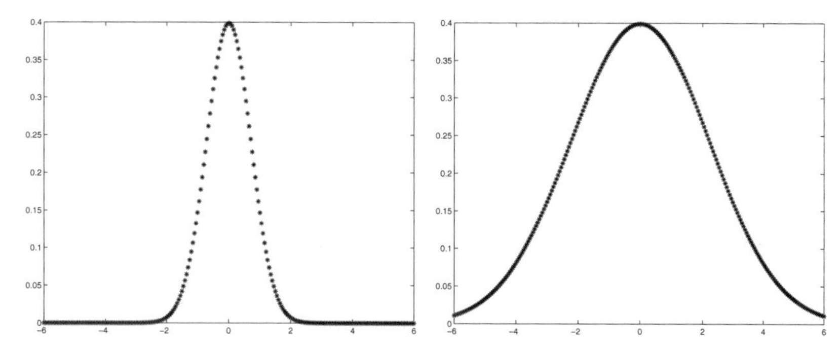

The first curve is narrower than the second and this is quantified using the variance which is given by

$$\frac{1}{N-1}\sum_{i=1}^{N}(x_i - \bar{x})^2.$$

We can interpret this as the "mean" of the sum of the squares of the distance from the mean. This can be written in terms of expectations as $E(X^2)-E(X)^2$. The variance can be determined using the simple code

```
N = length(x);
xbar = sum(x)/N;
var = sum((x-xbar).^2)/(N-1);
```

or unsurprisingly using the MATLAB command `var(x)`. Another related measure is the standard deviation which is merely the square root of the variance (determined using `std(x)`). This also is a measure of the width of the distribution but has the advantage that it has the same units as the data.

Example 9.2 *Calculate the variance and standard deviation of the data given in Example 9.1.*

This is done simply by augmenting the above code with the lines

```
var(x)
std(x)
```

There are various other measures, some of which are higher-order moments, the mean being the first, the variance the second, the skewness the third and the kurtosis the fourth (see Task 9.3).

We note that there are also measures which show how two sets of data might be related. These are the *covariance* and *correlation*. They are defined as:

Definition 9.1 (Covariance)

This quantity is defined as

$$\sigma_{XY} = E(XY) - E(X)E(Y),$$

and for a set of data is evaluated using

$$\begin{aligned} \sigma_{XY} &= \frac{1}{n}\sum_{i=1}^{n} x_i y_i - \frac{1}{n^2}\sum_{i=1}^{n} x_i \sum_{i=1}^{n} y_i \\ &= \frac{1}{n}\sum_{i=1}^{n}(x_i - \bar{x})(y_i - \bar{y}). \end{aligned}$$

Note that the variance of a random variable X is merely its covariance with itself. Here we have used a factor of n rather than $n-1$ (which is relevant to the covariance of a sample). The correlation is a measure of two variables' variability with each other when compared to their own "spread".

Definition 9.2 (Correlation)

The correlation of two random variables is defined as

$$\begin{aligned} r_{XY} &= \frac{\sigma_{XY}}{\sigma_X \sigma_Y} \\ &= \frac{\sum_{i=1}^{n}(x_i - \bar{x})(y_i - \bar{y})}{\sqrt{\sum_{i=1}^{n}(x_i - \bar{x})^2 \sum_{i=1}^{n}(y_i - \bar{y})^2}}. \end{aligned}$$

The correlation takes values between minus one and plus one.

A correlation of plus one means that the random variables are positively correlated, that is $Y = aX + b$ (where a is positive), and similarly for $r = -1$, in that they are negatively correlated.

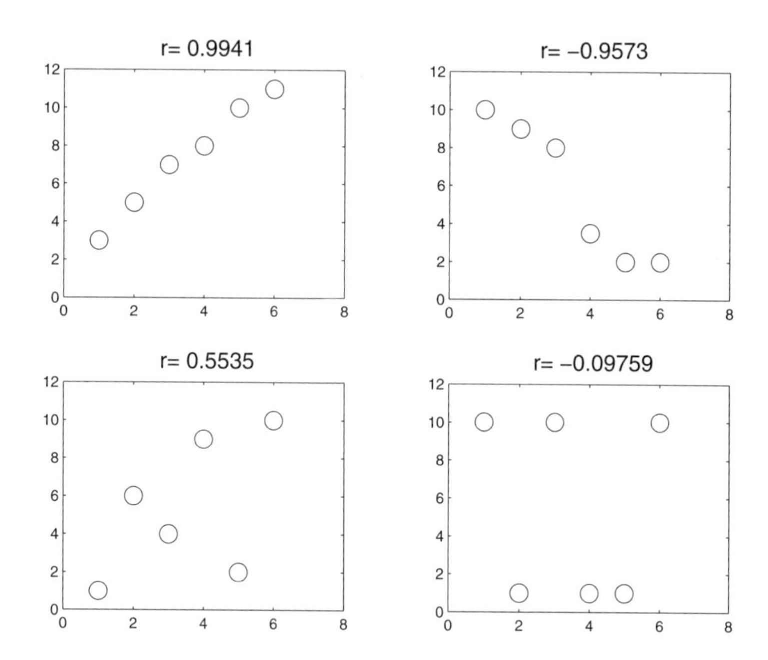

There are MATLAB commands `cov` and `corrcoef`. The first of these calculates the covariance between two vectors and returns a matrix including the variances so that `cov(x,y)` returns

$$\begin{pmatrix} \sigma_X & \sigma_{XY} \\ \sigma_{YX} & \sigma_Y \end{pmatrix}$$

and similarly the correlation command `corrcoef(x,y)` gives

$$\begin{pmatrix} 1 & r_{XY} \\ r_{YX} & 1 \end{pmatrix}.$$

The elements along the diagonal merely show that a random variable is correlated with itself (unsurprisingly!).

9.3 Random Numbers and Distributions

In order to generate random numbers we can use the various commands available within MATLAB. We shall not worry about how this is done and the technical aspects of the seeding. It is possible to get similar results from a program by always getting it to use the same set of random numbers but we will discourage this.

We shall start with the simple command `rand` which returns a random number between zero and one. Let us start with a simple dice program

```
%
% roll.m
%
function [number] = roll()
number = ceil(rand*6+eps);
```

which is called using

```
%
% mroll.m
%
for ii = 1:6000
     rr(ii) = roll;
end
for j = 1:6
     ii(j) = length(find(rr==j));
end
disp([' Mean ' num2str(mean(rr))])
```

We include eps in case the value of the random number is zero, which is improbable but possible.

This prints the mean of the sample which is defined as

$$\bar{x} = \frac{1}{N}\sum_{i=1}^{N} x_i,$$

which should be given by 7/2, derived from

$$\begin{aligned}\sum_{i=1}^{N_{outcomes}} ip(i) &= \frac{1}{6}1 + \frac{1}{6}2 + \cdots + \frac{1}{6}6 \\ &= \frac{1}{6}(1+2+3+4+5+6) = \frac{21}{6} = 3\frac{1}{2}.\end{aligned}$$

We can also determine the sample variance as

$$\text{var} = \frac{1}{N-1}\sum_{i=1}^{N}(x_i - \bar{x})^2,$$

and this can be determined for this case as

$$\begin{aligned}\overline{x^2} - \bar{x}^2 &= \sum_{i=1}^{6} i^2 p(i) - \bar{x}^2\\ &= \frac{1}{6}\sum_{i=1}^{6} i^2 - \frac{49}{4}\\ &= \frac{91}{6} - \frac{49}{4} = \frac{35}{12} \approx 2.9167.\end{aligned}$$

It also gives the distribution of the values in the array which can be plotted using `bar(ii)` to give

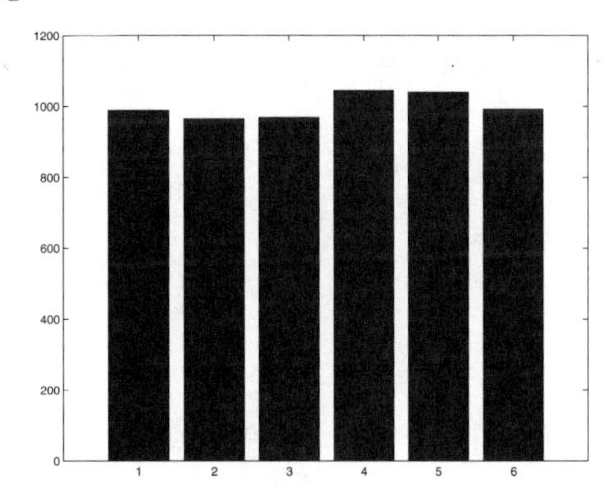

If we wanted the above experiment to investigate the effect of using two dice, we could modify the line within the `for` loop to read `rr(ii) = roll+roll;` which yields the plot

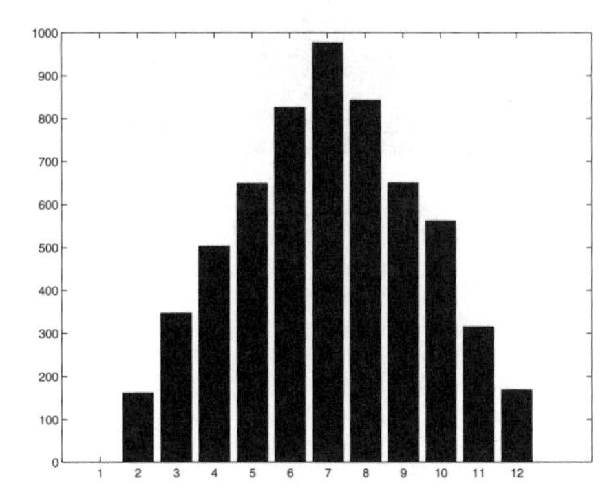

Example 9.3 *We can use this process for other "games", for instance tossing a coin:*

```
function [out] = toss()
A = ['heads';'tails'];
i = ceil(rand*2+eps);
out = A(i,:);
```

This can then be called just using **toss** *which generates either* ***heads*** *or* ***tails****.*

There are various other experiments we could try to analyse but let us discuss other random number generators, or at least variants of the above.

9.3.1 Normal Distribution

The normal distribution has two parameters associated with it: the mean and the variance. The MATLAB command **randn** generates random numbers which have a mean of zero and a variance of unity. We use the command `x = randn(1,500);` to set up the array and the command `hist(x,12)` to produce the plot

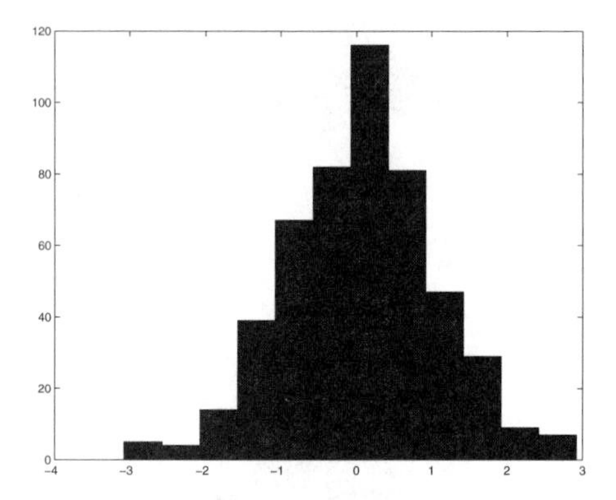

We can now use the commands **mean** and **var** to analyse how close to the advertised values the random number generator achieves. This gives

```
>> mean(x)

ans =

    0.0702

>> var(x)

ans =
```

```
0.9873
```

Usually we would have the data from an "experiment" but here we shall generate it from the internal generator.

Example 9.4 *Consider the distribution of salaries with mean of £20000 and standard deviation of £3000. How many people would you expect to have salaries in excess of £22000 in a group of 200 people?*

We use the code

```
no_experiments = 200;
for ii = 1:no_experiments
x = randn(1,200);
salaries = x*3000+20000;
number(ii) = length(find(salaries>22000));
end
mean(number)
```

which gave `mean(number)=50.94`*, and if it is repeated similar values are obtained. It is possible to determine this value from tables.*

9.3.2 Calculating Probabilities

We note that the probability that a random variable X is less than a value z, where the random variable is known to be distributed normally with mean μ and variance σ^2, and can be written as

$$P(X \leqslant z) = \frac{1}{\sigma\sqrt{2\pi}} \int_{q=-\infty}^{z} \mathrm{e}^{-\frac{1}{2}\left(\frac{q-\mu}{\sigma}\right)^2} \,\mathrm{d}q.$$

Of course we can calculate this quantity using MATLAB (relatively easily).

```
mu = 0.2;
sig = 1.5;
min_inf = -6*sig;
z = input('Enter z: ');
clear x n
x = linspace(min_inf,z,1000);
n = length(x);
h = x(2)-x(1);
f = exp(-((x-mu)/sig).^2/2);
val = (sum(f)-f(1)/2-f(n)/2)*h;
prob = val/sqrt(2*pi)/sig
```

In this code we have used the trapezium rule and the fact that the rate at which the integrand decays to zero is proportional to σ (and as such we set the lower end of the range of integration as -6σ). We can use this code to work out confidence intervals.

9.3.3 Permutations

MATLAB has many other commands for this kind of operation but it also provides other devices which can be used for randomisation of lists.

Example 9.5 *Let us suppose that we wish to produce a random ordering of the five names: Avril, Beryl, Carol, Daisy and Emily.*

```
a = ['Avril'; 'Beryl'; 'Carol'; 'Daisy'; 'Emily'];
r = randperm(5);
a(r,:)
```

9.4 Maps and White Noise

We have already met incidences of maps (for instance page 91) and these can be described as taking a value x_n and returning the next iterate. They can be written in a variety of ways but here we shall use the nomenclature:

$$x_{n+1} = f(x_n)$$

for a one-dimensional map. A fixed point of the map is defined to be $x_n = f(x_n)$. We can also consider repetitions of the map, for instance $f(f(x_n))$ where $f(\ .\ )$ operates on x_{n+1}. This is written as $f^2(x_n)$ and the mth iterate of the map would be written as $f^m(\ .\ )$. In the case where $f^m(x^*) = x^*$ we say that x^* is a fixed point of period m and the sequence of points $\{x^*, f(x^*), \cdots, f^{m-1}(x^*)\}$ is the corresponding orbit.

Example 9.6 (Tent Map) *We define the Tent map as* $f : \mathbb{R} \to \mathbb{R}$ *where*

$$f(x) = \begin{cases} ax & x \leqslant \frac{1}{2}, \\ a(1-x) & x > \frac{1}{2}. \end{cases}$$

This can be rewritten as

$$f(x) = a\left(\frac{1}{2} - \left|x - \frac{1}{2}\right|\right),$$

which is more convenient for use in MATLAB. We now use the code

```
function [out] = tent(a,x)
out = a*(0.5-abs(0.5-x));
```

If we plot the results of using $a = 4$:

```
x = linspace(-1,2);
plot(x,tent(4,x))
h = gca;
set(h,'YGrid','On')
```

we have

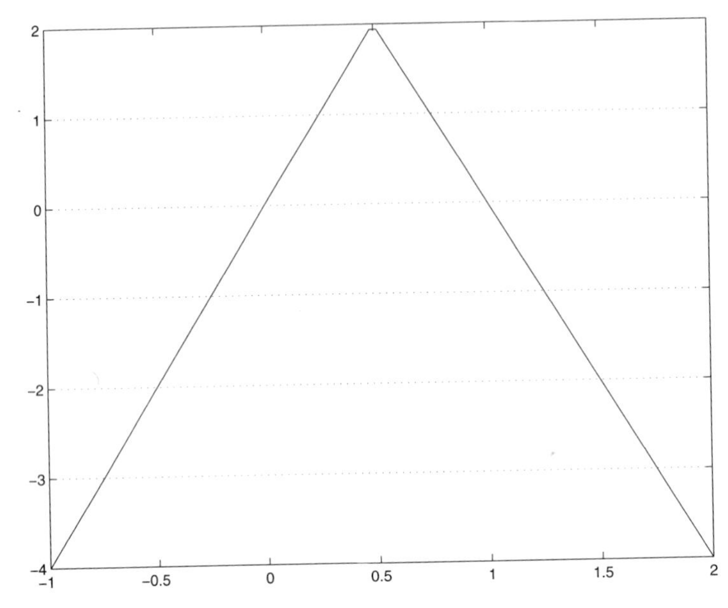

and producing the second iterate (`y = tent(4,x); plot(x,tent(4,y))`*) we have*

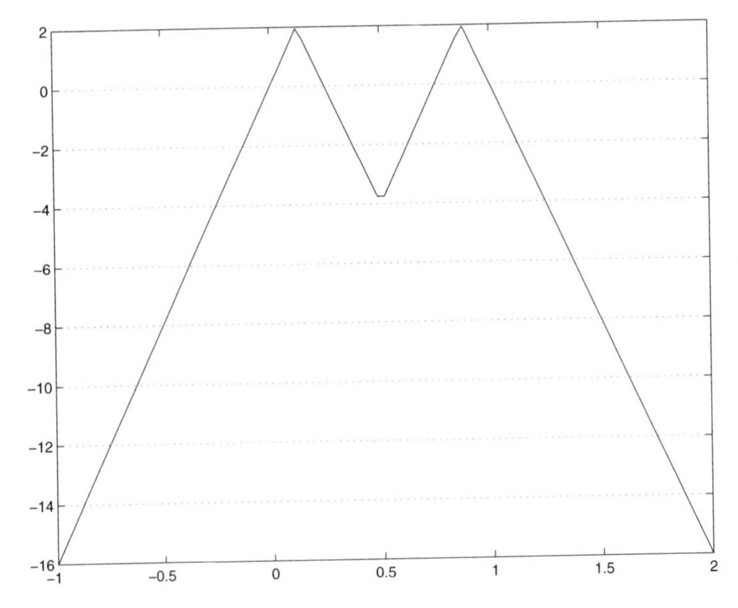

In fact most points will be expelled to minus infinity, but some points will remain. These are the fixed points. For instance points of period two will lie on the intersection of the lines shown in the previous figure and the line $y = x$.

This is a very rich topic for further investigation.

We shall return to one-dimensional maps, but shall start by deriving a simple two-dimensional map, which is written as

$$x_{n+1} = f(x_n, y_n), \qquad y_{n+1} = g(x_n, y_n).$$

This is a two-dimensional explicit map and can be used to simulate various behaviours depending on the forms of f and g. We shall start with the simple example of

$$\dot{x} = y, \qquad \dot{y} = -x,$$

where a dot denotes the derivative with respect to time. We shall use the Crank–Nicolson method to discretise this system and denote the values at the nth time level by the subscript n. The system becomes

$$\begin{aligned} \frac{x_{n+1} - x_n}{\Delta t} &= \frac{y_n + y_{n+1}}{2}, \\ \frac{y_{n+1} - y_n}{\Delta t} &= -\frac{x_n + x_{n+1}}{2}. \end{aligned}$$

We can write this as the matrix system

$$\begin{pmatrix} 1 & -\frac{\Delta t}{2} \\ \frac{\Delta t}{2} & 1 \end{pmatrix} \begin{pmatrix} x_{n+1} \\ y_{n+1} \end{pmatrix} = \begin{pmatrix} x_n + \frac{\Delta t}{2} y_n \\ y_n - \frac{\Delta t}{2} x_n \end{pmatrix}.$$

This can now be solved to yield the next point. Before we give the code for this purpose we solve the system analytically. Multiply the first equation by x and the second by y and then add them to obtain

$$x\dot{x} + y\dot{y} = 0$$

which can be written as

$$\frac{1}{2}\frac{\mathrm{d}}{\mathrm{d}t}\left(x^2 + y^2\right) = 0.$$

This can be integrated to give $x^2 + y^2 = C$, which are circles of radius $\sqrt{C}$. Hence the initial conditions will give the value of C and then the points should transcribe a circle. Notice if we use an Euler scheme the inaccuracies in the method soon become evident. However, using the following codes we can track a point.

```
%
% maps.m
%
function [f,g]=maps(x,y)
global delt
A = [ 1 -delt/2; delt/2 1];
rhs = [x+delt/2*y; -delt/2*x+y];
v = A\rhs;
f = v(1);
g = v(2);
```

and

```
%
% crmap.m
%
global delt
delt = 0.05;
x = 0.5; y = 0.4;
noise = 0.0;
nop = 500;
po(1) = x+i*y;
for it = 2:nop
    [xn,yn] = maps(x,y);
    xn = xn+ ...
      noise*sqrt(-2*log(rand(1)))*cos(2*pi*rand(1));
    yn = yn+ ...
      noise*sqrt(-2*log(rand(1)))*sin(2*pi*rand(1));
    po(it) = xn+i*yn;
    x = xn; y = yn;
end
plot(po,'.')
axis equal
```

This gives the result:

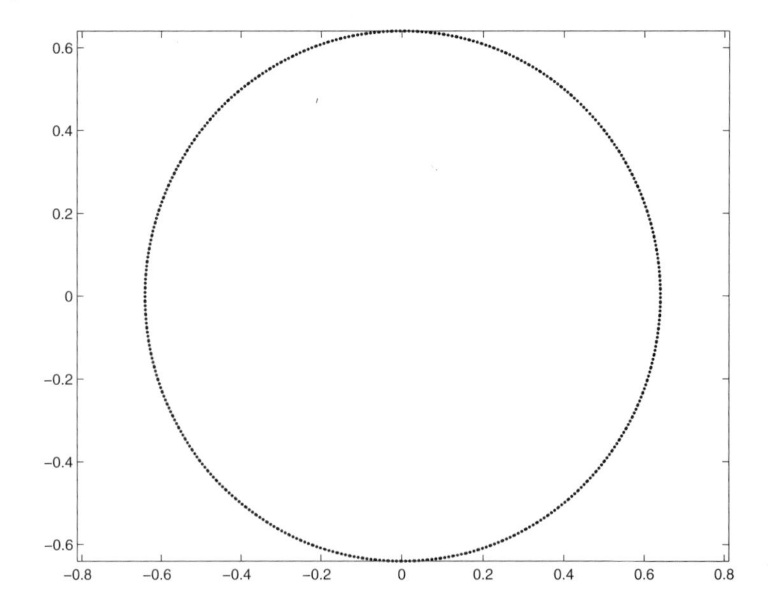

where we have used the array `po` as being complex. Hence it can be plotted in a straightforward manner using `plot`. The radius of the circle is $\sqrt{0.4^2 + 0.5^2} \sim 0.6403$ (and this can be verified by typing `max(real(po))`). In the above code we have introduced some noise (at the moment it is set to have an amplitude of zero). This is two-dimensional Gaussian noise and its effect is to add diffusion to the system. With `noise=0.01` we have:

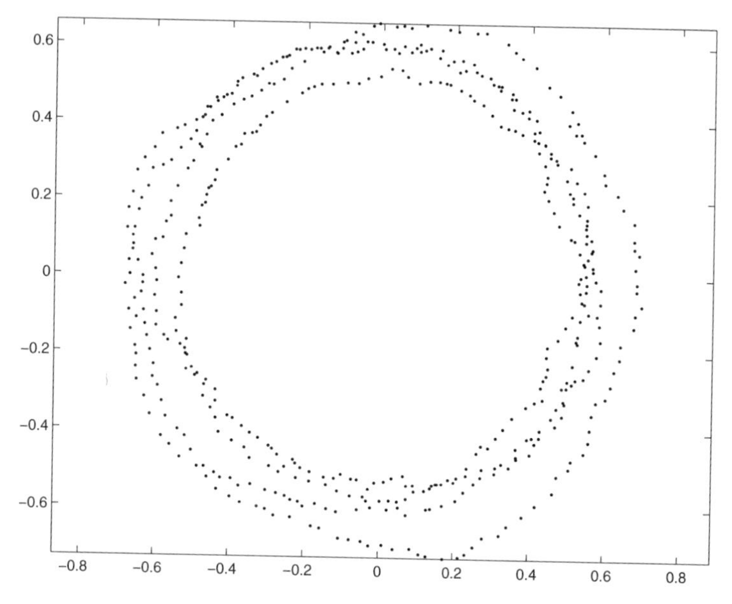

We could look at the noise purely in isolation. Here we consider a random walk using the simple code:

```
x= 0.0; y= 0.0;
noise = 0.05;
nop = 50;
po(1) = x+i*y;
for it = 2:nop
   y = y+noise*(rand(1)-0.5);
   x = x+0.1;
   po(it) = x+i*y;
end
plot(po)
```

which yields

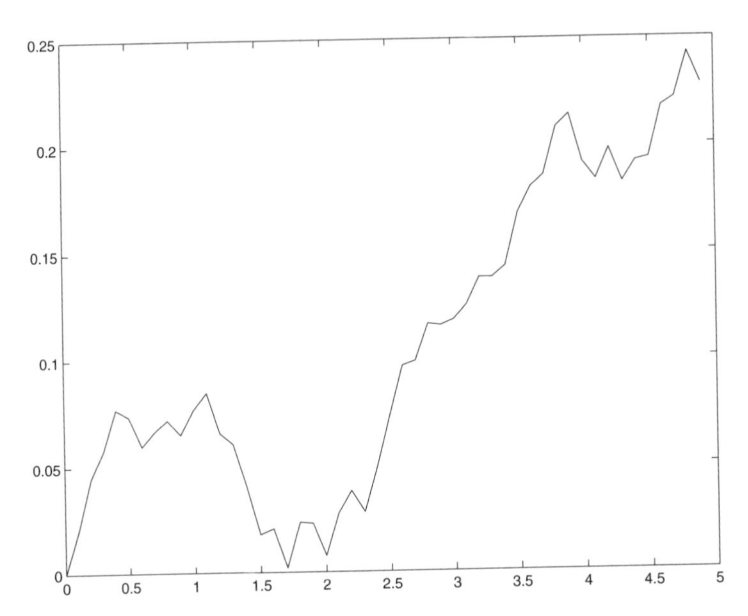

The above is a very simple map derived from the Crank–Nicolson scheme and does not exhibit any strange (that is chaotic) behaviour, but we could try the map

$$x_{n+1} = (x_n + y_n)|1, \qquad y_{n+1} = (x_n + 2y_n)|1,$$

which gives

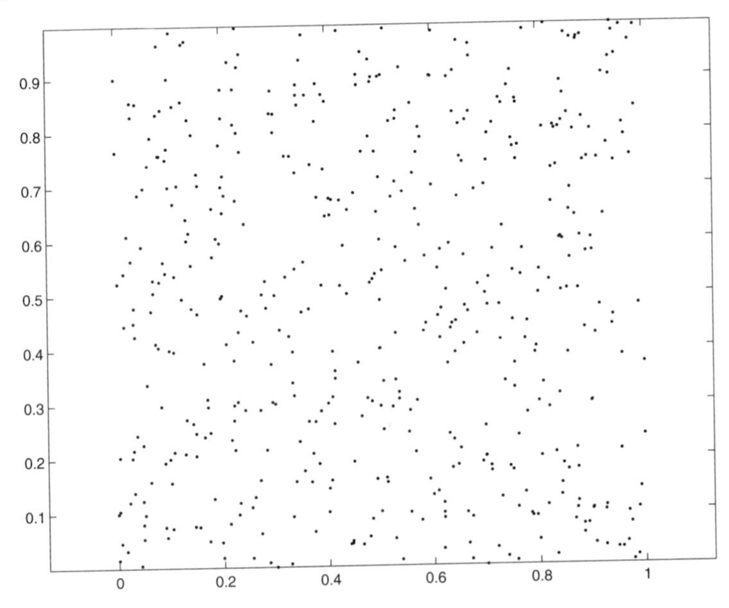

This plot is given with no noise as the map (which is called the Baker map (Drazin *Nonlinear Systems*)) is chaotic. This was coded using the MATLAB command `rem` so that the map part of the code is given by

```
% mapb.m
function [f,g] = mapb(x,y)
f = rem(x+y,1);
g = rem(x+2*y,1);
```

The function `rem` gives the remainder when dividing the first argument by the second.

9.4.1 Modelling Discrete Systems

There are many problems which require simulation: this is largely because it is not viable to actually run the physical tests. This covers things from experiments on aircrafts (which are too dangerous or expensive) to population studies and economic strategies.

We consider the example of two populations N and P. The evolution of these co-existing populations are governed by the difference equations:

$$\begin{aligned} N_{t+1} &= rN_t \mathrm{e}^{-aP_t} \\ P_{t+1} &= N_t \left(1 - \mathrm{e}^{-aP_t}\right). \end{aligned}$$

This is an example of a predator–prey model.

This is solved simply using the code

```
r = 2;
a = 1;
nt = 20;
t = 1:1:nt;
N = zeros(size(t));
P = zeros(size(t));
N(1) = r*log(r)/(a*(r-1))+0.01;
P(1) = log(r)/a;
for j = 1:(nt-1)
    N(j+1) = r*N(j)*exp(-a*P(j));
    P(j+1) = N(j)*(1-exp(-a*P(j)));
end
subplot(2,1,1)
plot(t,N,'o','MarkerSize',12)
title('N values')
subplot(2,1,2)
plot(t,P,'r+','MarkerSize',12)
title('P values')
```

In this code we have used the MATLAB command `subplot` which allows more than one graph to be added to a figure. The arguments are: the number of rows, the number of columns and the position within that system. For instance `subplot(2,3,5)` has two rows and three columns and makes the second figure on the second row the current figure. The figures are numbered across the rows, so that "1" is top left and "6" is bottom right.

The above calculation yields

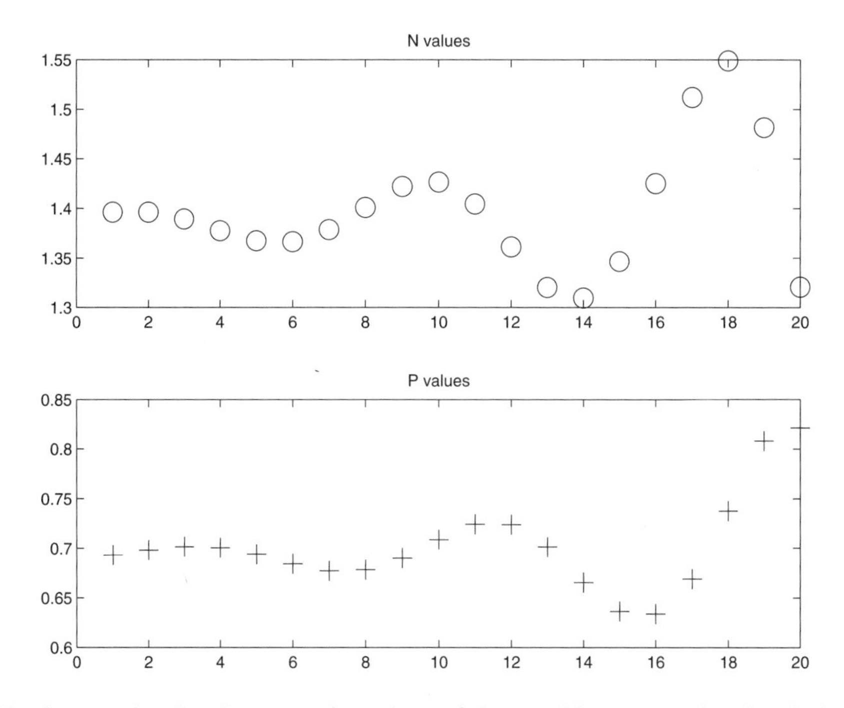

We observe that by changing the values of the equilibrium amplitudes slightly, the populations start to oscillate and eventually diverge from the values we start with. This would seem to suggest that this equilibrium is unstable.

9.4.2 Periodicity and Chaos

Here we discuss one of the most famous models of populations, which has an incredibly rich structure. Consider

$$N_{t+1} = N_t + rN_t(1 - N_t).$$

This difference equation has very different solutions depending on the values of r. The new population N_{t+1} is taken to be the old population N_t plus a term which is proportional to the term $N_t(1 - N_t)$, where unity is taken to be the optimal population. Notice that if $N_t > 1$ then the second term in this expression is negative, and the new population is reduced. This is supposed to represent the idea of finite resources. This model was posed by Verhulst and it is related to the logistic model. In the main we shall consider an initial population corresponding to $N = 1/10$.

```
str = 'Please enter the ';
r = input([str 'rate constant r :']);
N0 = input([str 'initial population N(0) :']);
n = input([str 'number of years n :']);
t = 1:1:n;
N = zeros(size(t));
N(1) = N0;
for j = 1:(n-1)
    N(j+1) = N(j) + r*N(j)*(1-N(j));
end

plot(t-1,N,'*','MarkerSize',10)
```

This gives

```
N =

  Columns 1 through 7

    0.1000    0.1900    0.3439    0.5695    0.8147    0.9657    0.9988

  Columns 8 through 10

    1.0000    1.0000    1.0000
```

and as observed in the other notes the population tends to unity, whereas when the rate constant $r = 2.1$ we find that

```
N =

  Columns 1 through 7

    0.1000    0.2890    0.7205    1.1434    0.7991    1.1362    0.8112

  Columns 8 through 10

    1.1328    0.8168    1.1310
```

So after an initial period the population tends to oscillate between two values, which can be seen by extending the simulation to 100 years

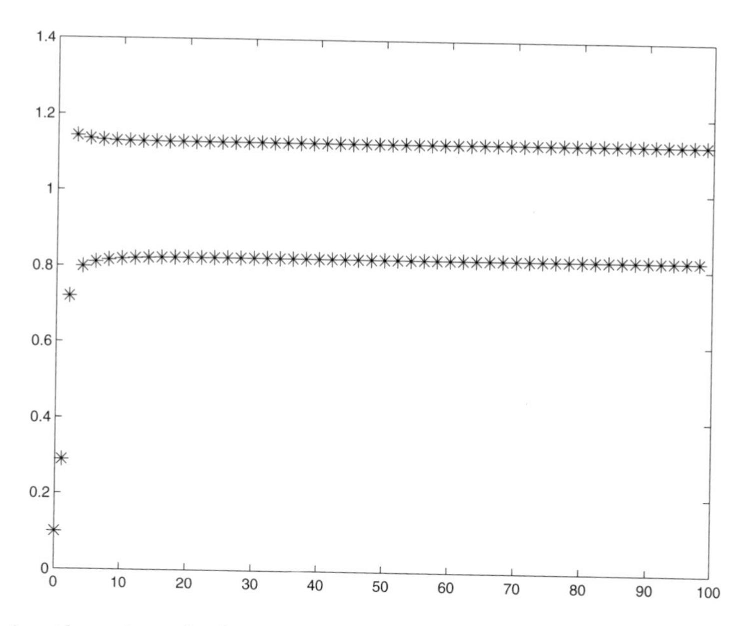

Increasing the value of r further leads to higher-order periodicity, for instance $r = 2.5$ gives

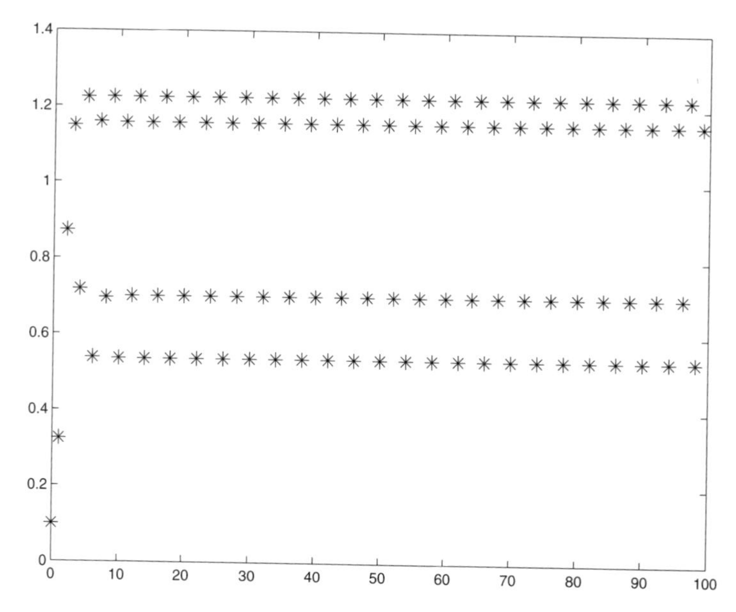

which is a period four structure. Increasing it further produces structures which don't seem to possess any periodic structures, for instance $r = 2.8$ gives

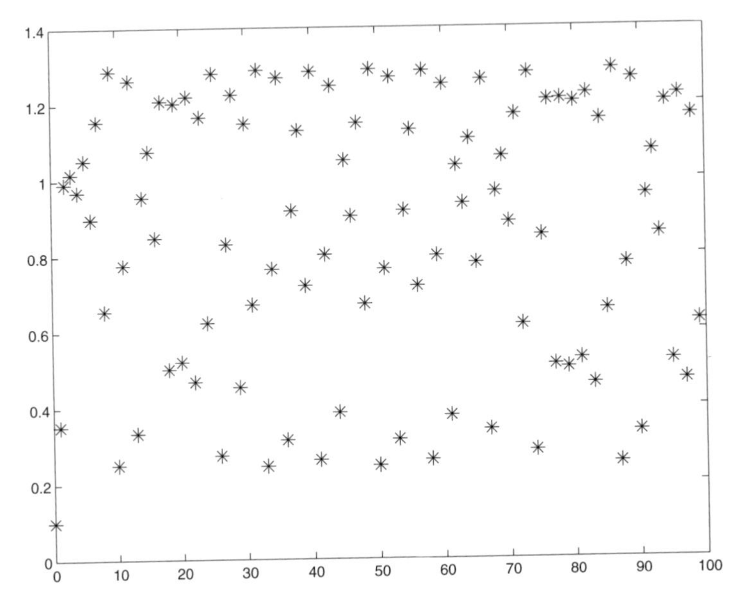

This has changed into a chaotic orbit: for further details refer to Drazin's book on Nonlinear systems or any popular science book on Chaos, for instance the book by Gleick.

We pause here and determine analytical results for certain values of the rate constant r. Firstly we determine the fixed point in which case $N_{t+1} = N_t$ and using the equation for N_{t+1} we find that N_t equals zero and one are fixed points (independent of the choice of r). If we seek points of period two, we require $N_{t+2} = N_t$ so that we have the coupled equations

$$\begin{aligned} N_{t+1} &= N_t + rN_t(1 - N_t) \\ N_t &= N_{t+1} + rN_{t+1}(1 - N_{t+1}). \end{aligned}$$

These can be combined to form a single quartic

$$0 = N_t(1 - N_t)\left\{1 + (1 + r(1 - N_t))\,(1 - N_t r)\right\},$$

where we have extracted factors of N_t and $(1 - N_t)$ since we know that the fixed points must also be solutions. The remaining quadratic has solutions

$$N_t = \frac{r + 2 \pm \sqrt{r^2 - 4}}{2r}.$$

This second pair of solutions is only real provided r is greater than or equal to two, and the system undergoes what we call a bifurcation. In fact at a higher value of r period four solutions come into existence and this phenomena is called *period doubling*. For the case $r = 2.1$ we find that the two points correspond approximately to 1.1286 and 0.8237.

As we see we can determine the values of the points within the orbit analytically (sometimes): however here we present a method for finding the values numerically using the Newton–Raphson method. Firstly we include a routine which returns the image of the point after n iterations of the map:

```
% maplog
function [new] = maplog(old,r,n)

for j=1:n
    new = old+r*old*(1-old);
    old = new;
end
```

and this is called from the main routine mapsearch which is

```
r = input('Please enter the value of r :');
starting_guess = input('Please enter starting guess ');
ms = 'Please enter the period of the orbit sought '
period = input(ms);
maxits = 200;
tolerance = 1e-10;
delta = 1e-4;
x = starting_guess;
for its = 1:maxits
    err = maplog(x,r,period)-x;
    if abs(err)<tolerance
       break
    end
    err_dx = maplog(x+delta,r,period)-(x+delta);
    x = x - delta*err/(err_dx-err);
end
if its==maxits
     disp(['No period ' int2str(period) ' point found '])
else
     disp('Orbit')
     for j = 1:period
          disp(x)
          x = maplog(x,r,1);
     end
end
```

The results of which for the two cases above are:

```
>> mapsearch
Please enter the value of r :2.1
Please enter starting guess 0.82
Please enter the period of the orbit sought 2
Orbit
    0.8237

    1.1286

>> mapsearch
Please enter the value of r :2.5
Please enter starting guess 0.58
Please enter the period of the orbit sought 4
Orbit
    0.6000

    1.2000

    0.6000

    1.2000
>> mapsearch
Please enter the value of r :2.5
Please enter starting guess 0.52
Please enter the period of the orbit sought 4
Orbit
    0.5359

    1.1577

    0.7012

    1.2250
```

Notice in the second case the program actually generated a period 2 point, and so an alternative initial guess was required.

We can actually plot a diagram of the fixed points of this map using

```
clf
for r = 0:0.01:4;
    x = rand(1);
% Remove transients
    for j = 1:100
        x = x+r*x*(1-x);
    end
    xout = [];
    for j = 1:400
        x = x+r*x*(1-x);
        xout = [xout x];
    end
    plot(r*ones(size(xout)),xout,'.','MarkerSize',3)
    hold on
end
axis([0 4 0 1])
```

This gives:

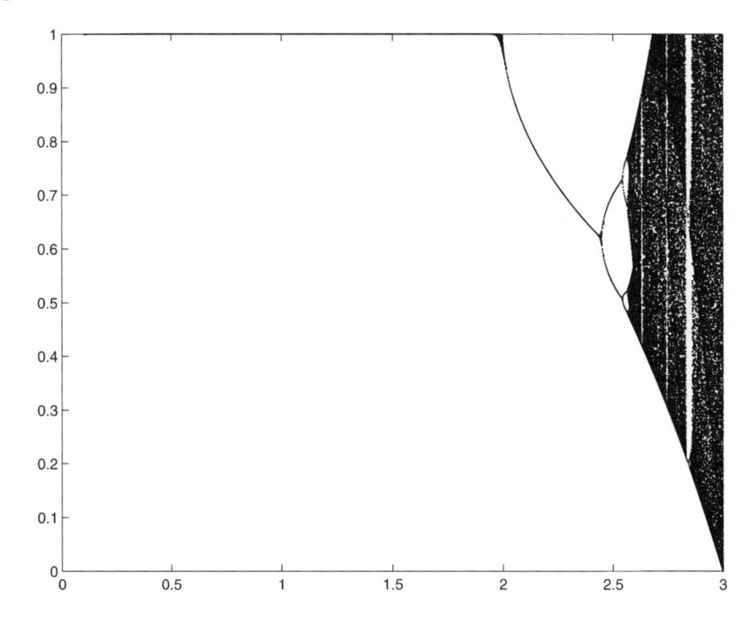

and zooming in

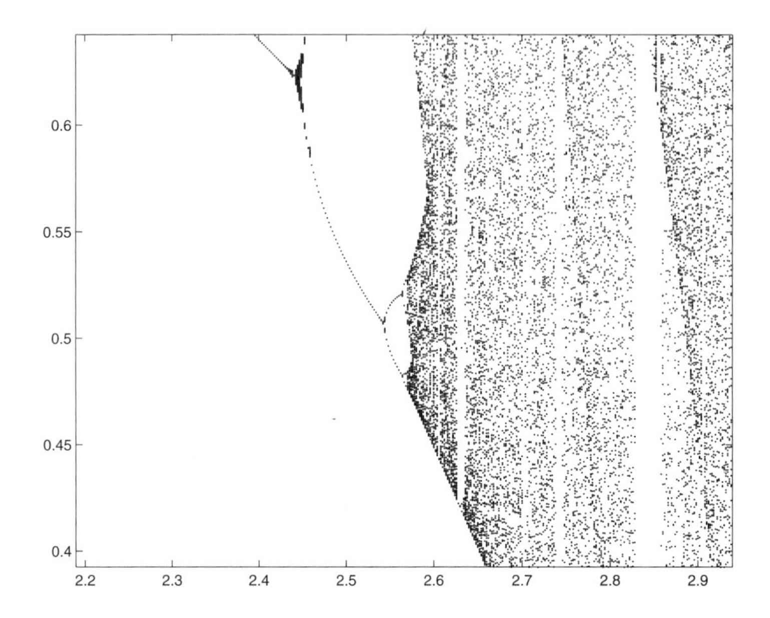

9.4.3 Random Motion

Let us now consider a particle in a box and ask the question of how likely it is to hit the east wall. We assume that the western wall repels the particles and that there is no vertical motion. In this case the position at a subsequent time is related to the current position by

$$x_{t+1} = x_t + f(x_t) + \text{Noise},$$

where $f(x) = k/x^2$ so that the force away from the western wall at $x = 0$ is proportional to the inverse of the distance squared. The noise is given by ϵ times a random number between minus one and one. The eastern wall is taken to be at $x = 5$ and the constant of proportionality for the repulsive force is $k = 0.5$, with a noise amplitude of 1.

```
x0 = 2.5;
noise = 1.;
east_wall = 5;
k = 0.5;
no_experiments = 200;
for j = 1:no_experiments
    x = x0;
    steps = 0;
    while x<east_wall
        steps=steps+1;
        x = x + k./x^2 + noise*(rand(1)-0.5)*2;
    end
    st_store(j) = steps;
end
```

The result of one simulation is given as

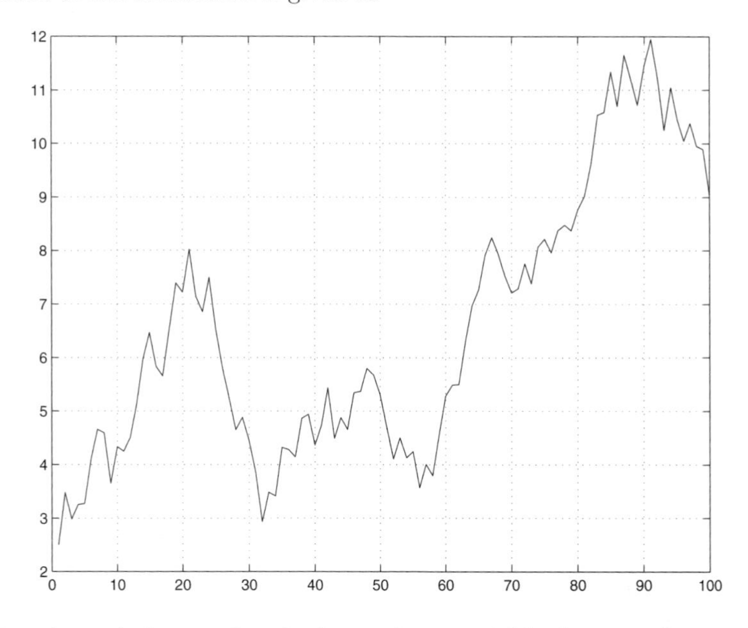

Running the code for two hundred experiments yields these results:

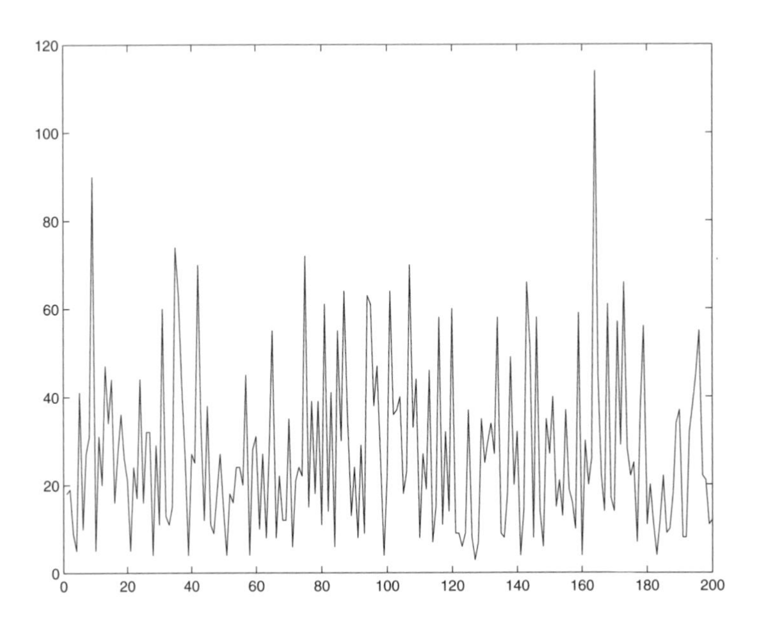

We can use the MATLAB commands **mean** and **std** to extract statistical information from the experiments. But perhaps the most helpful command is `hist(st_store)` which yields:

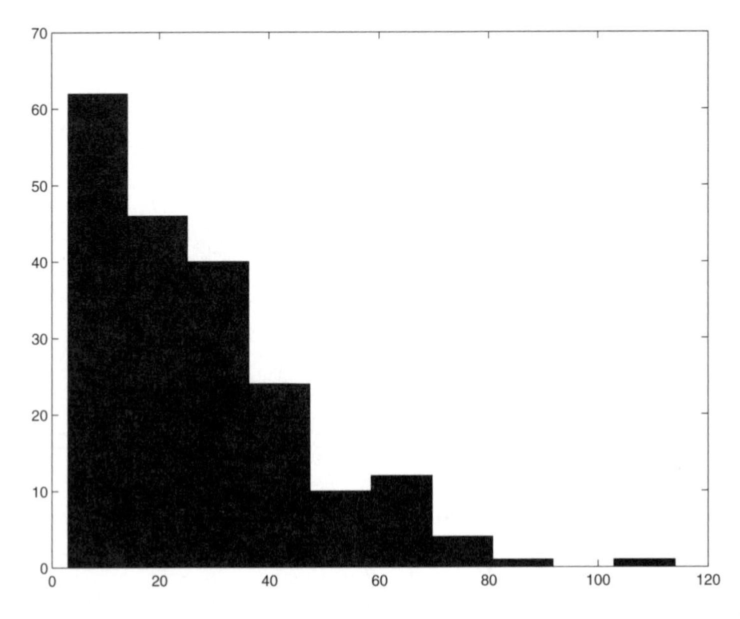

This shows how many of the runs took for instance above 100 steps to reach the eastern wall.

9.5 Tasks

Task 9.1 *Enter the data*

$$X = \{3, 2, 4, 5, 6, -1, 5, 6, 7, 8, 2\} \quad Y = \{2, -6, 3, 2, 0, 1, 4, 5, 6, 7, 8\}$$

and calculate the means and medians of both distributions, their variances and the covariance and correlation.

Task 9.2 *Show that the correlation r of two random variables who are related by the relation $Y = aX + b$ is given by sign(a).*

Task 9.3 (*) *A distribution can be quantified by higher-order moments. Write a code to calculate the skewness and the kurtosis, defined as*

$$\frac{1}{n}\sum_{i=1}^{n}\left(\frac{x_i - \bar{x}}{\sigma_x}\right)^3$$

and

$$\frac{1}{n}\sum_{i=1}^{n}\left(\frac{x_i - \bar{x}}{\sigma_x}\right)^4 - 3,$$

respectively. These represent the level of symmetry and sharpness of the distribution.

We now revisit the topic of maps. We shall start with a one-dimensional map $f : \mathbb{R} \to \mathbb{R}$ (which is read as: takes a real number and returns another one). We shall write the map as

$$x_{n+1} = f(x_n, \mu) \qquad n \in \mathbb{Z},$$

where x_n is the current value, x_{n+1} is the next one and μ is a parameter. We have also restricted the values of n to be integers. In general we will only consider natural numbers: however you should be aware that these maps can go backwards. The parameter μ can be passed through as an argument to the routine, `map(xn,mu)`; or as a global variable (this is preferable if the value is unlikely to change). The codes we will use for this are:

```
function [xn] = map(xo)
global mu
xn = mu*xo*(1-xo);
```

in conjunction with

```
global mu
mu = 0.4;
xold = 0.25;
xnew = map(xold);
```

This example has been given for the map $x_{n+1} = f(x_n, \mu) = \mu x_n(1 - x_n)$. The command `global` allows the program to know the value of variables globally. Any routine which is going to use (or set) a parameter must contain this statement. In the above example the variable is declared as global in `main.m` and then set, and when the routine `map.m` is run and the command `global mu` is found it sets `mu` to be the value set within the main program.

Task 9.4 *Rewrite the code* `main.m` *to evaluate the map nine times and hence starting from* $x_1 = 0.25$ *determine* x_{10}.

The fixed points of this map are where `xnew` is equal to `xold` and are given by solutions of

$$x = \mu x(1 - x) \qquad \Longrightarrow x = 0 \text{ or } x = 1 - \frac{1}{\mu}.$$

Sometimes we are interested in longer cycles, for instance period 2 points which are such that $x_{n+2} = x_n$, where

$$x_{n+1} = \mu x_n(1 - x_n) \tag{9.1}$$

$$x_{n+2} = \mu x_{n+1}(1 - x_{n+1}). \tag{9.2}$$

We can substitute x_{n+1} from (9.1) into (9.2) to give

$$x_n = \mu^2 x_n(1 - x_n)(1 - \mu x_n(1 - x_n)),$$

which can be manipulated to give the quartic

$$0 = -\mu^3 x_n^4 + 2\mu^3 x_n^3 - \mu^2(1 + \mu)x_n^2 + (\mu^2 - 1)x_n.$$

We can find the roots of this using MATLAB code:

```
mu = 0.4;
co = [-mu^3 2*mu^3 -mu^2*(1+mu) (mu^2-1) 0];
[r] = roots(co);
```

Unfortunately this quartic only has two real roots, which are in fact the points for which $x_n = x_{n+1}$ (i.e. $x = 0$ and $x = 1 - 1/\mu$). There are values of μ for which there are real period 2 cycles.

Task 9.5 *By experimenting with the value of* `mu` *in the above code identify a value for which there is a real period 2 cycle and then show that the points occur in pairs using the code* `map.m`.

In order to find roots of higher order it is obviously impractical to perform this manipulation; we can use a root finding technique associated with the function

$$x_n - f^m(x_n, \mu)$$

where the m denotes that the map has been applied m times. The zeros of this function occur at the points of period m.

We can also use these maps in higher dimensions $f : \mathbb{R}^2 \to \mathbb{R}^2$ (which takes two real values and returns two further values).

Let us consider the example

$$\begin{aligned} x_{n+1} &= x_n + \sin(x_n + y_n) \\ y_{n+1} &= \cos(x_n - y_n). \end{aligned}$$

We can evaluate the map using the code:

```
function [xn,yn] = map2(xo,yo)
xn = xo+sin(xo+yo);
yn = cos(xo-yo);
```

in conjunction with

```
x(1) = 0.2;
y(1) = 0.4;
for i = 2:20
    [x(i),y(i)] = map2(x(i-1),y(i-1));
end
```

Task 9.6 *Change the above code to evaluate the first 20 images of the points* $(1/2, 1/3)$ *and* $(1/5, 1/5)$ *under the influence of the map*

$$\begin{aligned} x_{n+1} &= (x_n + 2y_n)|1, \\ y_{n+1} &= (3x_n - 2y_n)|1. \end{aligned}$$

Task 9.7 *Consider the Hénon map,*

$$\begin{aligned} x_{n+1} =& x_n \cos\theta - y_n \sin\theta + x_n^2 \sin\theta, \\ y_{n+1} =& x_n \sin\theta + y_n \cos\theta - x_n^2 \cos\theta. \end{aligned}$$

for a variety of values of θ using the code:

```
cosa = 0.34;
sina = sqrt(1-cosa^2);
ii = 0;
for st = -0.5:0.05:0.5

  x = st; y = st;

  for its = 1:1000
      ii = ii+1;
      xn = cosa*x-sina*y+x^2*sina;
      yn = sina*x+cosa*y-x^2*cosa;
      x = xn; y = yn;
      po(ii) = x+i*y;
      if or(abs(x)>10 ,abs(y)>10)
         break
      end
  end
end
plot(po,'.','MarkerSize',4)
axis equal
axis([-1 1 -1 1])
```

This code sets up an initial rake of points running from $(-0.5,-0.5)$ *to* $(0.5,0.5)$. *Try using the* `zoom` *command to see the details.*

A
A Mathematical Introduction to Matrices

Matrices are objects which have special properties and there are a number of rules which must be adhered to in order to manipulate them in a consistent (and correct) manner. A matrix can most readily be defined as an $n \times m$ array of numbers which is comprised of n rows and m columns. For example, a two-by-one matrix (two rows and one column) has the general form

$$\begin{pmatrix} a \\ b \end{pmatrix}$$

whereas a three-by-two matrix (three rows and two columns) has the general form

$$\begin{pmatrix} a & b \\ c & d \\ e & f \end{pmatrix}.$$

In these examples $a, b, \ldots, f$ may be real or complex numbers. To refer to individual elements of the matrix we use the notation $a_{i,j}$ to denote the element in the ith row and the jth column. Using this notation the three-by-two example could be written in the general form

$$\begin{pmatrix} a_{1,1} & a_{1,2} \\ a_{2,1} & a_{2,2} \\ a_{3,1} & a_{3,2} \end{pmatrix}.$$

Matrices for which $n = m$ (so that the number of rows equal the number of columns) are referred to as *square* matrices. If $m = 1$ then the matrix is simply a *column vector* (as in the first example above). If $n = 1$ (the matrix has only one row) then we refer to it as a *row vector*. A scalar is simply a matrix in which both n and m are equal to one (that is, a one-by-one matrix). Throughout this text we will adopt the universal convection that both vectors and matrices are denoted by a bold font[1]. In general we shall use upper-case letters to denote matrices and lower-case for vectors.

Example A.1 *We show the rows and columns of a general three-by-three matrix.*

- *First row*

$$\begin{pmatrix} \boxed{a \quad b \quad c} \\ d \quad e \quad f \\ g \quad h \quad i \end{pmatrix}$$

- *Second row*

$$\begin{pmatrix} a \quad b \quad c \\ \boxed{d \quad e \quad f} \\ g \quad h \quad i \end{pmatrix}$$

- *Third row*

$$\begin{pmatrix} a \quad b \quad c \\ d \quad e \quad f \\ \boxed{g \quad h \quad i} \end{pmatrix}$$

- *First column*

$$\left(\begin{array}{|c|cc} a & b & c \\ d & e & f \\ g & h & i \end{array}\right)$$

- *Second column*

$$\left(\begin{array}{c|c|c} a & b & c \\ d & e & f \\ g & h & i \end{array}\right)$$

- *Third column*

$$\left(\begin{array}{cc|c|} a & b & c \\ d & e & f \\ g & h & i \end{array}\right)$$

[1] Of course, other notations do exist. For example, many textbooks employ the notation of a single underbar for vectors $\underline{x}$ and a double underbar for matrices $\underline{\underline{A}}$.

Of all the operations which can be performed on matrices one of the simplest is that of transposition (or taking the transpose of a matrix); this operation is usually denoted by a subscript T or a prime $'$. If $\mathbf{A}$ is n-by-m then $\mathbf{B} = \mathbf{A}^T$ is m-by-n, where the elements of $\mathbf{B}$ are defined by

$$b_{j,i} = a_{i,j} \qquad i = 1, \cdots, n; \quad j = 1, \cdots, m;$$

the transpose is thus obtained by interchanging the rows and columns of matrix $\mathbf{A}$ to give the matrix $\mathbf{B} = \mathbf{A}^T$. If the matrix $\mathbf{A}$ is square then the operation of taking the transpose is equivalent to a reflection in the leading diagonal (which runs from the top left corner to the bottom right). Matrices for which $\mathbf{A}^T = \mathbf{A}$ are referred to as *symmetric* and those for which $\mathbf{A}^T = -\mathbf{A}$ are *anti-symmetric*[2]. In the case of three-by-three matrices the general symmetric and anti-symmetric three-by-three matrices can be written as

$$\mathbf{A}_{\text{symm}} = \begin{pmatrix} a & b & c \\ b & d & e \\ c & e & f \end{pmatrix} \text{ and } \mathbf{A}_{\text{anti}} = \begin{pmatrix} 0 & b & c \\ -b & 0 & e \\ -c & -e & 0 \end{pmatrix}.$$

where $a, b, \ldots, f \in \mathbb{C}$. We remark that if the complex conjugate transpose of a matrix (with elements $a^*_{j,i}$) is equal to the matrix then it is called *Hermitian* and if it is equal to minus its complex conjugate transpose then it is referred to as *skew Hermitian.*

We pause here to state that $\mathbf{A} = \mathbf{B}$ implies that $a_{i,j} = b_{i,j}$ for all the elements of the matrices, whereas $\mathbf{A} \neq \mathbf{B}$ only requires that $a_{i,j} \neq b_{i,j}$ for one pair (i, j).

Example A.2 *Determine the transposes of the following matrices:*

$$\begin{pmatrix} 1 & 4 & 7 \\ -4 & -3 & 4 \end{pmatrix}, \qquad \begin{pmatrix} 4 & 2 & 3 \end{pmatrix}, \qquad \begin{pmatrix} 1 & 2 & 3 \\ 2 & 0 & 2 \\ 3 & 2 & 3 \end{pmatrix}.$$

The solutions are

$$\begin{pmatrix} 1 & -4 \\ 4 & -3 \\ 7 & 4 \end{pmatrix}, \quad \begin{pmatrix} 4 \\ 2 \\ 3 \end{pmatrix}, \quad \begin{pmatrix} 1 & 2 & 3 \\ 2 & 0 & 2 \\ 3 & 2 & 3 \end{pmatrix}.$$

Notice elements on the leading diagonal, that is elements of the form $a_{i,i}$, remain unchanged by transposition. In the final example the transpose is equal to the original matrix and therefore the matrix is symmetric.

[2] Note that the diagonal elements of anti-symmetric matrices must be equal to zero. This can be seen by setting $i = j$ in the relation $a_{i,j} = -a_{j,i}$, which is only true if $a_{i,i} = 0$).

The usual arithmetic operations of addition, subtraction and multiplication also apply to matrices. However, there are now several additional rules (or constraints) under which these operations can be performed on two (or more) matrices. These are outlined below.

Addition and subtraction: two matrices can only be added together if they are the same size (that is, they have the same number of rows **and** the same number of columns). In this case the operation of addition is performed element by element. For example, if $\mathbf{A}, \mathbf{B}$ are both n-by-m matrices then $\mathbf{C} = \mathbf{A} + \mathbf{B}$ is defined as the matrix with elements $c_{i,j} = a_{i,j} + b_{i,j}$. A similar rule holds for subtraction.

Scalar multiplication: matrices of any size can be multiplied by scalars. The multiplication is performed element by element so that $\mathbf{C} = \lambda\mathbf{A}$ where $c_{i,j} = \lambda a_{i,j}$.

Matrix multiplication: in order to multiply two matrices $\mathbf{A}$ and $\mathbf{B}$ together the number of columns of $\mathbf{A}$ must equal the number of rows of $\mathbf{B}$. To perform the multiplication we "multiply" the first row of $\mathbf{A}$ by the first column of $\mathbf{B}$, multiplying the first element of each together and then the second ones, etc, and finally adding up all the results. This gives the element in the top left hand corner. We then proceed to multiply the first row by the second column in the same manner (and put the result in the first row, second column). Mathematically this can be written as

$$c_{i,j} = \sum_{k=1}^{m} a_{i,k} b_{k,j} \qquad i = 1, \cdots, n; \qquad j = 1, \cdots, p,$$

where $\mathbf{A}$ is n-by-m and $\mathbf{B}$ is m-by-p. Then the answer $\mathbf{C}$ is n-by-p.

This rule for matrix multiplication highlights one of their important properties, namely that the order of multiplication is important. In this example, with $\mathbf{A}$ an n-by-m matrix and $\mathbf{B}$ an m-by-p matrix, the operation "$\mathbf{A}$ *times* $\mathbf{B}$" is defined. However, the operation "$\mathbf{B}$ *times* $\mathbf{A}$" (that is, pre-multiplying matrix $\mathbf{A}$ by matrix $\mathbf{B}$) is not defined **unless** $p = n$. Even if $p = n$, in general $\mathbf{AB} \neq \mathbf{BA}$. This is equivalent to saying that, unlike scalar multiplication, matrix multiplication is not commutative.

Example A.3 *We demonstrate these concepts by an example involving two two-by-two matrices, namely*

$$\mathbf{A} = \begin{pmatrix} a & b \\ c & d \end{pmatrix} \text{ and } \mathbf{B} = \begin{pmatrix} \alpha & \beta \\ \gamma & \delta \end{pmatrix}.$$

The sum $\mathbf{C} = \mathbf{A} + \mathbf{B}$ *is determined as follows. Firstly the element* $c_{1,1}$ *is obtained by adding the corresponding elements in* $\mathbf{A}$ *and* $\mathbf{B}$, *so that*

$$\begin{pmatrix} \boxed{a} & b \\ c & d \end{pmatrix} + \begin{pmatrix} \boxed{\alpha} & \beta \\ \gamma & \delta \end{pmatrix} = \begin{pmatrix} \boxed{a+\alpha} & \\ & \end{pmatrix}.$$

Now for the $c_{1,2}$ *entry (the top right element):*

$$\begin{pmatrix} a & \boxed{b} \\ c & d \end{pmatrix} + \begin{pmatrix} \alpha & \boxed{\beta} \\ \gamma & \delta \end{pmatrix} = \begin{pmatrix} a+\alpha & \boxed{b+\beta} \\ & \end{pmatrix}.$$

Similarly for $c_{2,1}$ *(the bottom left element)*

$$\begin{pmatrix} a & b \\ \boxed{c} & d \end{pmatrix} + \begin{pmatrix} \alpha & \beta \\ \boxed{\gamma} & \delta \end{pmatrix} = \begin{pmatrix} a+\alpha & b+\beta \\ \boxed{c+\gamma} & \end{pmatrix}$$

and finally for $c_{2,2}$ *(the bottom right element) we have*

$$\begin{pmatrix} a & b \\ c & \boxed{d} \end{pmatrix} + \begin{pmatrix} \alpha & \beta \\ \gamma & \boxed{\delta} \end{pmatrix} = \begin{pmatrix} a+\alpha & b+\beta \\ c+\gamma & \boxed{d+\delta} \end{pmatrix}.$$

Example A.4 *An example of multiplication of two two-by-two matrices will serve to highlight the differences between addition and multiplication of matrices. Consider the matrices*

$$\mathbf{A} = \begin{pmatrix} a & b \\ c & d \end{pmatrix} \text{ and } \mathbf{B} = \begin{pmatrix} \alpha & \beta \\ \gamma & \delta \end{pmatrix}.$$

By our earlier rule the product $\mathbf{C} = \mathbf{AB}$ *is defined and is determined as follows. We start with the top left entry, namely* $c_{1,1}$ *(formed by multiplying the first row of* $\mathbf{A}$ *by the first column of* $\mathbf{B}$*)*

$$\begin{pmatrix} \boxed{a \quad b} \\ c \quad d \end{pmatrix} \begin{pmatrix} \boxed{\begin{matrix} \alpha \\ \gamma \end{matrix}} & \begin{matrix} \beta \\ \delta \end{matrix} \end{pmatrix} = \begin{pmatrix} \boxed{a \times \alpha + b \times \gamma} & \\ & \end{pmatrix};$$

now the top right entry, namely $c_{1,2}$ *(which is formed by multiplying the first row of* $\mathbf{A}$ *by the second column of* $\mathbf{B}$*)*

$$\begin{pmatrix} \boxed{a \quad b} \\ c \quad d \end{pmatrix} \begin{pmatrix} \begin{matrix} \alpha \\ \gamma \end{matrix} & \boxed{\begin{matrix} \beta \\ \delta \end{matrix}} \end{pmatrix} = \begin{pmatrix} a \times \alpha + b \times \gamma & \boxed{a \times \beta + b \times \delta} \\ & \end{pmatrix};$$

next the bottom left entry, namely $c_{2,1}$ *(which is formed by multiplying the second row of* **A** *by the first column of* **B***)*

$$\begin{pmatrix} a & b \\ \boxed{c \quad d} \end{pmatrix} \begin{pmatrix} \alpha & \beta \\ \gamma & \delta \end{pmatrix} = \begin{pmatrix} a\times\alpha+b\times\gamma & a\times\beta+b\times\delta \\ \boxed{c\times\alpha+d\times\gamma} & \end{pmatrix};$$

and finally the bottom right entry, namely $c_{2,2}$ *(which is formed by multiplying the second row of* **A** *by the second column of* **B***)*

$$\begin{pmatrix} a & b \\ \boxed{c \quad d} \end{pmatrix} \begin{pmatrix} \alpha & \beta \\ \gamma & \delta \end{pmatrix} = \begin{pmatrix} a\times\alpha+b\times\gamma & a\times\beta+b\times\delta \\ c\times\alpha+d\times\gamma & \boxed{c\times\beta+d\times\delta} \end{pmatrix}.$$

In general, to calculate the $c_{i,j}$ *entry we multiply the* i^{th} *row of the first matrix* **A** *by the* j^{th} *column of the second matrix* **B** *term by term.*

Example A.5 *Calculate the product* **AB** *of the matrices*

$$\mathbf{A} = \begin{pmatrix} 1 & -1 \\ 0 & 3 \end{pmatrix} \text{ and } \mathbf{B} = \begin{pmatrix} -2 & 1 \\ 4 & -2 \end{pmatrix}.$$

Using the method given in the previous example

$$\begin{aligned} \mathbf{AB} &= \begin{pmatrix} 1 & -1 \\ 0 & 3 \end{pmatrix} \begin{pmatrix} -2 & 1 \\ 4 & -2 \end{pmatrix} \\ &= \begin{pmatrix} 1\times(-2)+(-1)\times 4 & 1\times 1+(-1)\times(-2) \\ 0\times(-2)+3\times(4) & 0\times 1+3\times(-2) \end{pmatrix} \\ &= \begin{pmatrix} -6 & 3 \\ 12 & -6 \end{pmatrix}. \end{aligned}$$

(It is worth practising these calculations; try calculating the product **BA** *yourself by hand.).*

We now turn our attention to matrix multiplication in which the matrices are not necessarily square.

Example A.6 *Consider the matrices*

$$\mathbf{A} = \begin{pmatrix} 3 & 0 & -1 \\ -4 & 2 & 2 \end{pmatrix}, \quad \mathbf{B} = \begin{pmatrix} -1 & 7 \\ 3 & 5 \\ -2 & 0 \end{pmatrix} \text{ and } \mathbf{C} = \begin{pmatrix} 2 & 0 \\ -1 & -3 \end{pmatrix}.$$

Calculate the quantities: $\mathbf{AB}$, $\mathbf{BA}$, $\mathbf{A}+\mathbf{B}^T$, $\mathbf{AC}$, $\mathbf{A}^T\mathbf{C}$, $3\mathbf{C}+2(\mathbf{AB})^T$, $(\mathbf{AB})\,\mathbf{C}$ *and finally* $\mathbf{A}\,(\mathbf{BC})$, *where possible (and if not, state the reason why the calculations cannot be performed).*

We shall start (for the first couple of examples) by providing full solutions and thereafter just give answers with a minimum of intermediate steps. So,

$$\begin{aligned}
\mathbf{AB} &= \begin{pmatrix} 3 & 0 & -1 \\ -4 & 2 & 2 \end{pmatrix} \begin{pmatrix} -1 & 7 \\ 3 & 5 \\ -2 & 0 \end{pmatrix} \\
&= \begin{pmatrix} 3\times(-1)+0\times 3+(-1)\times(-2) & 3\times 7+0\times 5+(-1)\times(0) \\ (-4)\times(-1)+2\times 3+2\times(-2) & (-4)\times 7+2\times 5+2\times(0) \end{pmatrix} \\
&= \begin{pmatrix} -1 & 21 \\ 6 & -18 \end{pmatrix}.
\end{aligned}$$

Similarly

$$\begin{aligned}
\mathbf{BA} &= \begin{pmatrix} -1 & 7 \\ 3 & 5 \\ -2 & 0 \end{pmatrix} \begin{pmatrix} 3 & 0 & -1 \\ -4 & 2 & 2 \end{pmatrix} \\
&= \begin{pmatrix} -1\times 3+7\times(-4) & -1\times 0+7\times 2 & -1\times(-1)+7\times 2 \\ 3\times 3+5\times(-4) & 3\times 0+5\times 2 & 3\times(-1)+5\times 2 \\ -2\times 3+0\times(-4) & -2\times 0+0\times 2 & -2\times(-1)+0\times 2 \end{pmatrix} \\
&= \begin{pmatrix} -31 & 14 & 15 \\ -11 & 10 & 7 \\ -6 & 0 & 2 \end{pmatrix}
\end{aligned}$$

$$\begin{aligned}
\mathbf{A}+\mathbf{B}^T &= \begin{pmatrix} 3 & 0 & -1 \\ -4 & 2 & 2 \end{pmatrix} + \begin{pmatrix} -1 & 3 & -2 \\ 7 & 5 & 0 \end{pmatrix} \\
&= \begin{pmatrix} 2 & 3 & -3 \\ 3 & 7 & 2 \end{pmatrix}.
\end{aligned}$$

It is not possible to pre-multiply matrix $\mathbf{A}$ *by* $\mathbf{C}$ *since* $\mathbf{A}$ *is two-by-three and* $\mathbf{C}$ *is two-by-two, so the inner dimensions do not agree (that is, the second dimension of the first matrix and the first dimension of the second matrix are different). For the next example* $\mathbf{A}^T\mathbf{C}$ *we observe that* $\mathbf{A}^T$ *is three-by-two so that the inner dimensions agree and hence the calculation is possible. We obtain*

$$\mathbf{A}^T\mathbf{C} = \begin{pmatrix} 3 & -4 \\ 0 & 2 \\ -1 & 2 \end{pmatrix} \begin{pmatrix} 2 & 0 \\ -1 & -3 \end{pmatrix} = \begin{pmatrix} 10 & 12 \\ -2 & -6 \\ -4 & -6 \end{pmatrix}.$$

The next example requires scalar multiplication and the use of the first of the results in this example; we have

$$\begin{aligned}
3\mathbf{C} + 2\left(\mathbf{AB}\right)^T &= 3\begin{pmatrix} 2 & 0 \\ -1 & -3 \end{pmatrix} + 2\begin{pmatrix} -1 & 6 \\ 21 & -18 \end{pmatrix}, \\
&= \begin{pmatrix} 6 & 0 \\ -3 & -9 \end{pmatrix} + \begin{pmatrix} -2 & 12 \\ 42 & -36 \end{pmatrix}, \\
&= \begin{pmatrix} 4 & 12 \\ 39 & -45 \end{pmatrix}.
\end{aligned}$$

The final two calculations serve to demonstrate that matrix multiplication is associative, that is for three matrices $\mathbf{A}, \mathbf{B}$ *and* $\mathbf{C}$, $\mathbf{A}(\mathbf{BC}) = (\mathbf{AB})\mathbf{C}$. *(Notice that this does not constitute a formal proof.)*

$$\left(\mathbf{AB}\right)\mathbf{C} = \begin{pmatrix} -1 & 21 \\ 6 & 18 \end{pmatrix}\begin{pmatrix} 2 & 0 \\ -1 & -3 \end{pmatrix} = \begin{pmatrix} -23 & -63 \\ 26 & 54 \end{pmatrix}.$$

And now the final example

$$\begin{aligned}
\mathbf{A}\left(\mathbf{BC}\right) &= \begin{pmatrix} 3 & 0 & -1 \\ -4 & 2 & 2 \end{pmatrix}\left\{\begin{pmatrix} -1 & 7 \\ 3 & 5 \\ -2 & 0 \end{pmatrix}\begin{pmatrix} 2 & 0 \\ -1 & -3 \end{pmatrix}\right\} \\
&= \begin{pmatrix} 3 & 0 & -1 \\ -4 & 2 & 2 \end{pmatrix}\begin{pmatrix} -9 & -21 \\ 1 & -15 \\ -4 & 0 \end{pmatrix} = \begin{pmatrix} -23 & -63 \\ 26 & 54 \end{pmatrix}.
\end{aligned}$$

In this example we see that the result of the multiplication $\mathbf{BA}$ *is a three-by-three matrix, further emphasising the fact that* $\mathbf{AB}$ *is not necessarily equal to* $\mathbf{BA}$. *In some cases it may not even be possible to perform this second multiplication. This example serves to demonstrate that matrix multiplication is associative (that is* $\mathbf{A}(\mathbf{BC}) \equiv (\mathbf{AB})\mathbf{C}$*), but it is not, in general, commutative. It is also a simple matter to show that matrix multiplication is distributive, that is* $\mathbf{A}(\mathbf{B} + \mathbf{C}) = \mathbf{AB} + \mathbf{AC}$ *for any three matrices* $\mathbf{A}, \mathbf{B}$ *and* $\mathbf{C}$ *for which the matrix multiplications are permitted.*

A.1 Special Matrices

There are two special matrices that we will make use of often within the text and we introduce them here. The first is the zero matrix, which we will denote by $\mathbf{0}$.

This is simply a matrix whose elements are all equal to zero[3]. Not surprisingly, the zero matrix has no effect when it is added to another matrix (of the same size). So $\mathbf{A} + \mathbf{0} = \mathbf{A} = \mathbf{0} + \mathbf{A}$. We will make use of this matrix to initialise matrices in preparation to assigning answers to a matrix.

The second important matrix we will have call to use often with the text is the *identity* (or *unit*) matrix, denoted by $\mathbf{I}$. The identity matrix $\mathbf{I}$ is a n-by-n matrix whose elements consist of 1s (ones) along the main diagonal and are zero everywhere else. For example, the three-by-three identity matrix is given by

$$\mathbf{I} = \begin{pmatrix} 1 & 0 & 0 \\ 0 & 1 & 0 \\ 0 & 0 & 1 \end{pmatrix}.$$

Multiplying a square n-by-n matrix $\mathbf{A}$ by $\mathbf{I}$ has no effect:

$$\mathbf{AI} = \mathbf{A} = \mathbf{IA}.$$

A.2 Inverses of Matrices

The inverse of a matrix, written as $\mathbf{A}^{-1}$, is defined as the matrix which when pre- and post-multiplied by the matrix $\mathbf{A}$ produces the identity matrix $\mathbf{I}$:

$$\mathbf{A}^{-1}\mathbf{A} = \mathbf{A}\mathbf{A}^{-1} = \mathbf{I}.$$

Only square matrices can have an inverse but it is only a subset of all square matrices for which the inverse exists. The existence of the inverse of a matrix (that is, whether the matrix is invertible or not) is intimately linked with the determinant of the matrix. We introduce this, and many other properties of matrices, in Chapter 6.

The utility of the inverse of a matrix is best seen when solving systems of equations. An example will serve by way of illustration.

Example A.7 *Consider the system of simultaneous equations:*

$$x_1 + x_2 = 3, \tag{A.1a}$$

$$x_1 + 2x_2 = 5. \tag{A.1b}$$

[3] Of course, we could include the dimensions of the matrix by writing $\mathbf{0}_{nm}$ to denote that it is an n-by-m matrix. We will adopt the convention in the text that when we refer to the zero matrix we are taking a matrix of the appropriate size required for the operation.

These equations can be solved using conventional means. To do this we first subtract (A.1a) from (A.1b) to give

$$x_1 + 2x_2 - (x_1 + x_2) = 5 - 3$$

or

$$x_2 = 2,$$

and now substituting back into either equation (let us use (A.1a)) we have

$$x_1 + 2 = 3$$

which gives

$$x_1 = 1.$$

We can just as easily write the system (A.1) as a matrix equation

$$\begin{pmatrix} 1 & 1 \\ 1 & 2 \end{pmatrix} \begin{pmatrix} x_1 \\ x_2 \end{pmatrix} = \begin{pmatrix} 3 \\ 5 \end{pmatrix}$$

or as

$$\mathbf{A}\mathbf{x} = \mathbf{b}.$$

Try multiplying out the matrix equation to check you get (A.1a) and (A.1b). Elementary linear algebra shows that the solution is given by $\mathbf{A}^{-1}\mathbf{b}$*, which can be written in MATLAB as* `inv(A)*b` *or* `A\b`*. The operator* \ *determines the effect of multiplying by the inverse of the first argument on the second, without ever constructing the inverse. The code for this example would be:*

```
>> A = [1 1; 1 2]; % Initialise the matrix A
>> b = [3; 5];      % Initialise the vector b
>> x = inv(A)*b     % Determine the solution vector x
```

Before we proceed we note in the previous three lines of MATLAB code everything after the percent sign % is taken by MATLAB to be a comment. Comments are a useful way of making your MATLAB code readable by both you and others.

We can check the answer from our matrix computation by typing `A*x`*, which should be equal to* `b`*.*

Example A.8 *Determine the vector* $\mathbf{x}$ *which satisfies the equation* $\mathbf{Ax} = \mathbf{b}$ *where*

$$\mathbf{A} = \begin{pmatrix} 1 & 2 & 3 & 4 \\ 4 & 3 & 2 & 1 \\ 1 & 0 & -1 & 0 \\ -1 & 1 & -1 & 1 \end{pmatrix} \text{ and } \mathbf{b} = \begin{pmatrix} 5 \\ 10 \\ 15 \\ 20 \end{pmatrix}.$$

We enter the matrix $\mathbf{A}$ *and vector* $\mathbf{b}$ *directly using*

```
>> A = [1 2 3 4; ...
   4 3 2 1; ...
   1 0 -1 0; ...
   -1 1 -1 1];
>> b = [5; 10; 15; 20];
```

(Note we have used the three dots ... (or ellipsis) to indicate to MATLAB that the input line continues. It is good practise to have a space before the dots at the end of the line.) These results can then be used to form $\mathbf{x}$*:*

```
>> inv(A)*b

ans =

    3.2500
    4.5000
  -11.7500
    7.0000
```

This solution was obtained by considering the equation $\mathbf{Ax} = \mathbf{b}$ *and pre-multiplying each side by the inverse of the matrix* $\mathbf{A}$ *(we have deliberately chosen* $\mathbf{A}$ *so that its inverse exists). This gives* $\mathbf{A}^{-1}\mathbf{Ax} = \mathbf{A}^{-1}\mathbf{b}$ *but we recall from the definition of the inverse that* $\mathbf{A}^{-1}\mathbf{A} = \mathbf{I}$ *and that* $\mathbf{Ix} = \mathbf{x}$*. Hence we have the solution* $\mathbf{x} = \mathbf{A}^{-1}\mathbf{b}$.

B
Glossary of Useful Terms

This appendix is provided purely as a guide. MATLAB has a very informative help feature `help command` which is supplemented with several other features `lookfor maths`. You can also access the help files on the web `helpdesk`.

This appendix is broken down into:

- arithmetic and logical operators
- symbols
- plotting commands
- general MATLAB commands

B.1 Arithmetic and Logical Operators

`+, -` Used to add or subtract variables of the same size together, whether they are matrices, vectors or scalars.

```
A = [1 2; 3 4];
B = ones(2);
A + B
A - B
5 + 0.5
7 - 4
```

It is also worth noting that these operations will add (or subtract) scalar quantities from matrices. For instance:

```
A = ones(3);
B = A + 2;
C = 3 - A;
```

This produces a three-by-three matrix full of threes in `B` and a three-by-three matrix full of twos in `C`. MATLAB will complain if these operations are not viable.

`*`, / Used to multiply or divide variables as long as the operation is mathematically viable:

```
4 * 3.2
4 / 2.3
A = [1 2; 3 4];
B = ones(2);
A * B
A / B
```

The last two operations give $\mathbf{AB}$ and $\mathbf{AB}^{-1}$. Notice that the multiplication operation is only viable if the inner dimensions agree: the number of columns of the first matrix must equal the number of rows of the second.

It can also multiply (or divide) matrices by scalars:

```
A = ones(3);
B = A/3;
C = A*4;
```

These commands give `B` as a three-by-three matrix full of 1/3's and `C` as a three-by-three matrix full of fours.

`.*` This binary operator allows one to perform multiplication calculations element by element on arrays of the same size. If we consider the vectors $\mathbf{x} = (x_1, x_2, x_3, x_4)$ and $\mathbf{y} = (y_1, y_2, y_3, y_4)$. Then the calculation **x'.*'y** gives $(x_1y_1, x_2y_2, x_3y_3, x_4y_4)$. Notice that both $\mathbf{x}$ and $\mathbf{y}$ were row vectors of length 4 and so is the answer. Let us consider the example:

```
A = [4 3; 2 1];
B = [1 2; 3 4];
C = A.*B;
```

This does the calculation element-wise and gives the result:

$$\texttt{C} = \begin{pmatrix} 4\times 1 & 3\times 2 \\ 2\times 3 & 1\times 4 \end{pmatrix} = \begin{pmatrix} 4 & 6 \\ 6 & 4 \end{pmatrix}.$$

This command can also be used on values which are scalars, so for instance `A = ones(2); B = A.*2;` gives `B` as a matrix full of twos.

The most common use of this operator is again in the construction of functions.

```
x = 1:5;
f = x.*sin(x);
g = (3*x+4).*(x+2);
```

This gives us `x = [1 2 3 4 5]` and then: $f = x\sin x$ evaluated at those points, i.e. `[sin(1) 2*sin(2) 3*sin(3) 4*sin(4) 5*sin(5)]`; and $g = (3x+4)(x+2)$ at the points. Notice it is not necessary to use the operator `.*` when calculating `3*x` since 3 is a scalar.

./ This binary operator allows one to perform division calculations element by element on arrays of the same size. If we consider the vectors $\mathbf{x} = (x_1, x_2, x_3, x_4)$ and $\mathbf{y} = (y_1, y_2, y_3, y_4)$, then the calculation $\mathbf{x}$'./'$\mathbf{y}$ gives $(x_1/y_1, x_2/y_2, x_3/y_3, x_4/y_4)$. Notice that both $\mathbf{x}$ and $\mathbf{y}$ were row vectors of length 4 and so is the answer. Let us consider the example:

```
A = [4 3; 2 1];
B = [1 2; 3 4];
C = A./B;
```

This does the calculation element-wise and gives the result:

$$\texttt{C} = \begin{pmatrix} 4/1 & 3/2 \\ 2/3 & 1/4 \end{pmatrix} = \begin{pmatrix} 4 & \frac{3}{2} \\ \frac{2}{3} & \frac{1}{4} \end{pmatrix}.$$

This command can also be used on values which are scalars, so for instance `A = ones(2); B = A./2;` gives `B` as a matrix full of halves.

The most common use of this operator is in the construction of functions.

```
x = 1:5;
f = x./sin(x);
g = (3*x+4)./(x+2);
```

This gives us `x = [1 2 3 4 5]` and then: $f = x/\sin x$ evaluated at those points, i.e. `[1/sin(1) 2/sin(2) 3/sin(3) 4/sin(4) 5/sin(5)]`; and $g = (3x+4)/(x+2)$ at the points.

We can use this command where either of its arguments are scalars (in fact, it is necessary for this example).

```
x = [1 2 3 4 5];
y = 2./x
```

This gives `[2/1 2/2 2/3 2/4 2/5]`. This construction is useful when working out functions of the form $f(x) = 2/x$.

.^ This binary operator allows one to perform exponentiation (raising to a power) calculations element by element on arrays of the same size. If we consider the vectors $\mathbf{x} = (x_1, x_2, x_3, x_4)$ and $\mathbf{y} = (y_1, y_2, y_3, y_4)$, then the calculation **x**'.^'**y** gives $(x_1^{y_1}, x_2^{y_2}, x_3^{y_3}, x_4^{y_4})$. Notice that both **x** and **y** were row vectors of length 4 and so is the answer. Let us consider the example:

```
A = [4 3; 2 1];
B = [1 2; 3 4];
C = A.^B;
```

This does the calculation element-wise and gives the result:

$$\mathtt{C} = \begin{pmatrix} 4^1 & 3^2 \\ 2^3 & 1^4 \end{pmatrix} = \begin{pmatrix} 4 & 9 \\ 8 & 1 \end{pmatrix}.$$

This command can also be used on values which are scalars, so for instance `A = ones(2); B = A.^2;` gives `B` as a matrix full of ones (that is one squared).

The most common use of this operator is in the construction of functions.

```
x = 1:5;
f = x.^(x+1);
g = (3*x+4).^x;
```

This gives us `x = [1 2 3 4 5]` and then: $f = x^{x+1}$ evaluated at those points, i.e. `[1^2 2^3 3^4 4^5 5^6]`; and $g = (3x+4)^x$ at the points.

As with the operator './' we can use this command when either of its arguments are scalars. For instance:

```
x = 1:5;
y = 2.^x;
z = x.^2;
```

This gives `y=[2^1 2^2 2^3 2^4 2^5]` and `z=[1^2 2^2 3^2 4^2 5^2]`.

\ This works out the effect of pre-multiplying by the inverse of the first argument on the second argument. So if we want to solve the set of linear equations represented in the matrix equation $\mathbf{Ax} = \mathbf{b}$, we need to construct $\mathbf{x} = \mathbf{A}^{-1}\mathbf{b}$.

```
A = [1 2; 3 4];
b = [2; 3];
x = A\b;
```

This solves the set of equations:

$$x_1 + 2x_2 = 2$$
$$3x_1 + 4x_2 = 3.$$

by defining

$$\mathbf{A} = \begin{pmatrix} 1 & 2 \\ 3 & 4 \end{pmatrix} \text{ and } \mathbf{b} = \begin{pmatrix} 2 \\ 3 \end{pmatrix},$$

with $\mathbf{x} = (x_1, x_2)^T$.

== Checks equality, rather than sets equal. This is usually exploited within logical statements with scalars:

```
if i==7
   disp(' i is seven ')
else
   disp(' i is not seven ')
end
```

This can be used in other forms: `x = 1:12; mod(x,3)==0`. This gives the output:

```
0  0  1  0  0  1  0  0  1  0  0  1
```

that is it is true provided the corresponding element of `x` is divisible by 3.

$\sim$= Checks not equal to. Again this is mainly used in logical statements with scalars:

```
if i~=7
   disp(' i is not seven ')
else
   disp(' i is equal to seven ')
end
```

This can be used inline `x = 1:12; mod(x,3)~=0`. This gives the output:

```
1  1  0  1  1  0  1  1  0  1  1  0
```

that is it is true provided the corresponding element of `x` is not divisible by 3. **Note**: great care is needed with this command since the simple order change to `y=~x` means set `y` equal to `not(x)`.

`>, >=` Checks greater than and greater than or equal to. Mainly used with scalars within logical statements:

```
if i>7
   disp(' i is greater than seven ')
end

if i>=7
   disp(' i is greater than or equal to seven ')
end
```

Can be used for an array `x = 1:12; x>7` which gives

```
0  0  0  0  0  0  0  1  1  1  1  1
```

whereas `x = 1:12; x>=7` gives

```
0    0    0    0    0    0    1    1    1    1    1    1
```

that is the second one is true for $x = 7$.

`<`, `<=` Checks less than and less than or equal to. Mainly used with scalars within logical statements:

```
if i<7
   disp(' i is less than seven ')
end

if i<=7
   disp(' i is less than or equal to seven ')
end
```

Can be used for an array `x = 1:12; x<7` gives

```
1    1    1    1    1    1    0    0    0    0    0    0
```

whereas `x = 1:12; x<=7`

```
1    1    1    1    1    1    1    0    0    0    0    0
```

that is the second one is true for $x = 7$.

`all` This returns a value of true if all of the arguments are true: `all(x.^2>0)` would be true (provided all the values of `x` are real); `x = 1:12; all(x>0)` would give true since all the elements of `x` are positive.

`and` Boolean operator for *and* can also use &. This is true only if both its arguments are true.

```
if and(x>7,x<9)
   disp(' x is between 7 and 9')
end
```

This can also be written as `x>7 & x<9` and can be applied to arrays, so that `x = 1:12; and(x>7,x<9)` gives:

```
0    0    0    0    0    0    0    1    0    0    0    0
```

`any` This returns a value of true if any of the arguments are true: `any(x<0)` would be true if any of the values in the vector `x` are strictly negative; `x = 1:12; any(x>10)` would be true since some elements of `x` are greater than 10.

`find` Provides a list of integers when the condition is true, for instance `x = 1:10; [i]=find(x.^2>26);` gives the locations `i=[6 7 8 9 10]`.

`not` Negates logical variables, can also use ~. This can be used to turn true into false and vice versa, so that

```
if not(x>=7)
   disp(' x is not greater than or equal to 7')
end
```

This could also be written as `~(x>=7)`. This can be applied to arrays, so that `x = 1:12; not(x>=7)` gives:

```
1    1    1    1    1    1    0    0    0    0    0    0
```

`or` Boolean operator for *or* can also use |. This is true provided one of its arguments is true.

```
if or(x<7,x>9)
   disp(' x is less than 7 or greater than 9')
end
```

This can also be written as `x<7 | x>9` and can be applied to arrays, so that `x = 1:12; or(x<7,x>9)` gives:

```
1    1    1    1    1    1    0    0    0    1    1    1
```

`xor` Exclusive or. This gives true if one of the values is true and false if they are both false or both true.

```
x = [0 0 1 1];
y = [0 1 0 1];
xor(x,y)
```

gives [0 1 1 0].

We note that this can also be done for or and and.

B.2 Symbols

... Used to link lines together (but cannot be used within a string):

```
x = 1:10;
f = x.^2+sqrt(x.^2+1) ...
     + cos(x)./x;
```

This gives the function $f = x^2 + \sqrt{x^2+1} + \cos x/x$ at the values from x equals 1 to 10.

% Means the rest of the line is a comment. It also has special meaning at the start of a code. It can be very useful for eliminating the execution of certain lines of a code.

```
% This code calculates the value
% of x times sin(x)
f = x.*sin(x);

% and the value of cos(x^2)

g = cos(x.^2); % where x can be a vector.
```

If this code was saved as mycode.m then typing help mycode would produce:

```
This code calculates the value
of x times sin(x)
```

; Used at the end of a phrase to suppress output. It can occur at the physical end of a line or at the end of a set of commands.

```
a = ones(2);
b = ones(2); c = ones(2);
d1 = ones(2); d2 = ones(2)
```

This code will set up the two-by-two matrices `a`, `b`, `c`, `d1` and `d2` full of ones. It will only report on the initiation of `d2` since this phrase is not concluded with a semicolon.

It can also be used to end lines within matrices (enclosed within pairs of square brackets)

```
A= [ 1 2; 3 4];
```

This gives the matrix

$$\mathbf{A} = \begin{pmatrix} 1 & 2 \\ 3 & 4 \end{pmatrix}.$$

(Note that the semicolon is used in both its incarnations here.) We note that both uses of the semicolon correspond to the end of the phrase: it is merely that the square brackets are unbalanced in the latter case so MATLAB knows that it is going to read the next line of the matrix.

: Used to delimit values when setting up a vector and also to refer to entire rows or columns of matrices. When setting a row vector there are two syntaxes:

```
x1 = 1:12
x2 = 1:2:13
```

The first of these commands sets up an array from 1 to 12 in steps of unity (and is equivalent to `1:1:12`) whereas the second array runs from 1 to 13 in steps of 2, thus it yields `[1 3 5 7 9 11 13]`. The syntaxes are `a:b` (an array from `a` to a value not exceeding `b` in steps of unity) and `a:h:b` (an array from `a` to a value not exceeding `b` in steps of `h`) (see also `linspace`). Note that `h` can be negative `11:(-2):1` gives `[11 9 7 5 3 1]`. We note that the length of an array set up using `a:h:b` is not $(b-a)/h$ (presuming that this is an integer), but $(b-a)/h+1$. For example `0:0.1:1` has eleven points `[0 0.1 0.2 0.3 0.4 0.5 0.6 0.7 0.8 0.9 1.0]` rather than ten. This dimension can be determined using `length`.

The second use of this symbol allows reference to all viable values of a row or column.

```
A = [11 12 13; 21 22 23; 31 32 33];

A(1,:)        % First row of A
A(:,2)        % Second column of A
A(:,:)        % Whole of A.
A(:,1:2)      % First and second columns of A
A(1:2:3,:)    % First and third rows of A
A(1:2,1:2)    % Top left two-by-two corner of A
```

, Can be used to delimit sets of commands where feedback is required:

```
a = 1, b = 2
```

and to separate elements of matrices on a particular line

```
A = [1,2,3,4;5,6,7,8];
```

It is also used to separate arguments of functions

```
x = linspace(0,1,200);
```

' ' Used to surround strings and for passing arguments to various functions:

```
a = 'Do robots dream of electronic sheep?'
x = 1:12;
y = x.^2;
plot(x,y,'LineWidth',2)
```

see the comments on `plot` below.

' Gives the complex conjugate transpose of a matrix. If the n-by-m matrix `A` has elements $a_{i,j}$, then `B = A'` is m-by-n and has elements $b_{i,j} = a^*_{j,i}$ (where $i \in \{1, \cdots, m\}$ and $j \in \{1, \cdots, n\}$).

```
A = [1+2*i 2-i; 3 4+i];
B = A';
```

This gives

$$\mathbf{B} = \begin{pmatrix} 1-2i & 3 \\ 2+i & 4-i \end{pmatrix}.$$

`.'` Gives the transpose of a matrix. If the n-by-m matrix `A` has elements $a_{i,j}$, then `B = A.'` is m-by-n and has elements $b_{i,j} = a_{j,i}$ (where $i \in \{1, \cdots, m\}$ and $j \in \{1, \cdots, n\}$).

```
A = [1+2*i 2-i; 3 4+i];
B = A.';
```

This gives

$$\mathbf{B} = \begin{pmatrix} 1+2i & 3 \\ 2-i & 4+i \end{pmatrix}.$$

It is better to use this form rather than `A'` unless you are concerned with issues of complex matrices.

`(space)` Can be used to separate elements of matrices, as for the comma; but cannot be used to separate commands. Hence we can use:

```
A = [1 2 3; 4 5 6];
B = [ 1:3 ; 3 2 1];
```

The reason that `(space)` cannot be used to delimit commands is that it can occur naturally within a command, for instance `a = 2` which is used rather than `a=2` merely to improve the readability of the code.

`.` decimal point – This can be used firstly as the mathematical decimal point: `3.145` or `567.3245`. It is also used to punctuate file names into two parts: an identifier (descriptive of the contents of the code, or its function) and then the file's type (`.m`, `.mat`, `.dat` or `.fig`). For this reason you should only have one '.' in each number or filename.

`[ ]` used to enclose elements of a vector or matrix:

```
A = [1 2 3 5 6 7];
B = [1 2 3;
     3 4 5];
ms = ['Vector array' ' of strings'];
```

These brackets should balance.

( ) used to surround mathematical expressions and lists of arguments for functions:

```
x= 1:12
y = 1./(x.^2+1);
z = x.*sin(y);
```

Again these brackets should balance.

B.3 Plotting Commands

Before we start this section we need to discuss the idea of a handle. This is essentially a variable which allows us to access properties of an object.

```
>> x = 1:12; y = x.^2;
>> h = plot(x,y)
>> get(h)
    Color = [0 0 1]
    EraseMode = normal
    LineStyle = -
    LineWidth = [0.5]
    Marker = none
    MarkerSize = [6]
    MarkerEdgeColor = auto
    MarkerFaceColor = none
    XData = [ (1 by 12) double array]
    YData = [ (1 by 12) double array]
    ZData = []

    ButtonDownFcn =
    Children = []
    Clipping = on
    CreateFcn =
    DeleteFcn =
    BusyAction = queue
    HandleVisibility = on
    Interruptible = on
    Parent = [3.00012]
```

```
Selected = off
SelectionHighlight = on
Tag =
Type = line
UserData = []
Visible = on
```

The variable `h` allows us to access all of these commands, which can be changed using the command `set`, for instance `set(h,'MarkerSize',10)`.

`axes` Used to initiate a set of axes

```
figure(1)
h = axes
figure(2)
h = axes('position',[0.2 0.2 0.6 0.6])
```

The first command yields a full window set of axes with default ranges. The second command gives a reduced set of axes:

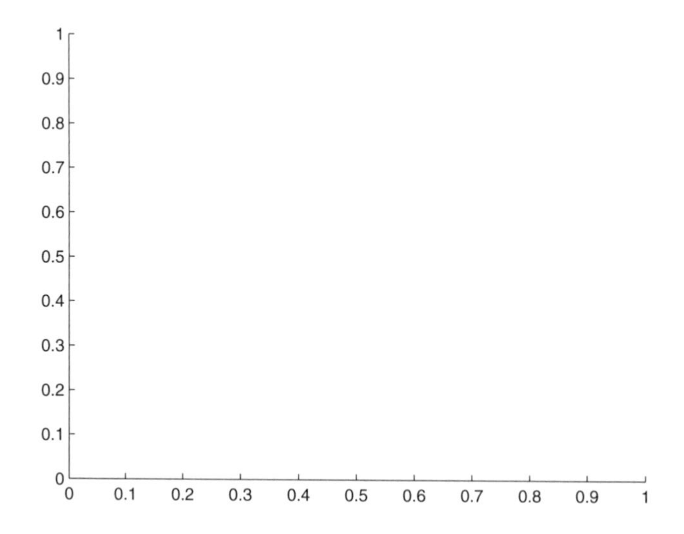

in the centre of the window of dimension 0.4 times the window width and height. The attributes of the handle can then be used by `get` and `set`.

`axis` In plots used to set the range of the plot, the argument is a row vector of length 4 (for two-dimensional plots) or of length 6 (for three-dimensional plots).

```
x = 0:pi/20:pi; y = sin(x);
plot(x,y)
axis([-pi/2 pi/2 -1 1])
```

The initial ranges for the axes are given by `[min(x) max(x) min(y) max(y)]` which gives:

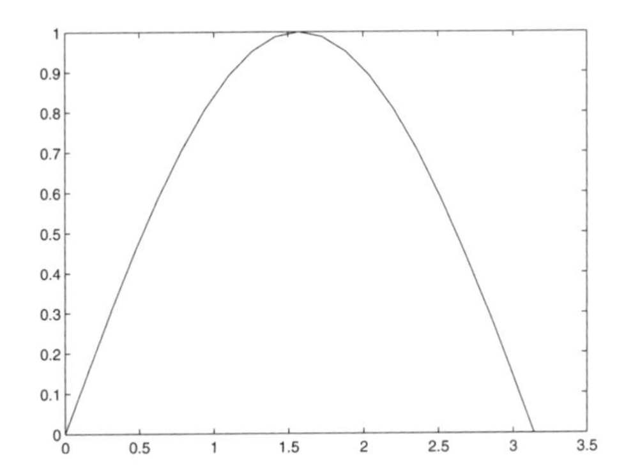

whereas typing the **axis** command above gives

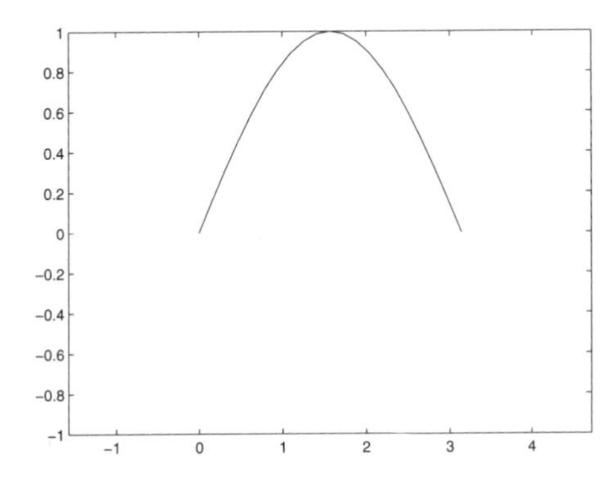

For three dimensions a similar structure applies with an extra pair of variables representing the minimum and maximum of the third variable.

```
t = linspace(0,2*pi);
x = cos(t);
y = sin(t);
z = t;
plot3(x,y,z)
grid on
axis([-1.5 1.5 -1.5 1.5 -pi/2 2*pi])
```

This gives the plot:

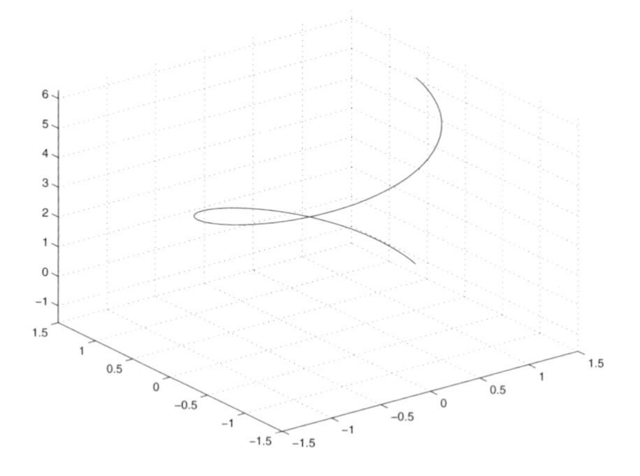

We have used `grid on` to add the dotted lines.

The command can also be used as:

```
axis off     % Removes the axis from the current figure

axis equal   % Sets the axis scaling to be equal.

axis square % Sets the axis to be square.

axis tight   % Uses the max and min of the
             % data for axis limits
```

for more variants see `help axis`.

`bar` Produces a bar chart of the data.

```
x = [1 2 3 4 5];
y = [13 12 9 8 15];
bar(x,y)
```

This gives:

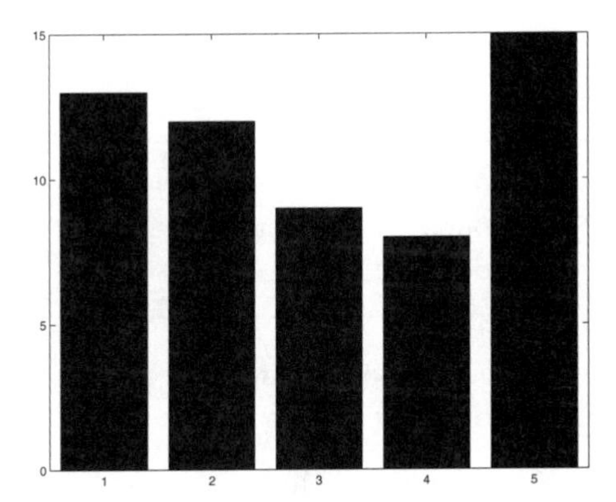

If only one argument is supplied it uses `x = 1:m` where `m` is the number of values of `y`.

`barh` Produces a horizontal bar chart of the data.

```
x = [1 2 3 4 5];
y = [13 12 9 8 15];
barh(x,y)
```

This gives:

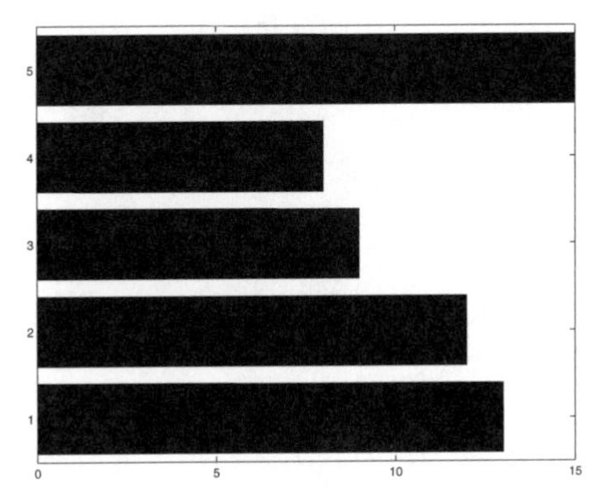

If only one argument is supplied it uses `x = 1:m` where `m` is the number of values of `y`.

`clf` This clears the current figure, in fact it removes all children with visible handles. It removes any axes from the current figure. It is useful to use this before we start plotting.

`close` This is used to close figures; `close all` closes all figures. There are basically three syntaxes:

```
close        % Closes the current figure
close(4)     % closes figure number 4
close all    % closes all figures.
```

This can be used for other windows by using their handle: `close(h)` closes the window with handle `h`.

`figure` This brings the requested figure to the fore, or creates it if it doesn't exist.

```
figure       % Creates a figure
figure(3)    % Ensures that Figure No. 3 is the current
             % figure and is at the fore.
figure(h)    % Ensures that the figure with the handle
             % h is the current figure
```

`gca` Returns the handle to the current axis; this allows various properties to be displayed (`get`) and modified (`set`).

There are many variables involved:

```
>> x = 1:12; y=x.^2;
>> plot(x,y)
>> h = gca
>> get(h)
        AmbientLightColor = [1 1 1]
        Box = on
        CameraPosition = [6 75 17.3205]
        CameraPositionMode = auto
        CameraTarget = [6 75 0]
        CameraTargetMode = auto
        CameraUpVector = [0 1 0]
        CameraUpVectorMode = auto
        CameraViewAngle = [6.60861]
```

```
CameraViewAngleMode = auto
CLim = [0 1]
CLimMode = auto
Color = [1 1 1]
CurrentPoint = [ (2 by 3) double array]
ColorOrder = [ (7 by 3) double array]
DataAspectRatio = [6 75 1]
DataAspectRatioMode = auto
DrawMode = normal
FontAngle = normal
FontName = Helvetica
FontSize = [10]
FontUnits = points
FontWeight = normal
GridLineStyle = :
Layer = bottom
LineStyleOrder = -
LineWidth = [0.5]
NextPlot = replace
PlotBoxAspectRatio = [1 1 1]
PlotBoxAspectRatioMode = auto
Projection = orthographic
Position = [0.13 0.11 0.775 0.815]
TickLength = [0.01 0.025]
TickDir = in
TickDirMode = auto
Title = [287.001]
Units = normalized
View = [0 90]
XColor = [0 0 0]
XDir = normal
XGrid = off
XLabel = [288]
XAxisLocation = bottom
XLim = [0 12]
XLimMode = auto
XScale = linear
XTick = [ (1 by 7) double array]
XTickLabel =
        0
        2
        4
        6
        8
        10
        12
XTickLabelMode = auto
XTickMode = auto
YColor = [0 0 0]
YDir = normal
YGrid = off
YLabel = [289]
YAxisLocation = left
```

```
YLim = [0 150]
YLimMode = auto
YScale = linear
YTick = [0 50 100 150]
YTickLabel =
        0
        50
        100
        150
YTickLabelMode = auto
YTickMode = auto
ZColor = [0 0 0]
ZDir = normal
ZGrid = off
ZLabel = [290]
ZLim = [-1 1]
ZLimMode = auto
ZScale = linear
ZTick = [-1 0 1]
ZTickLabel =
ZTickLabelMode = auto
ZTickMode = auto

ButtonDownFcn =
Children = [285]
Clipping = on
CreateFcn =
DeleteFcn =
BusyAction = queue
HandleVisibility = on
HitTest = on
Interruptible = on
Parent = [1]
Selected = off
SelectionHighlight = on
Tag =
Type = axes
UIContextMenu = []
UserData = []
Visible = on
```

`gcf` Returns the handle of the current figure. This allows various properties to be displayed (`get`) and modified (`set`).

```
x = 1:12; y = x.^2;
plot(x,y)
h = gcf
get(h)
```

returns the list printed on page 348.

`get` Extracts a particular attribute from a list, retrieved for example by `gca` or `gcf`.

```
>> x = 1:12; y = x.^2;
>> plot(x,y)
>> h = gcf
>> a = gca
>> get(h,'Color')

ans =

    0.8000    0.8000    0.8000

>> get(a,'YTick')

ans =

     0    50   100   150
```

The colour variable is returned as a triple (a one-by-three vector giving the RGB (Red-Green-Blue) value). The values associated with the axis and the figure can be changed using the `set` command.

`ginput` Returns coordinates of mouse clicks on the current figure in terms of axis units; very useful for obtaining initial estimates for roots (see also `zoom`).

```
x = 0:pi/20:4*pi;
y = sin(x);
plot(x,y)
[xp,yp]= ginput
```

Notice that control is passed over to the figure and the points are stored in the arrays `xp` and `yp` (until the return key is hit). The command can also be used as `n = 10; [xp,yp] = ginput(n)` to only get 10 points.

`gplot` Plot a graph using the vertices and the adjacency matrix. To plot the graph

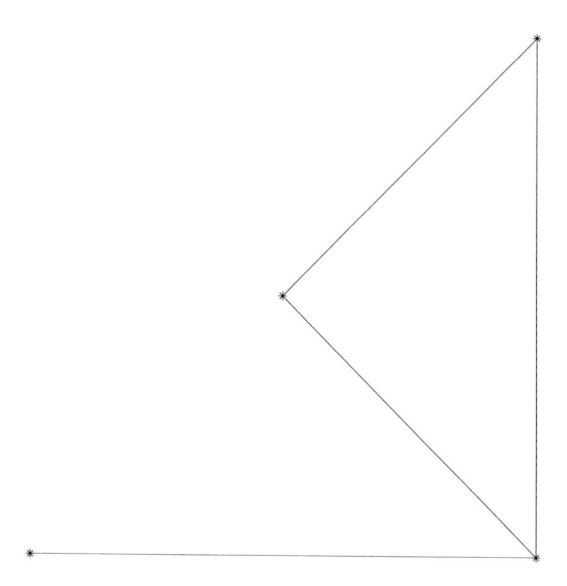

we use the code:

```
xy = [0 0;1 0;
      1 1; 0.5 0.5];
A = [0 1 0 0;
     1 0 1 1;
     0 1 0 1;
     0 1 1 0];
gplot(A,xy,'-*')
axis equal
axis off
```

`grid` Add a grid to a plot (or turn it off).

```
grid on    % Turns the grid on

grid off   % Turns the grid off

grid       % Toggles the grid state
```

`hist` Produces a histogram of the data.

```
>> y = [1 2 3 43 32 54 33 2 4 53 5 63 21 1 5 2];
>> hist(y)
```

gives

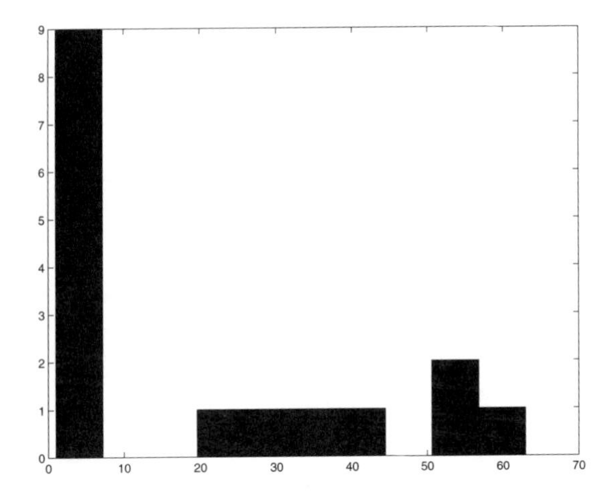

We can use a second argument for `hist` to define the number of bins.

`hold` Stops overwriting of current figure.

```
hold on    % Turns the hold on

hold off   % Turns the hold off

hold       % Toggles the hold state
```

This is useful for putting multiple lines on a plot.

```
x = 0:pi/20:pi;
y = sin(x);
z = cos(2*x);
clf
plot(x,y)
hold on
plot(x,z)
hold off
```

`legend` Used to add a “legend” to a figure, to describe the lines.

```
x = 1:10;
y = x.^2;
z = sin(x);
plot(x,y,x,z)
legend('x^2','sin x', 0)
```

This gives

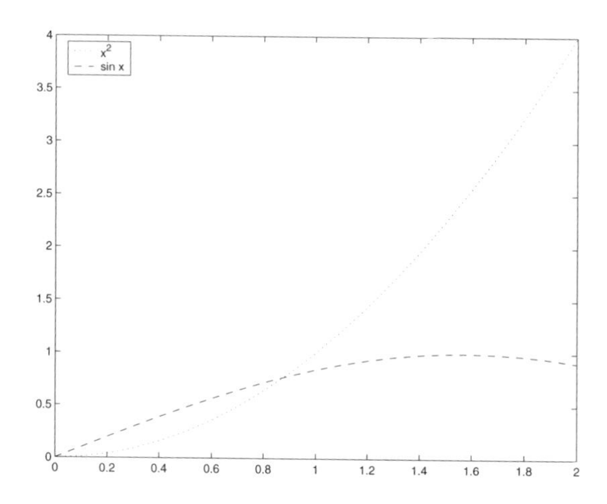

The arguments for `legend` consist of the labelling for the lines and a number which is chosen from:

0	=	Automatic "best" placement (least conflict with data)
1	=	Upper right hand corner (default)
2	=	Upper left hand corner
3	=	Lower left hand corner
4	=	Lower right hand corner
-1	=	To the right of the plot

`loglog` Produces a graph of natural log against natural log of the data.

```
x = [1e-3 1e-2 1e-1 1 1e1 1e2];
y = [10e-3 10e-2 10e-1 10 10e1 10e2];
loglog(x,y)
```

This gives:

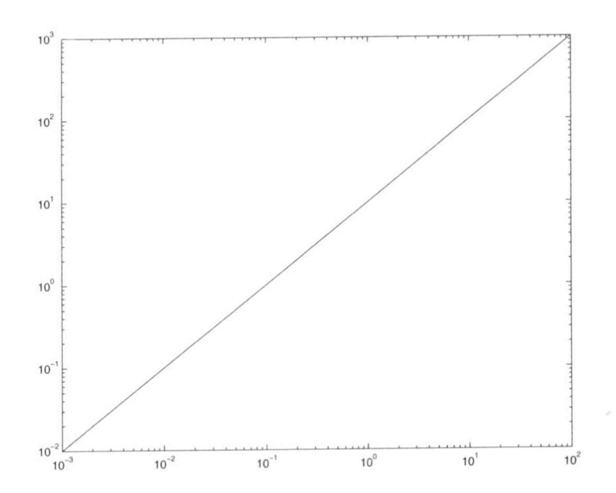

see also `semilogx` and `semilogy`.

`plot` Used to set up figures and plotting of data. The simplest form would be `plot(x,y)`. This command is extremely powerful and has many variants (typing `help plot` helps tremendously).

```
x = -1:0.1:1;
y = x.^3;
plot(x,y)         % Produces a simple plot of y=x^3
                  % using the default line colour

plot(x,y,'go')    % Produces a simple plot of y=x^3
                  % using green circles.
```

The arguments in `plot` either occur: in pairs in which they are pairs of coordinates for plotting or in triples in which case they are pairs of coordinates and the line style with which the data is to be plotted. You can also define variables associated with the plot in the statement:

```
plot(x,y,'LineWidth',2)
```

plots the line but with a thicker line.

`plot3` This is similar to `plot` but gives a line in three dimensions.

```
x = 0.:0.1:3.0;
y = x.^2;
z = 3*x+1;
plot3(x,y,z)
grid on
```

This gives

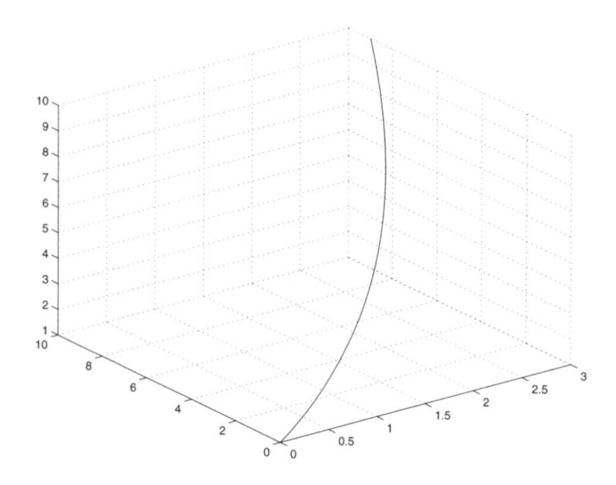

`print` Used to output the contents of figures to files, for Postscript use `print -dps2 output.ps`. For colour Postscript use `print -dpsc2 output.cps` or for JPEG format `printf -djpeg90 output.jpg`.

`semilogx` Produce a plot of y versus $\ln x$.

```
x = linspace(1,100,30);
semilogx(x,log(x))
```

gives:

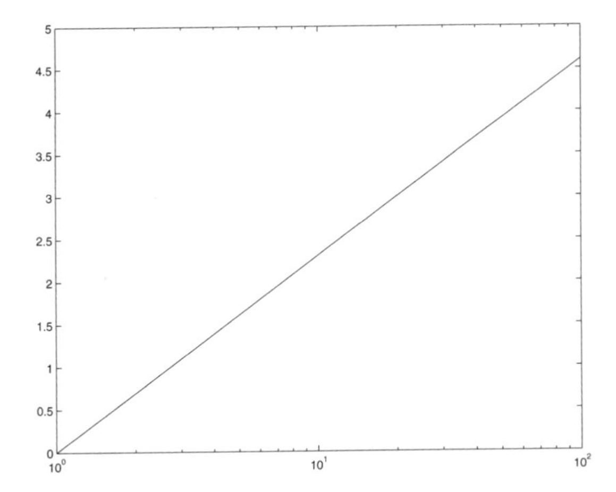

`semilogy` Produce a plot of $\ln y$ versus x.

```
x = linspace(1,100,30);
semilogy(log(x),x)
```

gives:

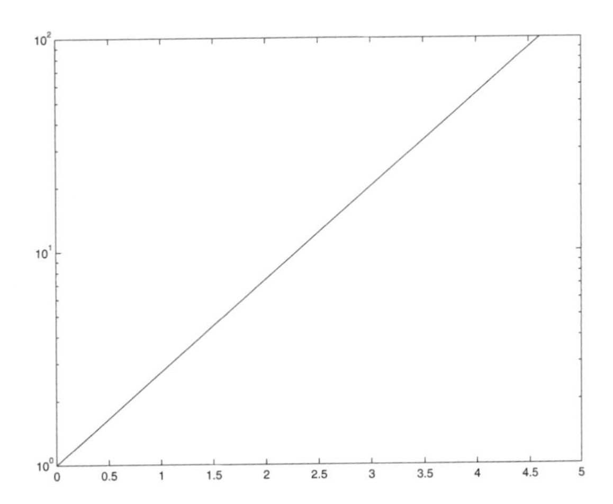

`set` Allows the definition of a particular attribute from a list, retrieved for example by `gca` and `gcf`, or directly from a `plot` command.

```
x = 1:12;
y = 1./x;
h = plot(x,y)
set(h,'LineWidth',4)
```

This gives

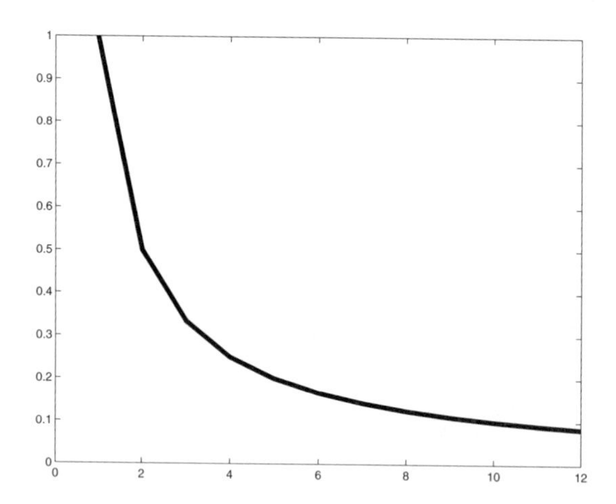

For a list of the attributes of an object (and their values) use `get(h)`, where `h` is defined directly using `gca` or `gcf`.

`subplot` Sets up sub-elements of a plot and points to which one is current. It takes three arguments:

```
subplot(4,2,3)  % Gives an array of figures 4-by-2
```

and sets the current axis to be the third figure, that is the left hand figure on the second row.

```
for j = 1:8
   subplot(4,2,j)
   text(0.5,0.5,int2str(j),'FontSize',24)
end
```

This gives eight subplots (which have been labelled with their corresponding numbers).

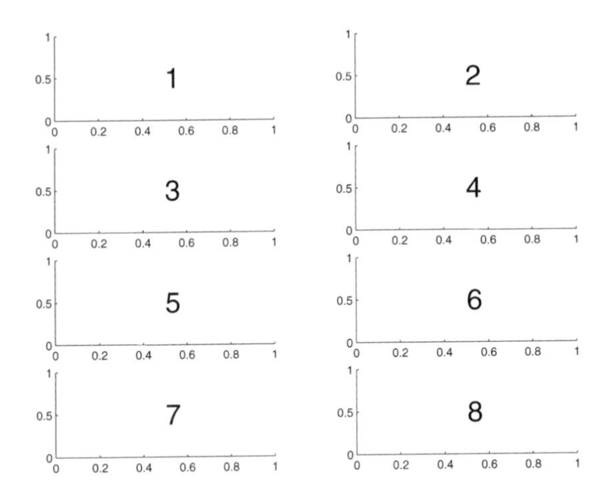

`text` Used to add useful labels to figures.

```
x = -1:0.1:1;
y = sin(x.^2);
plot(x,y)
h = text(0,0.5,'sin x^2')
set(h,'FontSize',18)
```

This gives:

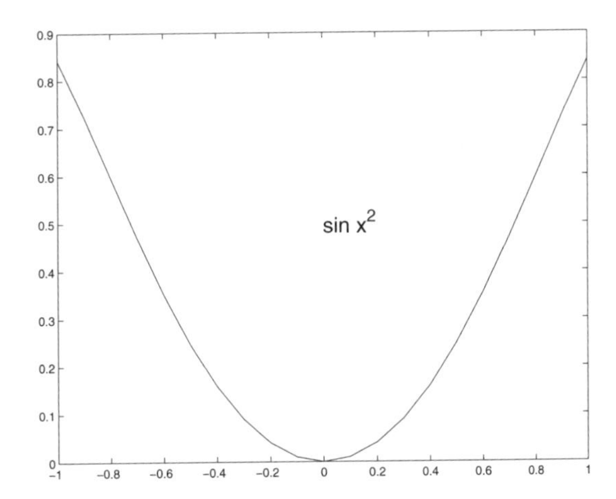

The syntax is `text(a,b,string)` where `(a,b)` is the coordinate in terms of the data and `string` is enclosed in single quotes. Notice here that `sin x^2` actually appears as sin x^2. These commands also recognise underscore for subscript and Greek letters in the form \omega, for instance.

`title` Used to set the title of a plot or subplot. This has quite simple syntax and attaches it to the current set of axes.

```
x = 0.0:0.1:5.0;
y = 1-exp(-x);
plot(x,y)
title('f(x) = 1-e^{-x}','FontSize',18)
```

which gives:

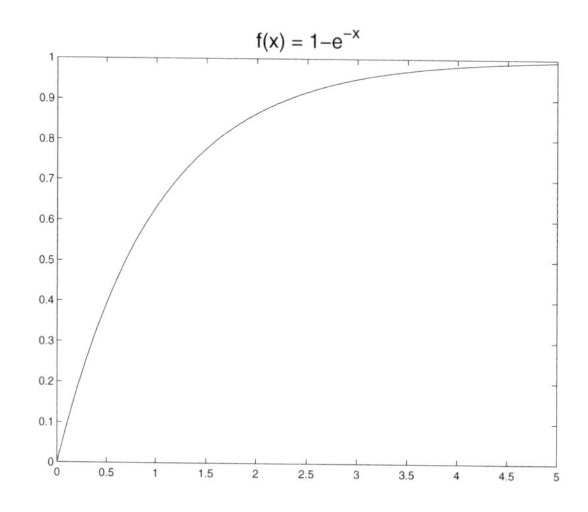

Notice we have used curly brackets to group the power of e for the title.

`xlabel, ylabel` Sets the text for the x and y axis.

```
x = linspace(-2,2,30);
y = (x - 3).^2;
plot(x,y)
hx = xlabel('x values')
hy = ylabel('y = (x-3)^2')
set(hx,'FontSize',16)
set(hy,'FontSize',16)
```

which gives:

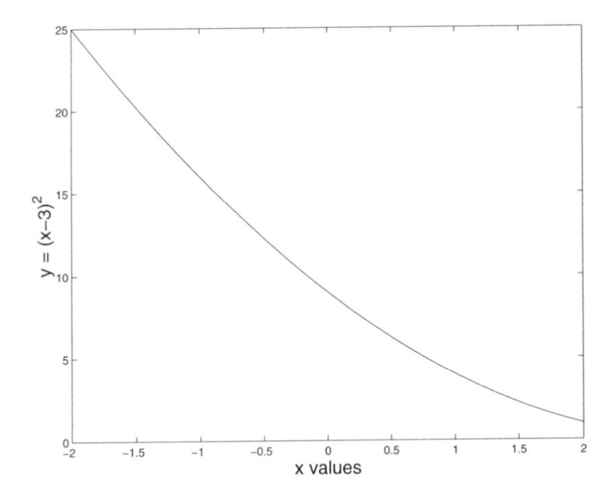

`zoom` Permits zooming into figures: right click enlarge, left click to reduce. This has the syntax:

```
zoom on    % Turns the zoom on

zoom off   % Turns the zoom off

zoom       % Toggles the zoom state
```

B.4 General MATLAB Commands

`abs` Returns the absolute value of a real number or the modulus of a complex one (actually both are the same thing).

```
abs(-1)        % Gives 1
abs(1+i)       % Gives sqrt(2)

x = [1 -2 3+3i];
abs(x)
```

This also works for vectors and matrices so that the final example gives `[1.0000 2.0000 4.2426]` (where the last value is $3\sqrt{2}$).

`angle` Returns the argument of a complex number.

```
angle(sqrt(-1))      % This gives pi/2
angle(2)             % This gives 0
angle(-2)            % This gives pi
```

This can be used for an array of values and returns a vector of the same size full of the corresponding arguments.

`atan2` Gives the arctangent with values between $(-\pi, \pi]$. (see also `tan,atan`). Rather than calculating y/x and then taking the arctangent this function takes account of which quadrant the value is in.

```
atan2(1,1)        % Gives pi/4
atan2(-1,-1)      % Gives -3pi/4
atan2(1,0)        % Gives pi/2
```

Note that the y value is given first. This command can be used to determine the argument of a complex number as `atan2(imag(z),real(z))` (which can be compared with `angle(z)`.

`besselj` Gives the solution $J_\nu(x)$ to Bessel's equation $x^2y''+xy'+(x^2-\nu^2)y = 0$.

```
x = linspace(0,6);
y = besselj(0.5,x);
plot(x,y)
```

This gives the graph:

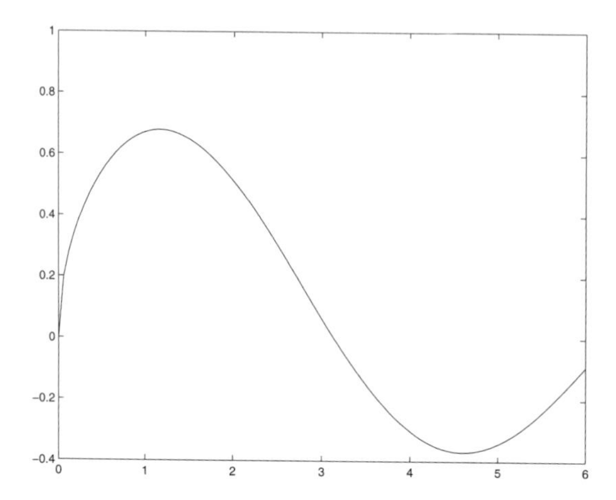

The parameter ν is set to be 1/2 here and in fact $J_{\frac{1}{2}}(x) = \sin x/\sqrt{x}$. There are other Bessel functions: `bessely(nu,x)`, `besseli(nu,x)` and `besselk(nu,x)`.

`break` Stop current level of execution and go back to the previous level, for instance exit a function.

```
function [sx] = takesqrt(x)
if x<0
   disp(' x is negative ')
   sx = NaN;
   break
end
sx = sqrt(x);
```

This routine finds the square root of positive quantities and `breaks` if x is negative.

`case` Elements of a `switch` list, plausible values which the argument can take (see `switch` entry for example).

`ceil` Rounds up to the integer above, has the syntax `ceil(x)` where `x` can be a matrix, vector or scalar.

```
>> x = [0.3 0.9; 1.01 -2.3];
>> ceil(x)
ans =

     1     1
     2    -2
```

`clear` Used to reset objects; `clear variables` removes all variables.

```
clear all           % Clears variables, globals, fns etc
clear variables     % Clear all variables
clear global        % Clear global variables
clear               % Same as clear variable

clear x             % Clear local variable x
clear x*            % Clear local variables starting with x.
```

`cond` Gives the condition number of a matrix, that is the ratio of its largest and smallest eigenvalues. This reflects the ease with which the matrix can be inverted, amongst other things.

```
A = [100 0; 0 0.1];
cond(A)
```

This matrix has eigenvalues of 100 and 0.1 and `cond(A)` returns 1000, that is 100/0.1. In particular the Hilbert matrix is particularly badly conditioned (ill-conditioned), see `hilb`.

`conj` Gives the conjugate of a complex number or an array of them.

```
x = [1 1+i -2-i 4+3i];
conj(x)
```

`corrcoef` Gives the correlation coefficient between two sets of data.

```
>> x = [ 1 2 3 4 5 6];
>> y = [ 3 4 2 1 4 5];
>> corrcoef(x,y)
ans =

    1.0000    0.3268
    0.3268    1.0000
```

This means that `x` and `y` are totally correlated with themselves and that the correlation coefficient between the vectors is `0.3268`. That is a slight positive correlation. The correlation coefficient between two random variables X and Y is given by

$$r = \frac{\text{cov}(X,Y)}{\sqrt{\text{var}(X)\text{var}(Y)}},$$

where the variances are defined in Equation (B.2) and the covariance in Equation (B.1).

`cos, acos` Cosine and arccosine. These functions need to be used with brackets `cos(x)` and `acos(x)` (without, it produces a bizarre result, for instance `cos pi` gives a one-by-two row vector with the elements cosine of the ASCII code for "p" followed by the cosine of the ASCII code for "i"). These functions can also be used for vectors and matrices.

```
x = 0:pi/20:pi;
y = cos(x)
z = acos(y)
```

`cosh` Hyperbolic cosine, equal to $(\mathrm{e}^x + \mathrm{e}^{-x})/2$.

`cov` Gives the covariance of two sets of data.

```
>> x = [ 1 2 3 4 5 6];
>> y = [ 3 4 2 1 4 5];
>> cov(x,y)
ans =

    3.5000    0.9000
    0.9000    2.1667
```

The top left and bottom right elements are the variances of `x` and `y` and the other elements are the covariances. The covariance is defined as

$$\sigma_{XY} = \frac{1}{N-1}\sum_{i=1}^{N}(x_i - \bar{x})(y_i - \bar{y}). \tag{B.1}$$

For the definition of the variance see Equation (B.2). This is normalised using $N - 1$ (rather than N) since this gives the best unbiased estimate of the covariance.

`cputime` Gives the current value of the CPU time. This can be used to time how long parts of the code take:

```
t = cputime;
A = rand(100);
B = inv(A);
t2 = cputime - t;
disp(['Took ' num2str(t2) ' seconds'])
```

`dec2hex` Converts a decimal number to a hexadecimal number. The output will be a string; `a=23456; dec2hex(a)`. See also `hex2dec`.

`demo` Demonstrates the features and capabilities of MATLAB.

`det` This gives the determinant of a matrix. `A = ones(10); det(A)`.

`diag` Sets one of the diagonals of a matrix. The diagonals are referred to as: 0 the leading diagonal, which runs from top left to bottom right. The super-diagonals 1, 2, etc are above the leading diagonal and the sub-diagonals -1, -2, etc are below the leading diagonal. Note that the nth diagonal is $|n|$ units longer than the leading one. If no diagonal is specified then the leading one is used.

```
x = 1:4;
A = diag(x);
B = diag(x,2) + diag(x,-2);
```

The matrix `A` is a four-by-four matrix with the elements of `x` (i.e. 1, 2, 3 and 4 down the leading diagonal), whereas `B` is the matrix

```
B =

     0     0     1     0     0     0
     0     0     0     2     0     0
     1     0     0     0     3     0
     0     2     0     0     0     4
     0     0     3     0     0     0
     0     0     0     4     0     0
```

The command can also be used in reverse, for instance `diag(A)` gives `[1 2 3 4]` and `diag(B,1)` gives `[0 0 0 0 0]`: here we have extracted diagonals.

As a further example we run the code

```
A = zeros(10);
for i = -9:9
    A = A+diag(ones(10-abs(i),1),i)*(i);
end
```

which gives

```
>> A

A =

     0     1     2     3     4     5     6     7     8     9
    -1     0     1     2     3     4     5     6     7     8
    -2    -1     0     1     2     3     4     5     6     7
```

```
    -3    -2    -1     0     1     2     3     4     5     6
    -4    -3    -2    -1     0     1     2     3     4     5
    -5    -4    -3    -2    -1     0     1     2     3     4
    -6    -5    -4    -3    -2    -1     0     1     2     3
    -7    -6    -5    -4    -3    -2    -1     0     1     2
    -8    -7    -6    -5    -4    -3    -2    -1     0     1
    -9    -8    -7    -6    -5    -4    -3    -2    -1     0
```

`diary` Records the user commands and output. This is useful to see which commands have been used. One can specify the file in which the output is stored:

```
diary('list.diary')
x = 1:10;
y = x.^2;
diary off
```

`diff` Gives the difference between successive elements of a vector (is one unit shorter than the original vector): `x=(1:10).^2; diff(x)`, where $d(j) = x(j+1) - x(j)$.

`disp` Displays its argument, which is usually a string: `a=10; disp(['The value of a is ' int2str(a)])`.

`edit` Invokes the MATLAB editor, which is very useful since it allows us to see the current values of variables and provides a very user friendly environment for developing MATLAB programs.

`eig` Returns all the eigenvalues and eigenvectors of a matrix.

```
A = [1 2; -1 2];
[V, D] = eig(A)
```

`V` is a two-by-two matrix with the eigenvectors and columns and `D` is a diagonal matrix containing the eigenvalues on the leading diagonal.

`eigs` Returns certain eigenvalues and eigenvectors of a matrix, specified by certain criteria. `[V,D] = eigs(A,3,'SM')` gives the three eigenvalues of **A** with the smallest magnitude (and the corresponding eigenvectors). There are many options for this command, see `help eigs`.

`else` If the argument for the preceding `if` statement is false execute these statements.

`elseif` Same as `else` but imposing an alternative constraint.

end This command ends all of the loop structures and for each starting argument there must be a corresponding **end** (used with **for**, **if**, **switch** and **while**).

eps The distance from 1.0 to the next largest floating point number. So MATLAB cannot tell the difference between `1` and `1+eps/2`, for instance.

error Causes the code to stop execution; `error('Broken!')` usually used within a conditional statement.

exist Checks to see whether an object exists:

```
if ~exist('a')
   disp(['The variable a ' ...
        'does not exist'])
```

This can be used beyond variables: see the help lines for the command.

exp Evaluate the expression e^x, can be used with vectors and matrices.

expm Evaluate the expression $e^{\mathbf{A}}$, where $\mathbf{A}$ is a matrix. This differs from `exp(A)`, which evaluates e^x for all the elements of `A` rather than $e^{\mathbf{A}}$ which is given by:

$$e^{\mathbf{A}} \equiv \mathbf{I} + \sum_{j=1}^{\infty} \frac{\mathbf{A}^j}{j!}.$$

eye Sets up the identity matrix: `eye(n)` gives $\mathbf{I}_n$.

factor Gives the prime factors of an integer.

factorial This calculates the factorial of an integer n: `factorial(n)` gives $n!$.

feval Evaluates a function, either user-defined or intrinsic. `feval('sin',pi)`.

fix Rounds to the nearest integer (closest to zero), also works for matrices.

fliplr Flips an object left to right. This has no effect on column vectors.

```
x = 1:6;
y = fliplr(x)
```

sets `y` to be `[6 5 4 3 2 1]`. This also works with matrices.

flipud Flips an object upside down. This has no effect on row vectors.

```
x = transpose(1:6);
y = flipud(x)
```

sets y to be `[6; 5; 4; 3; 2; 1]`. This also works with matrices.

`floor` Rounds down to the integer below, also works for matrices.

`fmins` This uses the Nelder–Mead simplex (direct search) method to find the minimum of a function. For instance to find the minimum of the function $f(x_1, x_2) = (x_1 + 2x_2 - 1)^2/(x_2^2 + 1)$ we use

```
function [f]=func(x)
f = (x(1)+2*x(2)-1)^2/(x(2)^2+1);
```

and the command `[x]=fmins('func',[0 0])`.

`for` Defines the start of a loop which runs over a list of objects.

```
N = 10
for j = 2:10
    disp(j)
end
```

displays the numbers from 2 to 10.

`format` Used to specify how MATLAB displays variables. The options can be retrieved using `help format`.

`full` Tells the programme to treat the matrix as `full`, rather than `sparse`.

```
A = [1 0 2; 0 -2 0; -1 0 0];
B = sparse(A);
C = full(B);
```

```
>> A

A =

     1     0     2
     0    -2     0
    -1     0     0
```

```
>> B

B =

   (1,1)        1
   (3,1)       -1
   (2,2)       -2
   (1,3)        2
```

and `C` is back to `A` again.

`function` Occurs at the start of a function definition;

`function [v1,v2]=testfn(in1,in2,in3)`.

`fzero` Determines one zero of the function passed as the first argument to `fzero`. This function takes many different arguments and these can be displayed using `help fzero`.

```
f = inline('sin(3*x)');
x = fzero(f,2);
```

finds a zero of the function $f(x) = \sin 3x$ near x equals two.

```
Zero found in the interval: [1.8869, 2.1131].
>> x

x =

    2.0944
```

`gcd` Gives the greatest common divisor of two integers. This is unity if they are coprime: used as `gcd(x,y)`.

`global` This enables variables to be accessed from other places in the code without being passed directly as an argument. There needs to be a `global` statement in the context in which the variable is defined and also one where it is used.

`help` Gives help on MATLAB commands and can be used to expand the material given in this glossary.

`helpbrowser` Launches a web browser help facility (MATLAB 6).

`helpdesk` Provides access to the web-based help facility.

`hex2dec` Converts a hexadecimal number to a decimal: the input needs to be a string; `a='FF0'; hex2dec(a)`. See also `dec2hex`.

`hilb(n)` This sets up the n-by-n Hilbert matrix with entries $(\mathbf{A})_{i,j} = 1/(i+j)$.

`i,j` Initially set to be the square root of minus 1. These can be used to set up complex numbers:

```
a = 3 + 2*i;
b = -4 + j;
```

(Note that once either of these variables has been used in another context they will not necessarily be equal to $\sqrt{-1}$.)

`if` Start of a conditional block: `if` the statement is true then execute its contents.

```
if x>1
   disp('x is greater than 1')
end
```

This statement uses the logical structures described in section B.1.

`imag` Gives the imaginary part of a complex number; `imag(1+i)` gives 1.

`Inf` This represents answers which are infinite, for instance 1/0.

`inline` Used to define functions which will be evaluated `inline`, see the help function; `g = inline('t^2')` gives $g = t^2$ and then it can be used in `feval` as `feval(g,5)`.

`input` Used for user entry of data `a=input('Enter a: ');`; can also be used to enter strings `name = input('Enter name ',s);`.

`int2str` This takes an integer and returns a string: `int2str(10)` gives '10'.

`inv` Works out the inverse of a matrix (if it exists).

```
a = [1 3; 2 -1];
b = inv(a);
a * b
```

This gives the two-by-two identity (which could be constructed using the command `eye(2)`).

`isempty` Checks whether a variable *is empty.*

```
if isempty(x)
   disp('x is empty')
end
```

This means that the array has either zero rows or zero columns but still exists; to check the existence of a variable use the `exist` command.

`isreal` Checks whether a variable *is real.*

```
isreal(exp(0))
isreal(exp(i))
```

Care is needed since this command checks to see if the imaginary part of the complex number is exactly zero and does not allow for computation errors: for instance $e^{i\pi} = -1$, but the command `isreal(exp(i*pi))` suggests that the quantity is complex.

`isprime` Checks whether a variable *is prime*: `isprime(24)` gives false (that is zero) whereas `isprime(3571)` gives true (that is one).

`length` Gives the length of a vector or alternatively the larger dimension of a matrix:

```
a = [1 2 3; 4 5 6];
b = 1:16
length(a)
length(b)
```

gives the values 3 and 16 respectively.

`linspace` Sets up a grid of one hundred points from the first argument to the second. If there are three arguments use this as the number of points in the grid.

```
x = linspace(0,1,6);
z = linspace(1,10);
```

This gives `x = [0 0.2 0.4 0.6 0.8 1]` and `z` being the array running from 1 to 10 in steps of $(10-1)/99$; since the endpoints are included the step length is not $(10-1)/100$ as you might initially expect.

`load` Reads in data, either directly into variable `load data` (which loads data.mat) or `load 'data.dat'` which returns a matrix `data`.

`log` Natural logarithms.

```
x = [1 exp(1) exp(2)]
y = log(x)
```

This would be written mathematically as $y = \ln x$.

`log10` Logarithm to base ten.

```
x = [1 10 10^2]
y = log10(x)
```

This would be written mathematically as $y = \log_{10} x$. We note that $\log_{10} x = \ln x / \ln 10$.

`lookfor` Allows one to search help files for any command which has a string in its specification `lookfor bessel`.

`lower` Converts the characters in a string to lower case:

```
name = 'Bob Roberts';
lower(name)
```

This gives `bob roberts` (the opposite command is `upper`).

`lu` This produces the LU decomposition of a matrix and can provide information concerning pivoting.

```
A = [1 2 3; -1 3 2; -1 0 1];
[L,U] = lu(A)
```

Gives

```
L =

    1.0000         0         0
   -1.0000    1.0000         0
   -1.0000    0.4000    1.0000
```

```
U =

     1     2     3
     0     5     5
     0     0     2
```

To obtain the pivoting information we would have used `[L,U,P] = lu(A)`. In this case `P` is the three-by-three identity.

`magic` Returns a magic square:

```
>> magic(4)

ans =

    16     2     3    13
     5    11    10     8
     9     7     6    12
     4    14    15     1
```

Notice that not only do the rows, columns and diagonals add up to 34, but so do the four numbers in each corner, the numbers in each two-by-two block in the corner and the central two-by-two block.

`max` This returns the maximum of a vector: if a matrix is supplied it returns a vector providing the maxima along the rows.

```
x = 0:pi/4:2*pi
max(sin(x))
```

gives 1, and

```
A = [1 2 3; 4 5 6];
max(A)
max(transpose(A))
```

gives `[4 5 6]` (that is the maxima of the columns) and `[3 6]` (the maxima of the rows), respectively. Instead of using the transpose we can use the syntax `max(A,[],2)`, which determines the maximum along the second dimension. To find the maximum of a two-dimensional array we use `max(max(A))`.

`mean` Calculates the mean of a set of data, $1/n \sum_{i=1}^{n} x_i$.

```
mean([1 2 3 4 5 6 7])
```

This can also be used on matrices:

```
a = [1 2 3; 4 5 6];
mean(a,1)
mean(a,2)
```

the former giving the averages of the columns and the latter the averages of the rows.

`median` Gives the median of a set of data, that is the one in the middle when the data is ordered. This works in the same way as `mean` on matrices.

`min` Similar to `max` but giving the minimum.

`mod` This gives the remainder when the first argument is divided by the second. If used in the context `mod(x,1)` this gives the fractional part of `x`. It is similar to `rem`.

`NaN` Not-a-Number, used to return quantities which are not assigned, for instance 0/0.

`norm` Gives the mathematical norm of a quantity, particularly useful for working out the length of vectors. For vectors we have

```
norm(v,p)
```

gives

$$\left(\sum_{i=1}^{n} |v_i|^p \right)^{1/p} .$$

If `p` equals two this is a "conventional" norm and the commands

```
norm(v,inf)
norm(v,-inf)
```

give `max(abs(V))` and `min(abs(V))`, respectively.

num2str Converts a number to a string with a specified number of digits; `num2str(pi,4)`. This can also be used without specifying the number of digits (which uses the default corresponding to four places after the decimal point).

ode23,ode45 , Hybrid Runge–Kutta routines to integrate functions, the former being a combination of second- and third-order schemes and the latter fourth and fifth.

We consider the solution of the differential equation

$$\frac{\mathrm{d}y}{\mathrm{d}t} = t^2 - y^2$$

subject to the initial conditions $y(0) = 1$. Firstly we need to set up a function to give y':

```
function [out] = func(t,y)
out = t.^2-y.^2;
```

(which we shall presume has been saved as `func.m`). This can now be called using:

```
trange = [0 1];
yinit = 1;
[t,y] = ode45('func',trange,yinit);
```

This gives the solution:

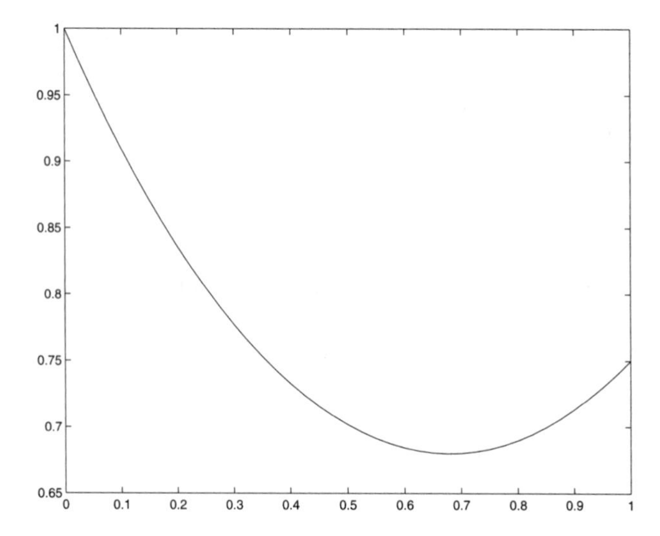

We can set tolerances, amongst other options. Consider the system of differential equations

$$\frac{\mathrm{d}x}{\mathrm{d}t} = t - y$$
$$\frac{\mathrm{d}y}{\mathrm{d}t} = x,$$

subject to the initial conditions $x(0) = y(0) = 0$. We introduce the vector $\mathbf{y} = (x(t), y(t))^T$, and as such we modify the code `func.m` to be:

```
function [out] = func(t,in)
% in(1) is x(t) and in(2) is y(t).
out = zeros(2,1);
out(1) = t-in(2);
out(2) = in(1);
```

This is called using the code:

```
trange = [0 1];
yinit = [0; 0];
options = odeset('RelTol',1e-4,'AbsTol',[1e-4 1e-4]);
ode45('func',trange,yinit,options);
```

This is for the above version of `func.m` for the coupled first-order systems, which sets the relative tolerance to be `1e-4` and the absolute tolerances to be `[1e-4 1e-5]`. This gives:

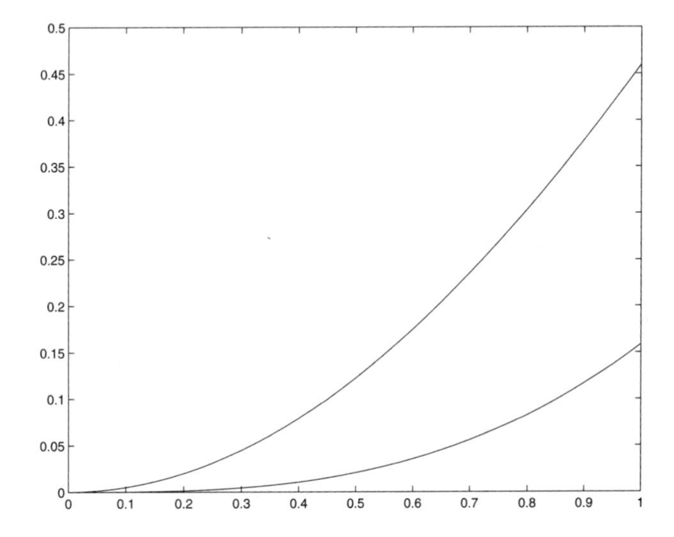

The upper line is $x(t)$ and the lower one is $y(t)$.

`ones` Sets up a matrix full of ones. `ones(n,m)` gives **A** which is an n-by-m matrix for which $a_{i,j} = 1$. `ones(n)` gives a square matrix (n-by-n).

`otherwise` If none of the `cases` correspond to the argument of the `switch` command then these statements are executed.

`path` A list of places MATLAB looks for files, also serves as a command to alter this variable. This command varies on different platforms and as such you should look at `help path`.

`pause` Causes the programme to wait a specified time, or can be used to wait until the user touches a key.

```
x = linspace(0,10);
for its = 1:5
    y = besselj(its/2,x);
    clf
    plot(x,y)
    pause
end
```

This programme runs through the functions $J_{n/2}(x)$ for $n = 1$, 2, 3, 4 and 5 as the user presses a key.

`pi` π - this is `4*atan(1)` or `imag(log(-1))`.

`poly` This returns the characteristic polynomial of a matrix.

```
a = [1 2 3;
     -1 2 0;
     -1 1 1];
poly(a)
```

This gives `[1.0000 -4.0000 10.0000 -7.0000]`, which represents $|\mathbf{A} - \lambda\mathbf{I}| = \lambda^3 - 4\lambda^2 + 10\lambda - 7$. The eigenvalues of the matrix can then be found using `roots`.

`polyfit` Tries to fit a polynomial of best fit, using a least squares idea. If there are n points in `x` (with no repeats) and `y`, and the user requests a polynomial of degree $n-1$ then the fit is "exact":

```
x = [1 2 3];
y = [4 5 -2];
p = polyfit(x,y,2);
```

Note that the coefficients are returned with the one corresponding to the largest power first. This gives the quadratic $-4x^2 + 13x - 5$: however `polyfit(x,y,1)` on the same data gives the straight line $-3x+25/3$ (which is the line of best fit). It is possible to get information about the level of the fit using the form `[p,s] = polyfit(x,y,1)`. The object `s` contains information, for instance the covariance `s.R`.

`polyval` Evaluates a polynomial specified by its coefficients at a set of data points `y=polyval(c,x)`.

```
x = [1 2 3];
c = [-4 13 -5];
y = polyval(c,x);
```

This reconstructs the data used in the example for `polyfit`. Notice that the coefficients are given with the one corresponding to the largest power first.

`primes` Lists the primes up to and including the argument. The syntax is simply `n=20; primes(n)`.

`prod` Similar to `sum` and gives the product of the elements of the vector `x`, so that `prod(x)` returns $\prod_{i=1}^{n} x_i$.

```
x = 1:10;
prod(x) % Gives 10!
z = [1 4 5 6 -2];
prod(z)
```

The factorial could also be found using `factorial`.

`rand` Generates random numbers between zero and one. This can be used to form a matrix of random numbers as well. There are many versions of the argument for this command, see `help rand`.

`randn` Generates normally distributed real numbers: again can be used to generate matrices `randn(n,m)`.

`randperm` Generates a random permutation of a list of objects: To rearrange the letters of our names:

```
s = 'otto denier';
for its = 1:20
   l = randperm(11);
   s(l)
   pause
end
```

`rank` This yields the rank of a set of linear equations and can be used to see whether the system has no solutions, a unique solution or an infinite number of solutions (yielding information about the degrees of freedom).

```
A = [-1 1 1 2;
     3 -1 1 1;
     0 0 1 2];
rank(A)
```

`real` Gives the real part of a complex number `real(z)`, where `z` can be a scalar, vector or matrix.

`realmax` This is the largest floating point number representable on the computer.

`realmin` This is the smallest floating point number representable on the computer.

`rem` This is the remainder attained by dividing the first argument by the second one: `rem(3.32,1.1)` gives `0.02`.

```
rem([3 4 5],2) % Gives 1 0 1
rem(5,[1 2 3]) % Gives 0 1 2
rem([3 4 5],[1 2 3])
               % Gives 0 0 2
```

`reshape` This simply reshapes a matrix into a new shape. This is used most commonly to make a vector into a matrix, but can be used to `reshape` matrices.

```
s = rand(100,1);
a = reshape(s,10,10);
b = [1 3 4; 2 3 4];
d = reshape(b,1,6);
```

This gives `d = [1 2 3 4 4]`; the elements of `b` are read column-wise.

`roots` This gives the roots of the polynomial which is passed to the routine. For instance to find the roots of the cubic $x^3 + 4x^2 + 7x + 2$ we use

```
co = [1 4 7 2];
roots(co)
```

Notice again that the coefficients are listed with the one corresponding to the largest power first (similarly for `polyfit` and `polyval`).

`round` Rounds to the nearest integer.

`save` Saves values of variables to a .mat file.

`sin, asin` Sine and arcsine. These functions need to be used with brackets `sin(x)` and `asin(x)` (without, it produces bizarre results, for instance `sin pi` gives a one-by-two row vector with the elements sine of the ASCII code for "p" followed by the sine of the ASCII code for "i"). These functions can also be used for vectors and matrices.

```
x = -pi/2:pi/20:pi/2;
y = sin(x)
z = asin(y)
```

`sinh` Hyperbolic sine.

`size` Returns the dimensions of a matrix: `[rows,cols]=size(A)`.

`sort` This returns a list of numbers sorted into ascending order, together with a map from their original position to that in the revised list.

`sparse` Defines the matrix as `sparse` so that the computer only operates on non-zero entries: this can dramatically reduce the time spent doing computations (see `full`).

`spline` Fits cubic splines through a set of data points (x, f) and evaluates them at a further set of points z; `y=spline(x,f,z)`.

`sqrt` This finds the square root of a matrix element by element. If necessary the answer may be returned as a complex number.

`std` Calculates the standard deviation of a vector.

`str2mat` As the name suggests, this takes a string and returns a matrix.

`sum` This sums the contents of a vector, or the rows of a matrix. It can also be called with a second argument which defines which dimension needs to be summed.

`switch` This defines the start of a group of statements, the argument for which is a variable, the likely values of which are listed in the `cases`.

`tan,atan` Tangent and arctangent. These functions need to be used with brackets `tan(x)` and `atan(x)` (without, it produces bizarre results, for instance `tan pi` gives a one-by-two row vector with the elements tangent of the ASCII code for "p" followed by the tangent of the ASCII code for "i"). These functions can also be used for vectors and matrices.

```
x = 0:pi/20:pi/4;
y = tan(x)
z = atan(y)
```

(see also `atan2`).

`tour` Gives a tour of the facilities of MATLAB.

`transpose` Returns the transpose of a matrix: can also be done using `A.'`. (Note that `A'` transposes and also takes the conjugate.)

`type` Prints out the contents of a MATLAB script.

`upper` Converts the characters in a string to upper case:

```
name = 'Bob Roberts';
upper(name)
```

This gives `BOB ROBERTS` (the opposite command is `lower`).

`var` Gives the variance of a set of data. The covariance is defined as

$$\sigma_X = \frac{1}{N-1}\sum_{i=1}^{N}(x_i - \bar{x})^2. \tag{B.2}$$

`warning` Allows codes to issue warnings when there may be problems. Is also used to affect how the system issues warnings.

`which` Tells a user where a MATLAB file can be found, and which version is going to be run.

`while` Defines the start of a loop which is continued while a certain condition is fulfilled.

`whos` List of all variables (with details): this can be restricted using things like `whos a*`.

`zeros` Sets up a matrix full of zeros: `zeros(n,m)` gives $\mathbf{A}$ which is an n-by-m matrix for which $a_{i,j} = 0$. `zeros(n)` gives $\mathbf{0}_n$.

C
Solutions to Tasks

Please note that these solutions are given for guidance only and are by no means unique. At the outset we shall give MATLAB output: however subsequently we shall merely give the commands which can be used to solve the problems.

C.1 Solutions for Tasks from Chapter 1

Solution 1.1 *The MATLAB code to solve these problems is:*

```
x = 1.3;
p = x^2+3*x+1

x = 30/180*pi;
y = sin(x);

x = 1;
f = atan(x);

x = sqrt(3)/2;
h = acos(x);
g = sin(h)
```

The values this returns are: `6.5900`, `0.5000`, `0.7854` *(which is* $\pi/4$*) and* `0.5000`*.*

Solution 1.2 *To calculate the function* $y(x) = |x| \sin x^2$ *we use the code:*

```
x = pi/3;
y = abs(x)*sin(x^2);
```

and similarly for $x = \pi/6$*. Notice care is needed with the brackets and the syntax.*

Solution 1.3 *The MATLAB commands to determine these quantities are:*

`sin(pi/2);`

`cos(pi/3);`

`tan(60/180*pi);`

`x=0.5; log(x+sqrt(x^2+1))` *(and with* `x=1`*);*

`x=0; x/((x^2+1)*sin(x))`

and finally `x=pi/4; x/((x^2+1)*sin(x))`*.*

Notice the penultimate part of this task generates `NaN`*: this is because MATLAB does not know how to evaluate zero divided by zero.*

Solution 1.4 *This is a matter of either typing out all the values or exploiting the fact that MATLAB can operate on vectors. We can use the code:*

```
x = [0.3 1/3 0.5 1/2 1.65 -1.34];
round(x)
ceil(x)
floor(x)
fix(x)
```

This example is included to help understand the rôle of various MATLAB commands which can be used to return different roundings to appropriate integers.

Solution 1.5 *The difference between* `rem` *and* `mod` *can be illustrated using the* `y` *value of 4. Thus we have*

```
>> x=[3 4 5]

x =

     3     4     5

>> rem(x,4)

ans =

     3     0     1

>> mod(x,4)

ans =

     3     0     1

>> rem(x,-4)

ans =

     3     0     1

>> mod(x,-4)

ans =

    -1     0    -3

>>
```

So for the second argument being positive we have that `rem` *and* `mod` *are equivalent, whereas for negative values the remainder is signed (thus it shows whether it is positive or negative).*

Solution 1.6 *The MATLAB code is:*

```
x = 1:0.1:2;

% Part 1
y = x.^3+3*x.^2+1;

% Part 2
y = sin(x.^2);

% Part 3
y = (sin(x)).^2;

% Part 4
y = sin(2*x)+x.*cos(4*x);

% Part 5
y = x./(x.^2+1);

% Part 6
y = cos(x)./(1+sin(x));

% Part 7
y = 1./x+x.^3./(x.^4+5*x.*sin(x));
```

Solution 1.7

```
x = 3:0.01:5;
y = x./(x+1./x.^2);
```

Solution 1.8

```
x = -2:0.1:-1;
f = 1./x;
y = f.^3+f.^2+3*f;
```

Solution 1.9 *The code should read*

```
clear all
x = linspace(0,1,200);
g = x.^3+1;
h = x+2;
z = x.^2;
y = cos(x*pi);
f = y.*z./(g.*h);
f(200)
```

The errors were: the second line should have been first, else this cleared out the contents of x*. The default number of points for* `linspace` *is 100, so this needed to be specified as being 200. There were dots missing from the definition of* `g` *and the calculation of* f*. MATLAB distinguishes between upper and lower case in variable names so we need to use* `h` *rather than* `H`*. The command to calculate* y *needs brackets around the argument of the cosine function and an asterisk between* x *and* `pi`*. As mentioned above the dots were missing from* f *and the denominator of the fraction needed to be contained within brackets. Finally the answer needs to be printed (which could have been done at the prompt).*

Solution 1.10 *The code should read*

```
x = linspace(-2,2,20);
c = [1 0 0 0 -1];
y = polyval(c,x);
plot(x,y)
```

The errors were: in the first line the vector as defined would contain 21 entries (try typing `length(x)` *after running the original code). A quartic actually has 5 coefficients, so there was a zero missing in the definition of* `c` *and finally the* x *and* y *needed to be transposed in the plotting command.*

This gives the figure

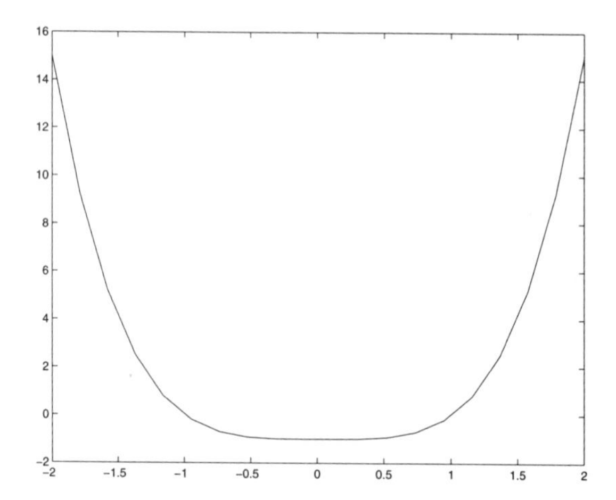

Solution 1.11 *The corrected code should be*

```
x = 0:0.1:3;
f = x.^3.*cos(x+1);
% x = 2
f(21)
% x = 3
f(end)
```

C.2 Solutions for Tasks from Chapter 2

Solution 2.1 *This is hopefully just a matter of typing the code and saving the answer correctly.*

Solution 2.2 *The revised code could be:*

```
a = input('Enter a : ');
b = input('Enter b : ');
res = a^b;
str1 = 'The answer is ';
str2 = '  when ' ;
str3 = ' is raised to the power ';
disp([str1 num2str(res) str2 ...
  num2str(a) str3 num2str(b)]);
```

Solution 2.3 *The function for this purpose is:*

```
function [out] = twox(x)
out = 2.^x;
```

Note the use of the dot so that it can be called with a vector (or even matrix). Try it with `x=1:8`*.*

Solution 2.4 *We can do this calculation using only one function*

```
function [out1,out2] = func(x,y)
out1 = x.^2 - y.^2;
out2 = sin(x+y);
```

or using the two codes:

```
function [out] = func1(x,y)
out = x.^2 - y.^2;
```

and

```
function [out] = func2(x,y)
out = sin(x+y);
```

and then using the plot commands in the example. To extend the range to $[0, 2\pi]$ *the first line would need to be changed to* `x=0:pi/10:2*pi;`*.*

Solution 2.5 *The code should be modified to:*

```
function [out1,out2,out3] = xfuncs(x)
out1 = sin(x);
out2 = cos(x);
out3 = out1.^2 + out2.^2;
```

where we have used the variables `out1` *and* `out2` *to set the value of* `out3` *(which is actually always going to be 1).*

Solution 2.6 *This code takes two inputs and returns two outputs:*

```
function [outx,outy] = mapcode(inx,iny)
outx = mod(inx+iny,1);
outy = mod(inx+2*iny,1);
```

Solution 2.7 *This requires the modification of the second line to read* `y = x.^3+3*x`. *The change in the limit requires the modification of the first line to* `x = -4:1/4:6`. *It is perhaps a good idea to clear out the variables, which is done using* `clear all`.

Solution 2.8 *This question can be solved by noting that this is in fact an equation which is quadratic in* x^2 *and as such can be solved using the formula to have roots:*

$$x^2 = \frac{-1 \pm \sqrt{1-4a}}{2}.$$

In order to have real roots we require $1 - 4a \geqslant 0$ *so that* $1/4 \geqslant a$ *and the quantity* x^2 *must be positive so that* $\sqrt{1-4a} > 1$, *which means that* a *has to be less than or equal to zero. This condition is more restrictive than the previous one so consequently we require* $a \leqslant 0$.

Solution 2.9 *In the question the step is not specified: however we shall use* $1/10$ *since this gives relatively smooth functions:*

```
x = -1:0.1:1;
f = x+3;
g = x.^2+1;
fg = f.*g;
f_over_g = f./g;
clf
plot(x,[f; g; fg; f_over_g])
legend('f','g','f*g','f/g',0)
```

This gives

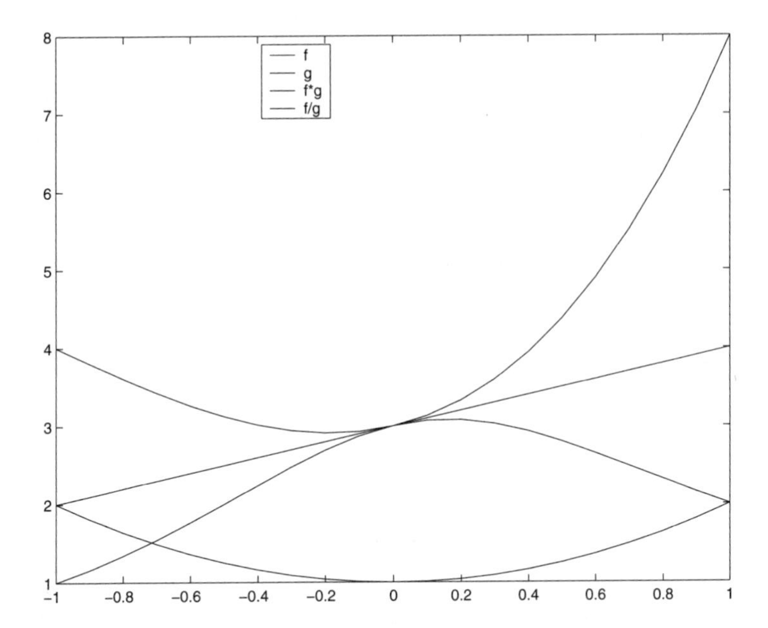

We have added a legend which uses the line styles to show which lines we have plotted. We could have used different style lines. We have used the additional argument at the end of the `legend` *command* `,0` *to tell MATLAB to put the legend in the "best" position.*

Solution 2.10 *Here we shall provide details of how these codes can be improved (or in some cases actually run).*

1. *In the first line we just add a semicolon on the end to suppress output. The second line contains a terrible error, although this is fine as a mathematical equation. In MATLAB you cannot set* `x+2` *equal to* `y`*, we need to set* `y` *equal to* `x+2`*. In the third line: the MATLAB variable for π has a lower case "p" and we are missing an asterisk to denote multiplication and brackets around the denominator of the fraction. The corrected code is:*

```
x = 4;
y = x+2;
z = 1/(y^2*pi);
```

2. *In the first line we are missing a single quote and a semicolon. In the definition of the* `for` *loop we have introduced a variable* `n`*, which should be* `N`*. We have the loop variable as* `i` *whereas it is used as* `j` *within the loop. On the next line we have the brackets missing which should surround the*

denominator of the fraction, forcing it to be evaluated first. In the display line the first square bracket is missing and the answer needs to be converted to a string, which should be `sum` *not* `s`*. On top of this we have failed to set* `sum` *to be zero outside the loop and the command within the loop merely gives the final value rather than calculating the cumulative sum.*

The corrected code is:

```
N = input('Enter N ');
sum = 0;
for j = 1:N
    sum = sum + 1/j+ 1/((j+2)*(j+3));
end

disp(['The answer is ' num2str(sum)])
```

3. *In the first line we have used two equals signs where we should only have one. Two equals signs ask if* x *is equal to the right-hand side, rather than setting* x *equal to it. The rest of the errors are in the second line. In the numerator of the fraction we are missing an asterisk and brackets to show that we are taking the cosine of* x*. In the denominator we have unbalanced brackets (an extra round bracket needs to be added at the end). We have also used a pair of square brackets which should be round. There is an extra asterisk after the division sign. Since we are operating on a vector all of the operators should be preceded with a dot. The corrected code is:*

```
x = 0.0:0.1:1.0;
f = x.*cos(x)./((x.^2+1).*(x+2))
```

4. *The first line of this code gives us a nine-by-nine matrix rather than a row vector. The third line is fine. In the* `for` *loop we have missed out all the asterisks and the* `end` *which terminates the loop. The colons at the end of the lines in the loops need to be changed to semicolons.*

   The corrected code is:

```
w = ones(1,9);

w(1) = 1;

for j = 1:4
    w(2*j) = 3;
    w(2*j+1) = 2*j+1;
end
```

Solution 2.11

(a) Here the conversion factor will be worked out by remembering that there are 1760 yards in a mile, a yard is 36 inches and one inch is 2.54cm:

```
s = 'Enter speed in mph ';
sp_mph = input(s);

sp_yards_ph = sp_mph*1760;
sp_inch_ph = sp_yards_ph*36;
sp_cm_ph = sp_inch_ph*2.54;
sp_m_ph = sp_cm_ph/100;
sp_km_ph = sp_m_ph/1000;
disp(['Speed in km/h is ' ...
      num2str(sp_km_ph) ])
```

(b) This is essentially the reverse of the example above:

```
s = 'Enter speed in m/s ';
sp_mps = input(s);
sp_cmps = sp_mps*100;
sp_inch_ps = sp_cmps/2.54;
sp_yard_ps = sp_inch_ps/36;
sp_miles_ps = sp_yard_ps/1760;
sp_mph = sp_miles_ps *3600;
disp(['Speed in mph is ' ...
      num2str(sp_mph)])
```

(c) We now take the solution above and convert it to be a function, so we have

```
function [output] = change(input);
sp_mph = input;
sp_yards_ph = sp_mph*1760;
sp_inch_ph = sp_yards_ph*36;
sp_cm_ph = sp_inch_ph*2.54;
sp_m_ph = sp_cm_ph/100;
sp_km_ph = sp_m_ph/1000;
output = sp_km_ph;
```

(d) This shows that a sprinter will run at 22 mph on average since they do 100 metres in 10 seconds, that is 10 metres per second.

Solution 2.12 *This is accomplished using the code:*

```
x = linspace(-pi/2,pi/2);
f = x./(1+x.^2);
g = tan(x);
fg = g./(1+g.^2);
gf = f./(1+f.^2);
plot(x,fg,x,gf)
```

Solution 2.13 *This is done using the code:*

```
a = input('Coefficient of x squared: );
b = input('Coefficient of x:');
c = input('The constant:');
y = linspace(0,pi);
x = sin(y);
q = a*x.^2+b*x+c;
plot(y,q)
```

C.3 Solutions for Tasks from Chapter 3

Solution 3.1 *The required code is:*

```
s = 0;
for i = 1:100
    s = s+1/i^2;
end
disp(['Required value is ' num2str(s)])
```

Notice that we have changed the command `int2str` *to* `num2str`*: this is because the answer is no longer an integer.*

Solution 3.2 *This only requires modification of the* `for` *line to* `for i=1:2:100`*, which gives a vector which increases in twos.*

Solution 3.3 *The code* `f.m` *needs to be changed to*

```
function [value] = f(i)

value = sin(i*pi/2)/(i^2+1);
```

and then use the same code.

Solution 3.4 *This is accomplished using the code:*

```
x = 0:pi/4:pi;
f = x.^2+1;
```

Solution 3.5 *The code requires very minor modification to:*

```
v = 0.:0.25:0.75;
cosx = zeros(size(v));
N = 10; range = 0:N;
ints = 2*range;
for n = range
    cosx = cosx + ...
    (-1)^n*v.^ints(n)/factorial(ints(n));
end
```

This gives very accurate answers:

```
cos(v)-cosx

ans =

     0     0     0     0
```

This means that the difference between the MATLAB and the series answers are less than `eps`.

Solution 3.6 *We can use code which performs the summations separately for different values of N or note that $S_{N+1} = S_N + 1/(N+1)^2$ where $S_1 = 1$.*

This leads to the simple code:

```
s(1) = 1;
for n = 1:2000
    s(n+1) = s(n)+1/(n+1)^2;
end
```

This gives a value which when divided by π^2 allows one to appreciate that $c \sim 6$. In fact

$$\sum_{i=1}^{\infty} \frac{1}{n^2} = \frac{\pi^2}{6}.$$

Solution 3.7 *We solve this task by using a nested loop structure*

```
for p = 1:4
    sum = 0;
    for j = 1:(p+1)
        sum = sum + j^p;
    end
    disp([' sum for p=' ...
     int2str(p) ' is ' int2str(sum)])
end
```

Solution 3.8 *The codes for this task are:*

```
sumln(1) = -1;
for n = 1:1000
    sumln(n+1) = sumln(n)+(-1)^(n+1)/(n+1);
end

sum2(1) = 1/2;
for n = 1:1000
    sum2(n+1) = sum2(n)+1/((n+1)*(n+2));
end
```

where we have used the terms in the summations evaluated at the $(N+1)^{\text{th}}$ *place.*

Solution 3.9 *The simplest way of doing this would be* `x>2&x<4`*, although there are other ways* `~xor(x>=2,x=<4)` *for instance. The second example can be done with* `xor`*using* `xor(x>3,x<-1)` *or with* `or` *using* `x>3|x<-1` *(you can use both since the sets are disjoint).*

Solution 3.10 `mod(n,2)==0` *tells us that the remainder when dividing by two is zero (that is n is even). In order to ensure that this is only true for values of* $n > 20$ *we need the statement* `mod(n,2)==0 & n>20`*.*

Solution 3.11 *This requires us to work out* $\tan(73\pi/4)$ *but since* $\tan$ *is periodic this is equal to* $\tan(\pi/4)$ *(which is one). Hence the first condition is true so* x *is set to 2 and as this is an integer* `floor(x)` *is equal to* `x` *and so* `x` *is set to 10, which is not prime so* `x` *is returned as the string "False". This changes with the initial value of* x*, try for instance* $x = 3$.

Solution 3.12 *This uses*

```
start = 1/7;
next = mod(5*start,1)
while floor(next*7) ~= floor(start*7)
      next = mod(5*next,1)
end
```

Here we have used quite a complicated structure to deal with the rounding errors intrinsic to MATLAB (noting that the only possible answers are $n/7$ *where* $n \in \mathbb{N}$*).*

Solution 3.13 *We have*

```
n = 1:50;
f = n.^3-n.^2+40;
ii = find(f > 1000 & mod(n,3) ~= 0);
n(ii)
```

Solution 3.14 *The key here is to start with the string of the first ten letters, namely "abcdefghij". We ask the user to enter the first value of* n *outside the* `while` *loop: this avoids the need for the* `first` *flag. The required code is then:*

```
str = 'abcdefghij';
msg = 'Enter an integer from 1 to 10: ';
n = input(msg);

while (round(n)~=n) | (n<1 | n>10)
    warning(' Not valid ')
    n = input(msg);
end
str(1:n)
```

Solution 3.15 *Although this problem can be solved in one code it is preferable to use a couple of functions. The first one checks whether a character is a letter (lower or upper case):*

```
function [val] = isletter(charac)

lchar = lower(charac(1));
if lchar>='a' & lchar <='z'
   val = 1;
else
   val = 0;
end
```

Firstly the first character is extracted and converted to lower case. Then a check is made to see if it is a character in the lower case alphabet: if it is the variable `val` *is set to be true (that is 1). A similar code checks for whether a character is a digit:*

```
function [val] = isdigit(charac)

lchar = charac(1);
if lchar>='0' & lchar <='9'
   val = 1;
else
   val = 0;
end
```

We can now use the functions:

```
msg = 'Please enter a letter and a digit ';

str = input(msg,'s');

while ~isletter(str(1)) | ~isdigit(str(2))
      warning('This is not valid')
      str = input(msg,'s');
end
```

The argument of the `while` *loop checks to see if either of the conditions isn't satisfied (and as such uses* `or`*, that is the vertical line).*

Solution 3.16 *This can be done with the code:*

```
x = linspace(-3,5,100);
for i = 1:length(x)
    if x(i) >= -1 & x(i) <= 1
       f(i) = x(i)^2;
    elseif x(i) > 1 & x(i) < 4
       f(i) = 1;
    else
       f(i) = 0;
    end
end
plot(x,f)
axis([-3 5 -0.5 1.5])
```

The final command is added purely so that the curve can be distinguished from the axis. This gives

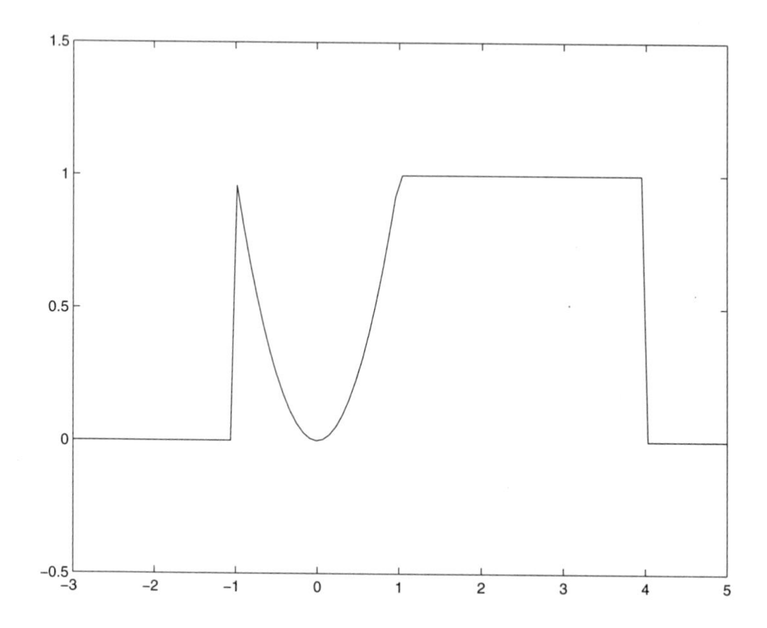

Solution 3.17 *Here we use the code*

```
x = -3:0.1:3;
g = cos(pi*x);
izero = find(abs(g)<=1e-15);
ii = find(abs(g)>=1e-15);
f(izero) = NaN;
f(ii) = 1./g(ii);
plot(x,f)
```

which gives

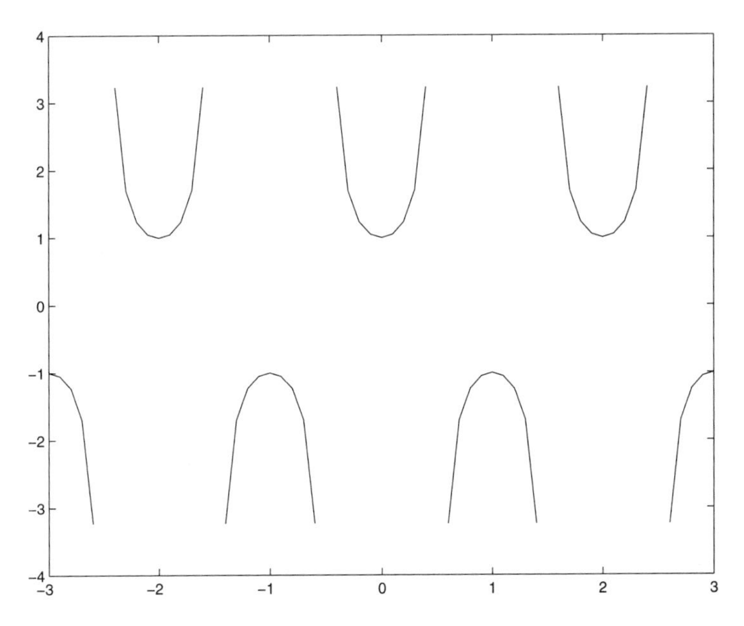

and the choice of 10^{-15} *is in a sense arbitrary, but reflects how close we come to the singularities.*

Solution 3.18 *The first line contains a spelling mistake: the command should be* `linspace`*. The second line is missing brackets around* x *and a semicolon. In the definition of the* `for` *loop we should have used a colon rather than a semicolon. In both the logical expressions on the* `if` *and the* `elseif` *lines the second reference to the array* x *uses* `i` *rather than* `j`*. The first logical expression should use an ampersand (&) rather than the word* `and`*; similarly the second one should also have an ampersand rather than the word* `or`*. The first part of the second expression* `x(j)` *should be checked to be greater than one or equal to one (although this change is academic). The following line should refer to* `x(j)` *rather than the entire vector* x*, and should be finished off with a semicolon. The variable* `zero` *is used without definition. Finally we are missing an* `end` *statement to balance with the* `for`*; a fact which would be clear if the correct indentation was used.*

The corrected code is:

```
x = linspace(-4,4);
N = length(x);

for j = 1:N
    if x(j) >= 0 & x(j) <= 1
       f(j) = x(j);
    elseif x(j) >= 1 & x(j) < 2
       f(j) = 2 - x(j);
    else
       f(j) = 0;
    end
end
```

C.4 Solutions for Tasks from Chapter 4

Solution 4.1 *The solution to this task involves noting that the zeros of a function* $f(x) = g(x)h(x)$ *will occur at the zeros of the functions* $g(x)$ *and* $h(x)$*, provided neither of the functions are singular. In this case* $g(x) = x$ *is zero at* $x = 0$ *and* $h(x) = \sin x$ *is zero at* 0, π, 2π *and* 3π *(within the range* $[0, 10]$*).*

Solution 4.2 *We note that* $\cosh x = (\mathrm{e}^x + \mathrm{e}^{-x})/2$ *so in order for* $\cosh x$ *to be zero we require that* $\mathrm{e}^x = -\mathrm{e}^{-x}$ *or multiplying through by* e^x *that* $\mathrm{e}^{2x} = -1$*, but since* $\mathrm{e}^x > 0$ *for all* x*, this can never be so. Consequently* $\cosh x$ *is never zero. Similarly for* $\sinh x$ *we find that* $\mathrm{e}^{2x} = 1$ *which is only true when* $x = 0$*, which is the single isolated zero of* $\sinh x$*. The zeros of the function of* $f(x)$ *occur at the zeros of* $\cosh^m x$ *and* $\sinh^n x$*, which are only at* $x = 0$ *(in which case it is an n-fold zero).*

Solution 4.3 *We consider the discriminant of the equation, which is* $b^2 - 4$*. For two real roots we have* $b^2 > 4$*, in which case* $|b| > 2$*, for one real root the discriminant is zero, so that* $b = 2$ *and finally for complex roots the discriminant is negative so that* $|b| < 2$*. This can be verified graphically using:*

```
b = input('Value of b ');
x = -10:0.1:10;
f = x.^2+b*x+1;
plot(x,f)
```

Notice that we have chosen the range $[-10, 10]$ *but we could have used knowledge of the structure of the function to make sure that both roots were on the image (where* $|b| > 2$*).*

Solution 4.4 *We shall try to write the equation in the form*

$$\frac{\cos\theta\sin x + \sin\theta\cos x}{\cos\theta}$$

in which case $\tan\theta = \beta$*, in which case* $\sin\theta = \beta/\sqrt{1+\beta^2}$ *and* $\cos\theta = 1/\sqrt{1+\beta^2}$*. Now the function* $f(x)$ *can be written as*

$$f(x) = \frac{\sin(\theta + x)}{\cos\theta},$$

which has zeros at $\sin(\theta+x) = 0$*, hence* $x = n\pi - \theta$ *where* $\theta = \sin^{-1}(\beta/\sqrt{1+\beta^2})$ *(which is evaluated in MATLAB using* `asin`*). If* $\beta = 0$ *then* $\theta = 0$ *and if* $\beta = 1$ *then* $\theta = \pi/4$*.*

Solution 4.5 *The equation* $f(x) = 0$ *can be rewritten in many forms but we choose* $x = \cos^{-1}(\sin x/2)$ *so that a fixed point scheme would be* $x_{n+1} = \cos^{-1}((\sin x_n)/2)$ *and the corresponding code is:*

```
function g = eqn(x)
g = acos(sin(x)/2);
```

The roots of this function can be determined analytically using a similar method to the previous example and are found to be $n\pi + \theta$ *where* $\theta = \sin^{-1}(2/\sqrt{5})$*.*

Solution 4.6 *The roots of this equation occur at*

$$x_b = \frac{-b \pm \sqrt{b^2 - 4}}{2}.$$

Firstly considering the option $g(x) = -(x^2 + 1)/b$ *we have that* $g' = -2x/b$ *which at the roots is*

$$g'(x_b) = 1 \mp \sqrt{1 - \frac{4}{b^2}}.$$

Considering $|b| > 2$ the modulus of this function is greater than one for the root corresponding to the negative sign and less than one for the root with the positive sign. For the other option we find that

$$g'(x) = \frac{b}{2\sqrt{-(bx+1)}}.$$

When the roots are substituted in we find that the above situation is reversed. Using the code

```
function g = eqn(x)
b = 3;
g = -(x^2+1)/b;
```

or with the alternative final line `g = -sqrt(-(b*x+1));` *we find starting with an initial guess of -1 we get different roots depending on which fixed point scheme we use.*

Solution 4.7 *We change the file* `func.m` *to be*

```
function [f] = func(x)
f = 2*x.^2-x.^3+sin(x);
```

Using the routine we produce the plot

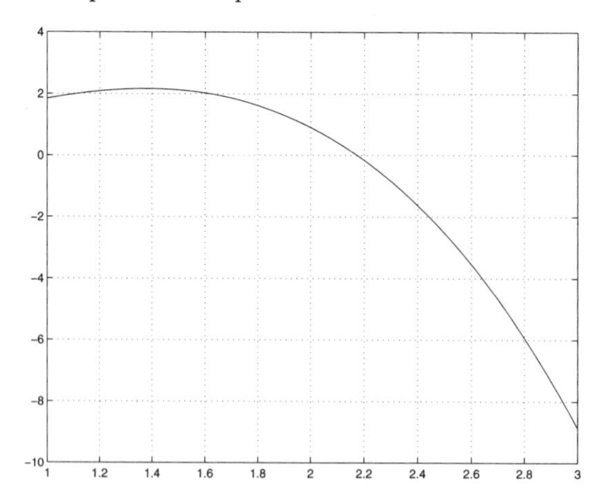

Using 2 and 3 as the ends of the range we obtain

```
>> mbisect
 Root = 2.1741 found in 14 iterations
```

with a tolerance of 1×10^{-4}.

Solution 4.8 *The function* $f(x) = \cos 3x$ *has three zeros in the range 0 to* π. *Using the full range, we encounter the left root* $\pi/6$ *(in fact we should check whether* $f((b+a)/2)$ *is smaller than the tolerance, which it is in this case. For the other two ranges we can select the lower or upper root. We note that the scheme still works for an odd number of roots (since the scheme eliminates them in pairs).*

Solution 4.9 *This is merely a matter of setting up the routines*

```
function [f] = func(x)
f = x.*cos(x)-sin(x);
```

```
function [fp] = func_prime(x)
fp = -x.*sin(x);
```

```
function [f] = func(x)
f = (x.^3-x).*sin(x);
```

```
function [fp] = func_prime(x)
fp = (3*x.^2-1).*sin(x) ...
       +(x.^3-x).*cos(x);
```

The roots of the first function are at $x = 0$ *and* $x \approx \pm 4.41$ *and many other roots which tend to the zeros of* $\cos x$ *(as* x *increases).*

The other function has zeros at $x = 0$ *and 1, and then at* $n\pi$ *where* $n \in \mathbb{Z}$.

Solution 4.10 *The zeros of this function occur where* $x - x^3 = n\pi$. *In order to obtain initial estimates for the range we plot the function*

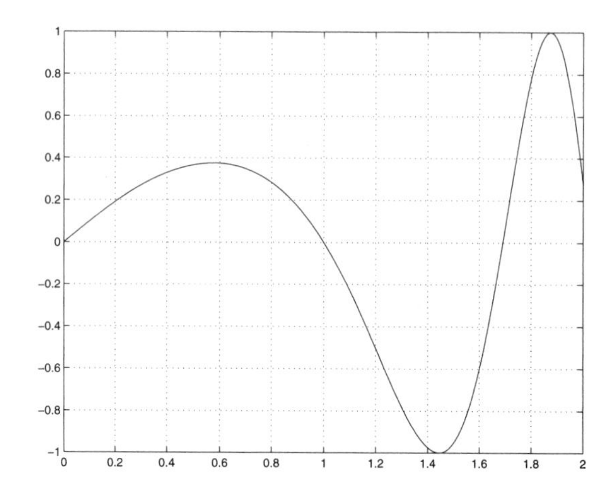

Now using the code ***False_Position.m****:*

```
>> False_Position
Starting guess point 1 :0.8
Starting guess point 2 :1.2
 Root = 0.9999999982 found in 12 iterations.
>> False_Position
Starting guess point 1 :1.6
Starting guess point 2 :1.8
 Root = 1.690631797 found in 4 iterations.
```

The first root corresponds to $n = 0$ above. We note that the function gets very oscillatory as x increases and may pose problems as more roots are required, in which case the roots of the cubic $x - x^3 = n\pi$ can be sought, using for instance `roots`*.*

Solution 4.11 *We use the code on page 124 which calls:*

```
function [f,fp,fpp] = fun2(x);
f = x.^3-4*x.^2+5*x+2;
fp = 3*x.^2-8*x+5;
fpp = 6*x.^2-8;
```

The roots are at 1 (twice) and 2.

Solution 4.12 *The roots can be calculated using the inline code* `roots([1 1 1 1])` *and* `roots([1 0 0 1 -2 -4])`*.*

Solution 4.13 *This can be done using the code*

```
function [f]=func1(x)
f = x.*sin(x)+cos(x);
```

and then use the code `fzero('ff',3)` *which gives the root* ≈ 2.7984. *There are many others. For the other cases the roots are:* $\sin x = 0$ *or* $\sin x = \pm 1$, *so that* $x = n\pi/2$ $(n \in \mathbb{Z})$*; in this case* `fzero` *fails and returns a root at* $x = 1$. *The code has mistaken the fact that the function changes sign for a root.*

Solution 4.14 *The function* $J_{1/2}(x)$ *is actually* $\sin x/\sqrt{x}$ *and consequently the roots are* $n\pi$. *The code needed for* `fzero` *is*

```
function [f] = ourbess(x)
f = besselj(1/2,x);
```

and then `fzero('ourbess',3)`.

Solution 4.15 *The function we want to find the zeros of is* $f(x) = x\sin x - x^2\cos x - 1$ *which has derivative* $f'(x) = \sin x + x\cos x - 2x\cos x + x^2\sin x = (1+x^2)\sin x - x\cos x$. *The correct definitions for* $f(x)$ *and* $f'(x)$ *are:*

f.m

```
function [out] = f(in)
out = in.*sin(in)-in.^2.*cos(in)-1;
```

fp.m

```
function [out] = fp(in)
out = (1+in.^2).*sin(in)-in.*cos(in);
```

and the code to use these would be

```
x = 0;
for j = 1:10
    x = x -f(x)/fp(x);
end
```

(notice that the second term in this expression was the wrong way up in the question). This code could be written far more eloquently.

C.5 Solutions for Tasks from Chapter 5

Solution 5.1 *This code inputs four values which represent the points* (x_1, y_1) *and* (x_2, y_2)*. These are then made into vectors* $\mathbf{x}$ *and* $\mathbf{y}$*, where the former contains the* x *coordinates and the latter the* y *coordinates. The command* **`polyfit`** *fits a straight line through the points and returns* $y = mx + c$*, where the gradient* m *is the first element of* **p** *and the intercept* c *is the second. Then the final command displays the result. If the user enters both values of* y*, the same equation is just returned as* $m = 0$ *and* c *equals that value. On the other hand if two values of* x *are the same then MATLAB tries to give the line an infinite gradient, since it should be of the form* $x = d$*.*

Solution 5.2 *We write the quadratic as* $y = a + b(x - x_0) + c(x - x_0)(x - x_1)$ *where we choose* x_0 *as the* x *coordinate of one of the points. For convenience we shall take the origin as the first point so that* $x_0 = 0$ *and we see that* $a = 0$*. Using the point* $(2, -1)$ *as the next point we note that the quadratic is* $y = bx + cx(x - 2)$ *and* $-1 = 2b$*. Finally the condition that the curve goes through the final point yields the equation* $5 = -5/2 + 15c$ *so that* $c = 1/2$*. Hence the quadratic is*

$$
\begin{aligned}
y &= -\frac{1}{2}x + \frac{1}{2}x(x-2) \\
&= \frac{x(x-3)}{2}.
\end{aligned}
$$

Solution 5.3 *We use the code:*

```
x = 0:10;
co = [1 3 2];
y = x.^2+3*x+2;
for i = 1:3
    xv = i-0.5;
    p = polyfit(x(i:(i+1)),y(i:(i+1)),1);
    err(i) = polyval(p,xv)-polyval(co,xv);
end
```

This gives the same error in each case, namely $1/4$ *(which we would expect from understanding the error associated with this method of interpolation (consider the second derivative)).*

Solution 5.4 *This quadratic will be of the form* $cx(\pi - x)$ *(since it is zero at*

0 and π). The value of c can be determined from the condition that the curve goes through the final point which gives

$$y = \frac{4x}{\pi^2}(\pi - x).$$

Solution 5.5 *Since the cubic is zero at the points 0 and π we know that it is of the form $x(\pi - x)(a(x - \pi/2) + b)$. The values of a and b can be determined from the other points: $(\pi/2, 1)$ gives $b = 4/\pi^2$; finally $(-\pi/2, -1)$ gives $a = 16/(3\pi^2)$. Hence the cubic is*

$$\begin{aligned} y &= x(\pi - x)\left(\frac{16}{3\pi^3}(x - \pi/2) + \frac{4}{\pi^2}\right) \\ &= x(\pi - x)\left(\frac{16x}{3\pi^3}x + \frac{4}{3\pi^2}\right). \end{aligned}$$

Solution 5.6 *For this we shall use the MATLAB command* **spline** *so that*

```
x = -pi:(pi/2):pi;
y = [0 -1 0 1 0];
z = -pi:(pi/20):pi;
f = spline(x,y,z);
plot(z,f,x,y,'o','MarkerSize',14)
```

This gives:

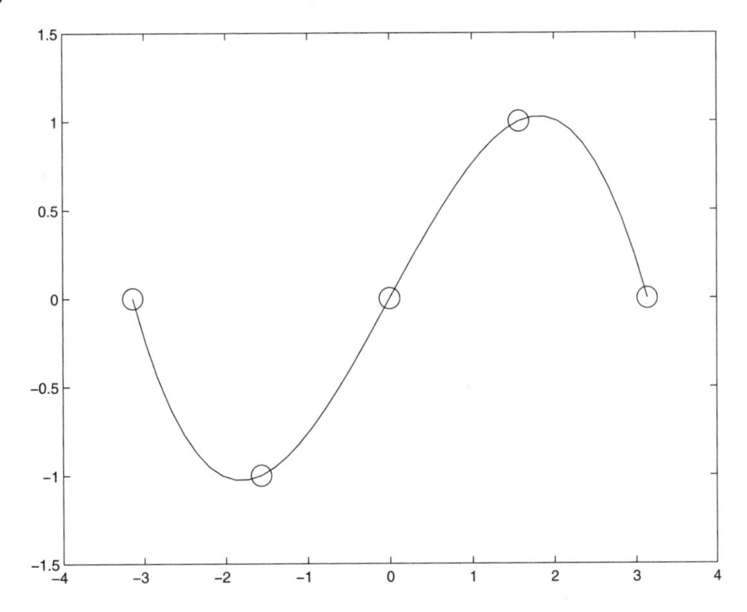

Solution 5.7 *Here we shall use the method of least squares. As such we shall produce details of the formulation. We calculate the sum of the squares of the errors:*

$$e = \sum_{i=1}^{n} (a \sin x_i + b \cos x_i - f_i)^2$$

and partially differentiating with respect to a and b we have

$$\frac{\partial e}{\partial a} = \sum_{i=1}^{n} \sin x_i (a \sin x_i + b \cos x_i - f_i)$$

$$\frac{\partial e}{\partial b} = \sum_{i=1}^{n} \cos x_i (a \sin x_i + b \cos x_i - f_i).$$

These equations can be rewritten in matrix form as:

$$\begin{pmatrix} \sum_{i=1}^{n} \sin^2 x_i & \sum_{i=1}^{n} \sin x_i \cos x_i \\ \sum_{i=1}^{n} \cos x_i \sin x_i & \sum_{i=1}^{n} \cos^2 x_i \end{pmatrix} \begin{pmatrix} a \\ b \end{pmatrix} = \begin{pmatrix} \sum_{i=1}^{n} f_i \sin x_i \\ \sum_{i=1}^{n} f_i \cos x_i \end{pmatrix}.$$

This can be solved using the code:

```
x = 0:0.1:1.0;
f = [3.16 3.01 2.73 2.47 2.13 1.82 ...
   1.52 1.21 0.76 0.43 0.03];
A = [sum(sin(x).^2) sum(cos(x).*sin(x)); ...
   sum(cos(x).*sin(x)) sum(cos(x).^2)];
r = [sum(f.*sin(x)); sum(f.*cos(x))];
sol = A\r;
```

This gives

```
sol =

   -1.9941
    3.1892
```

The original data was actually generated with -2 *and* 3.2, *and then noise was added.*

Solution 5.8 *This can be done by hand but in fact MATLAB will actually return the required coefficients.*

```
x = -pi:(pi/2):pi;
y = [0 -1 0 1 0];
z = -pi:(pi/10):pi;
pp = spline(x,y);
f = spline(x,y,z);
plot(z,f,x,y,'o','MarkerSize',14)
true = sin(z);
err = sum((true-f).^2);
```

This gives

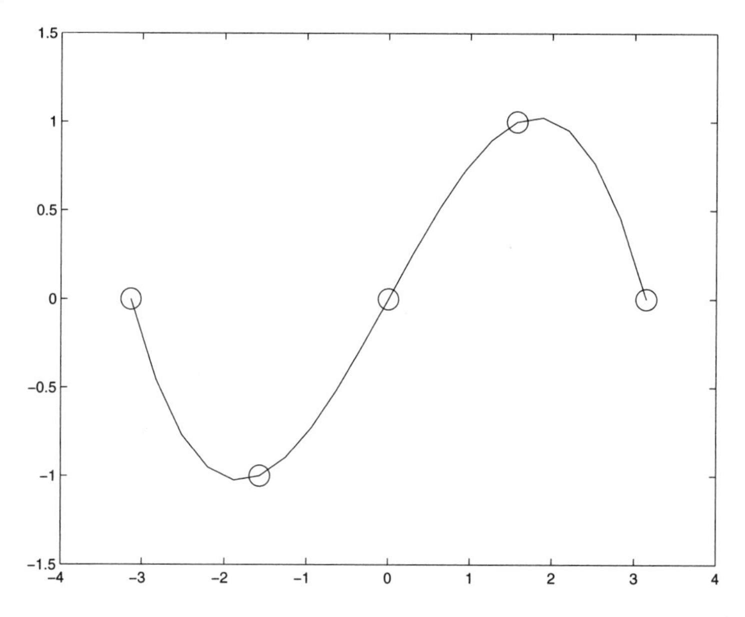

and

```
>> pp.coefs

ans =

   -0.0860    0.8106   -1.6977         0
   -0.0860    0.4053    0.2122   -1.0000
   -0.0860    0.0000    0.8488         0
   -0.0860   -0.4053    0.2122    1.0000
```

which are the four cubic equations (given with the coefficients of x^3 *first). The total sum of the errors was* 0.1944.

Solution 5.9 *The corrected code is:*

```
x = 2:11;
f = polyval([1 0 0 -1],x) + sin(x);
% x = 4.5
r = 3:4;
c = polyfit(x(r),f(r),1);

yy = polyval(c,4.5)

% x = 15 (extrapolation)
r = length(x)-1:length(x);
c = polyfit(x(r),f(r),1);
yy = polyval(c,15);
```

C.6 Solutions for Tasks from Chapter 6

Solution 6.1 *These calculations can be repeated using the code*

```
A = [3 0 -1; -4 2 2];
B = [-1 7; 3 5; -2 0];
C = [2 0; -1 -3];
A*B
B*A
A+transpose(B)
A*C
A*transpose(C)
3*C+2*transpose(A*B)
(A*B)*C
A*(B*C)
```

Solution 6.2 *Notice that here we set up the matrix* `A` *before changing the elements and this is unnecessary but good practice. Also it is not necessary to set up* `r` *and this can be defined inline. The definition of* `r` *allows for more versatile code.*

```
r = 1:4;
A = zeros(4);
A(1,r) = r;
A(r,4) = flipud(r');
```

Note that the final command could also be replaced by `A(flipud(r),1)=r';`*; you should try to understand this command.*

Solution 6.3 *Here we rely on the fact that MATLAB knows that a matrix with a super- or sub-diagonal of length 9 is a ten-by-ten matrix.*

```
a = diag(ones(1,9),1)+diag(-ones(1,9),-1);
```

Solution 6.4 *This is just a matter of typing the commands: however you should be able to decide which ones are viable before doing this.*

Solution 6.5 *The code to solve these problems is given by:*

```
A = [ 1 2; 3 4];
B = [3 4; -1 2];
A*B
C = [3 5; 6 -2];
D = [-1 0; 2 1];
2*C-4*D
E = [1 3 5];
F = [2 -1; -1 0; 7 -2];
E*F
```

This gives the answers

```
ans =

     1     8
     5    20


ans =
```

```
    10    10
     4    -8


ans =

    34   -11
```

Solution 6.6 *Consider a general matrix*

$$\mathbf{A} = \begin{pmatrix} a & b \\ c & d \end{pmatrix}$$

multiplied by the matrix

$$\mathbf{X} = \begin{pmatrix} \alpha & \beta \\ \beta & \alpha \end{pmatrix}.$$

Now

$$\mathbf{XA} = \begin{pmatrix} \alpha a + \beta c & \alpha b + \beta d \\ \beta a + \alpha c & \beta b + \alpha d \end{pmatrix}$$

and

$$\mathbf{AX} = \begin{pmatrix} \alpha a + b\beta & a\beta + b\alpha \\ c\alpha + d\beta & c\beta + d\alpha \end{pmatrix}.$$

Now comparing **XA** *and* **AX** *we find that it is necessary for* $a = d$ *and* $b = c$ *(provided* $\beta \neq 0$*). Hence the only matrices which commute with matrices of the form* **X** *are those of the same form.*

```
stl = 'Top left element of matrix ';
sbl = 'Bottom left element of matrix ';
for j = 1:2
        a(j) = input([stl num2str(j) ': ']);
        b(j) = input([sbl num2str(j) ': ']);
end
A = [a(1) b(1); b(1) a(1)];
B = [a(2) b(2); b(2) a(2)];
disp(A*B)
disp(B*A)
```

Notice that the answer is also of the form **X**.

Solution 6.7 *The $(i,j)^{\text{th}}$ element of* $\mathbf{B}$ *is* $a_{i,j}+a_{j,i}$ *and the* $(j,i)^{\text{th}}$ *element is* $a_{j,i}+a_{i,j}$ *(which are equal so* $\mathbf{B}$ *is symmetric). Similarly the* $(i,j)^{\text{th}}$ *element of* $\mathbf{C}$ *is* $a_{i,j}-a_{j,i}$*, whereas the* $(j,i)^{\text{th}}$ *element is minus this, namely* $a_{j,i}-a_{i,j}$ *so that* $\mathbf{C}$ *is anti-symmetric.*

Solution 6.8 *These matrices can be constructed in MATLAB*

```
theta = 0;
A0 = [cos(theta) sin(theta); -sin(theta) cos(theta)];
theta = pi/2;
A1 = [cos(theta) sin(theta); -sin(theta) cos(theta)];
theta = pi;
A2 = [cos(theta) sin(theta); -sin(theta) cos(theta)];
```

or mathematically and we find

$$A|_{\theta=0} = \begin{pmatrix} 1 & 0 \\ 0 & 1 \end{pmatrix},$$

$$A|_{\theta=\pi/2} = \begin{pmatrix} 0 & 1 \\ -1 & 0 \end{pmatrix},$$

and finally

$$A|_{\theta=\pi} = \begin{pmatrix} -1 & 0 \\ 0 & -1 \end{pmatrix}.$$

Working through the cases one at a time

$\theta = 0$ *This gives us the identity matrix, so multiplying leaves all points unchanged.*

$\theta = \pi/2$ *Here if we start with* $\mathbf{x} = (x,y)^T$ $\mathbf{Ax}$ *is* $(y,-x)^T$*. This moves the point round the origin by (unsurprisingly)* $\pi/2$*.*

$\theta = \pi$ *Now the action is to return* $(-x,-y)^T$*, which is a reflection in the origin (or in fact a rotation of* π*).*

In general the action of multiplying by this matrix is to rotate by θ *radians.*

We can work out the inverse of this matrix by noting that its determinant is unity and switching the terms on the leading diagonal and then multiplying the off diagonal terms by minus one. However, we could also exploit the fact that in order to invert the operation of rotating by an angle θ *we merely rotate by* θ *in the other sense (or more specifically by* $-\theta$*). The inverse is given by*

$$\mathbf{A}^{-1} = \begin{pmatrix} \cos(-\theta) & \sin(-\theta) \\ -\sin(-\theta) & \cos(-\theta) \end{pmatrix} = \begin{pmatrix} \cos\theta & -\sin\theta \\ \sin\theta & \cos\theta \end{pmatrix}.$$

This is easily verified by performing the multiplication of the matrices

$$\begin{pmatrix} \cos\theta & -\sin\theta \\ \sin\theta & \cos\theta \end{pmatrix}\begin{pmatrix} \cos\theta & \sin\theta \\ -\sin\theta & \cos\theta \end{pmatrix}$$
$$= \begin{pmatrix} \cos^2\theta + \sin^2\theta & \cos\theta\sin\theta - \sin\theta\cos\theta \\ \sin\theta\cos\theta - \cos\theta\sin\theta & \sin^2\theta + \cos^2\theta \end{pmatrix}$$
$$= \begin{pmatrix} 1 & 0 \\ 0 & 1 \end{pmatrix} = \mathbf{I}.$$

Solution 6.9 *This is solved using the code:*

```
A = [3 4; -1 2];
b = [2 ; 0];
x = A\b;
```

which gives $x = 2/5$ *and* $y = 1/5$.

Solution 6.10 *This is solved using the code:*

```
A = [1 1 2; 1 -1 -3; ...
   -2 -5 1];
b = [1; 0 ; 4];
x = A\b
```

This gives $x = 4/5$, $y = -1$ *and* $z = 3/5$.

Solution 6.11 *We are able to add* **A** *and* **B** *since they are of the same size, namely they both have three rows and two columns.*

$$\begin{pmatrix} 1 & -1 \\ 0 & 2 \\ 3 & 2 \end{pmatrix} + \begin{pmatrix} 2 & -1 \\ -1 & 0 \\ 3 & 2 \end{pmatrix} = \begin{pmatrix} 3 & -2 \\ -1 & 2 \\ 6 & 4 \end{pmatrix}.$$

The matrices **A** *and* **C** *can be multiplied together since the number of columns of* **A** *(two) matches the number of rows of* **C**.

$$\begin{pmatrix} 1 & -1 \\ 0 & 2 \\ 3 & 2 \end{pmatrix}\begin{pmatrix} -1 & 0 \\ 2 & 1 \end{pmatrix} = \begin{pmatrix} 1\times(-1)+(-1)\times 2 & 1\times 0+(-1)\times 1 \\ 0\times(-1)+2\times 2 & 0\times 0+2\times 1 \\ 3\times(-1)+2\times 2 & 3\times 0+2\times 1 \end{pmatrix}$$
$$= \begin{pmatrix} -3 & -1 \\ 4 & 2 \\ 1 & 2 \end{pmatrix}.$$

The multiplication of $\mathbf{C}$ *times* $\mathbf{B}$ *is not possible since the number of columns of* $\mathbf{C}$ *(two) is not equal to the number of rows of* $\mathbf{B}$ *(three).*

First we calculate $\mathbf{A}-\mathbf{B}$ *(which is possible since both matrices are the same size). This gives another matrix of the same size (again with three rows and two columns), which can now multiply* $\mathbf{C}$ *since this has two rows. The answer is*

$$\begin{pmatrix} 1 & 0 \\ 3 & 2 \\ 0 & 0 \end{pmatrix}$$

The final calculation should give the same answer. The MATLAB code for these calculations is:

```
>> a = [1 -1; 0 2; 3 2];
>> b = [2 -1; -1 0; 3 2];
>> c = [-1 0; 2 1];
>> a+b

ans =

     3    -2
    -1     2
     6     4

>> a*c

ans =

    -3    -1
     4     2
     1     2

>> (a-b)*c

ans =

     1     0
     3     2
     0     0

>> a*c-b*c

ans =
```

```
     1     0
     3     2
     0     0
```

Solution 6.12 *These calculations can both be performed and the solutions are*

$$\begin{pmatrix} 3 \\ 10 \end{pmatrix} \text{ and } \begin{pmatrix} 16 & -2 & 9 & -3 \\ 0 & 2 & -5 & -1 \end{pmatrix}$$

which can be checked using MATLAB

```
>> [1 -1 2; 3 0 1]*[3; 2; 1]

ans =

     3
    10

>> [5 -2;-1 2]*[4 0 1 -1; 2 1 -2 -1]

ans =

    16    -2     9    -3
     0     2    -5    -1
```

Solution 6.13 *The results of both calculations merely returns the matrix unchanged. This is the effect of multiplying by the identity.*

Solution 6.14 *Firstly, we reflect in the leading diagonal to give* $\mathbf{A}^T$ *so that*

$$\mathbf{A}^T = \begin{pmatrix} 3 & 0 \\ 2 & -1 \\ -1 & -2 \end{pmatrix}$$

The results of the multiplications are

$$\begin{pmatrix} 14 & 0 \\ 0 & 5 \end{pmatrix} \text{ and } \begin{pmatrix} 9 & 6 & -3 \\ 6 & 5 & 0 \\ -3 & 0 & 5 \end{pmatrix},$$

which can also be done using MATLAB code:

```
>> A = [3 2 -1; 0 -1 -2];
>> A*transpose(A)

ans =

    14     0
     0     5

>> transpose(A)*A

ans =

     9     6    -3
     6     5     0
    -3     0     5
```

Solution 6.15 *We assume that a general row vector is of the form*

$$(x_1, x_2, \cdots, x_N)$$

and consequently its transpose is the column vector

$$\mathbf{x}^T = \begin{pmatrix} x_1 \\ x_2 \\ \vdots \\ x_N \end{pmatrix}$$

Hence

$$\mathbf{x}\mathbf{x}^T = \begin{pmatrix} x_1 & x_2 & \cdots & x_N \end{pmatrix} \begin{pmatrix} x_1 \\ x_2 \\ \vdots \\ x_N \end{pmatrix} = x_1^2 + x_2^2 + \cdots + x_N^2.$$

This is a scalar which is positive since it is merely the sum of squares.

Solution 6.16 *The matrix equation can be expanded to give*

$$x + 4y = 1$$
$$-2x + 3y = -2$$

and the three simultaneous equations can be written as the single matrix equation

$$\begin{pmatrix} 1 & 1 & 1 \\ 1 & -2 & -1 \\ -1 & 3 & -1 \end{pmatrix} \begin{pmatrix} x \\ y \\ z \end{pmatrix} = \begin{pmatrix} 0 \\ 2 \\ -1 \end{pmatrix}.$$

Solution 6.17 *We simply present the code which can be used to determine the character of the systems (this exploits the code* `solns.m` *given on page 188)*

```
a = [3 2; 3 -2]; b=[7; 7];
solns(a,b)
a = ones(6);
for r = 2:6
  a(r,r) = -1;
end
b = ones(6,1);
solns(a,b)
```

This returns the comments:

```
There are 2 equations
with 2 variables
There is a unique solution
```

and for the second case

```
There are 6 equations
with 6 variables
There is a unique solution
```

Solution 6.18 *This can be accomplished using*

```
A = [1 0 0 -1; ...
    -1 2 -1 0; ...
    0 -1 2 -1;
    0 0 0 1];
r = [0 1;0 0; 0 0; 1 0];
sols = A\r;
```

This gives

```
sols =
```

```
    1.0000    1.0000
    1.0000    0.6667
    1.0000    0.3333
    1.0000         0
```

where we have solved both systems at once to give $(1,1,1,1)$ *and* $(1,\frac{2}{3},\frac{1}{3},0)$.

Solution 6.19 *We use the code*

```
s = pi:pi/3:(2*pi);
ns = length(s);
for j = 1:ns
    ss = s(j);
    A = [0 1 ss; ...
          ss 0 1; ...
          1 ss 0];
    z(j) = det(A);
end
c = polyfit(s,z,3)
```

which gives $c = (1,0,0,1)$ *so that the determinant of the matrix is* s^3+1.

Solution 6.20 *We note that*

$$\mathbf{B}^2 = \begin{pmatrix} 0 & 1 \\ -1 & 0 \end{pmatrix}\begin{pmatrix} 0 & 1 \\ -1 & 0 \end{pmatrix} = -\mathbf{I}.$$

As such we find that $\mathbf{B}^3 = \mathbf{B}\mathbf{B}^2 = -\mathbf{B}\mathbf{I} = -\mathbf{B}$ *and that* $\mathbf{B}^4 = \mathbf{B}^2\mathbf{B}^2 = (-\mathbf{I})(-\mathbf{I}) = \mathbf{I}$. *Hence we have the code*

```
n = input('What power :');
b = [0 1; -1 0];
switch mod(n,4)
      case 0
        bn = eye(2);
      case 1
        bn = b;
      case 2
        bn = -eye(2);
      case 3
        bn = -b;
end
```

Solution 6.21 *The eigenvalues can be determined using the code*

```
a = [1 0 0 -1; ...
     0 1 0 0; ...
     0 0 1 0; ...
     -1 0 0 1];
eig(a)
```

which gives 1 *(twice), 0 and 2.*

Solution 6.22 *We start with* $n = 1$ *which is merely the definition, that is* $\mathbf{A} = \mathbf{PDP}^{-1}$. *And we assume that our conjecture is true for* n, *that is* $\mathbf{A}^n = \mathbf{PD}^n\mathbf{P}^{-1}$. *Now premultiply by* $\mathbf{A}$

$$\begin{aligned}\mathbf{AA}^n &= \mathbf{A}(\mathbf{PD}^n\mathbf{P}^{-1})\\ \mathbf{A}^{n+1} &= \mathbf{PDP}^{-1}(\mathbf{PD}^n\mathbf{P}^{-1})\\ &= \mathbf{PD}(\mathbf{P}^{-1}\mathbf{P})\mathbf{D}^n\mathbf{P}^{-1}\\ &= \mathbf{PDD}^n\mathbf{P}^{-1}\\ &= \mathbf{PD}^{n+1}\mathbf{P}^{-1}.\end{aligned}$$

This is merely the statement of our initial conjecture for $n + 1$. *Thus we have shown by induction that* $\mathbf{A}^n = \mathbf{PD}^n\mathbf{P}^{-1}$.

Solution 6.23 *This gives*

```
>> co = charpoly(a);
>> roots(co)

ans =

   2.0000
   1.0000 + 0.0000i
   1.0000 - 0.0000i
   0.0000
```

This confirms the results above.

Solution 6.24 *The eigenvalues of this equation are* $(3 \pm \sqrt{5})/2$. *Consequently using the general form on page 216, we have*

$$\mathbf{x}(t) = \frac{1}{\sqrt{5}}\left\{\frac{1}{2}\left((3+\sqrt{5})\mathrm{e}^{(3-\sqrt{5})t/2} - (3-\sqrt{5})\mathrm{e}^{(3+\sqrt{5})t/2}\right)\mathbf{I} + \left(\mathrm{e}^{(3-\sqrt{5})t/2} - \mathrm{e}^{(3+\sqrt{5})t/2}\right)\mathbf{A}\right\}\begin{pmatrix} 1 \\ -1 \end{pmatrix}.$$

C.7 Solutions for Tasks from Chapter 7

Solution 7.1 *Please try it yourself first but this is the answer (or one of them):*

```
for i = 1:12
    switch mod(i,3)
         case 0
            f(i) = 1;
         case 1
            f(i) = 2;
         case 2
            f(i) = 3;
    end
end
```

There are many alternatives, for instance `f = mod(1:12,3)+1;`.

Solution 7.2 *For the one third case, we can work through the code with* $N = 9$, *so that*

rodd=1:2:N gives [1 3 5 7 9]
reven=2:2:(N-1) gives [2 4 6 8]
weights(rodd=2) gives [2 0 2 0 2 0 2 0 2]
weights(1)=1 gives [1 0 2 0 2 0 2 0 2]
weights(N)=1 gives [1 0 2 0 2 0 2 0 1]
weights(reven)=4 gives = [1 4 2 4 2 4 2 4 1]

and for the three eighths rule with $N = 10$ we have

m=(N-1)/3	*gives 3*
rdiff=3(1:(m-1))+1*	*gives [3*(1:2)+1] that is [4 7]*
*weights=3*ones(1,N)*	*gives [3 3 3 3 3 3 3 3 3 3]*
weights(1)	*gives [1 3 3 3 3 3 3 3 3 3]*
weights(N)	*gives [1 3 3 3 3 3 3 3 3 1]*
weights(rdiff)=2	*gives [1 3 3 2 3 3 2 3 3 1]*

Solution 7.3 *This is done using the code:*

```
function [val] = fn(x)
val = log(x+sqrt(x.^2+1));
```

Solution 7.4 *We shall use forty points (which should be more than enough) and note that the exact answer is*

$$\int_{x=1}^{3} x^2 - 3x + 2\,\mathrm{d}x = \left[\frac{x^3}{3} - \frac{3x^2}{2} + 2x\right]_1^3 = \frac{2}{3}.$$

The code is

```
N = 40;
x = linspace(1,3,N);
f = x.^2-3*x+2;
h = x(2)-x(1);
integral = (sum(f)-f(1)/2-f(N)/2)*h;
```

This gives the value 0.6675*, which is within* 8.7×10^{-4} *of the exact answer. Notice that by using either of Simpson's rules we could have retrieved the exact answer, since the original curve is a quadratic.*

Solution 7.5 *We use* $N = 11$ *and modify Simpson's 1/3 rule code on page 233, so that we have*

```
x = linspace(0,1,11);
h = x(2)-x(1);
N = length(x);
rodd = 1:2:N;
reven = 2:2:(N-1);
weights(rodd) = 2; weights(1) = 1;
weights(N) = 1; weights(reven) = 4;
f = x.^3-x+1;
integral = h/3*sum(weights.*f);
disp([integral])
```

This gives an answer of 0.75. *The exact answer is*

$$\int_0^1 x^3 - x + 1 \,\mathrm{d}x = \left[\frac{x^4}{4} - \frac{x^2}{2} + x\right]_0^1 = \frac{3}{4}.$$

So the scheme does exceedingly well and the error is of the order 10^{-16}. *This is unsurprising since the error is proportional to the fourth derivative, which is identically zero for a cubic.*

Solution 7.6 *In this task we produce a minor modification of the previous solution: the first line needs to read* `x=linspace(0,pi,N);` *and the line defining* $f(x)$ *needs modifying to* `f=sin(x)`. *This now allows us to try different values of* N, *which we do with a loop structure:*

```
Ns = 5:2:19;
for N = Ns
   clear rodd reven weights f x
   x = linspace(0,pi,N);
   h = x(2)-x(1);
   rodd = 1:2:N;
   reven = 2:2:(N-1);
   weights(rodd) = 2;
   weights(1) = 1; weights(N) = 1;
   weights(reven) =  4;
   f = sin(x);
   integral(N) = h/3*sum(weights.*f);
end
plot(Ns,abs(integral-2))
```

This gives

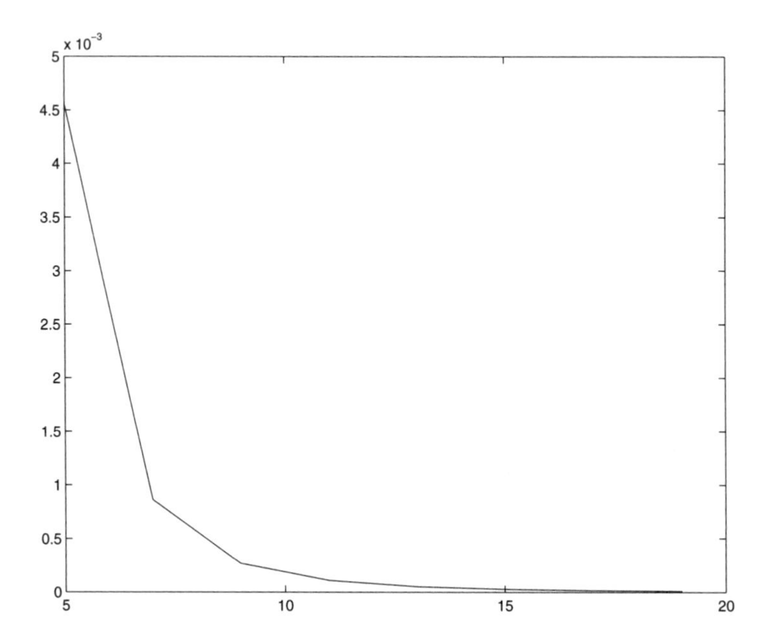

where we have plotted the errors versus the number of points. We have used the exact answer which is

$$\int_0^\pi \sin x \, \mathrm{d}x = [-\cos x]_0^\pi = 2.$$

As we can see the errors tend to zero very rapidly. Of course $\sin x$ *is a very smooth function over this interval and if we had a more oscillatory function more points would be needed.*

Solution 7.7 *We shall use the trapezium rule for simplicity. We also note that the value of this integral over the truncated domain is:*

$$\int_0^a \frac{1}{\sqrt{x^2+1}} \, \mathrm{d}x = \sinh^{-1}(a).$$

We note that $\sinh^{-1}(a) = \ln(a + \sqrt{a^2+1})$ *so in fact the value of the integral diverges, but very slowly.*

We use the code:

```
X = input('Truncate at:');
N = ceil(X)*3;
x = linspace(0,X,N);
h = x(2)-x(1);
f = 1./sqrt(x.^2+1);
int = (sum(f)-f(1)/2-f(N)/2)*h
```

The second line ensures that the step sizes will be smaller than $1/3$. *This gives:*

```
>> diverge
Truncate at:10

int =

    2.9981

>> diverge
Truncate at:100

int =

    5.2983

>> diverge
Truncate at:1000

int =

    7.6009
```

which as we see increases as the truncation point increases. (The corresponding values of arcsinh are 2.9982, 5.2983 and 7.6009; so that the integration does a good job).

Solution 7.8 *We can use the code:*

```
theta = 0:pi/20:(pi/2-pi/20);
N = 20;
for it = 1:length(theta);
   theta1 = theta(it);
   clear grid f
   grid = linspace(theta1,pi-theta1,N);
   f = sqrt(1+cos(grid).^2);
   h = grid(2)-grid(1);
   arclen(it) = (sum(f)-f(1)/2-f(N)/2)*h;
end
plot(theta/pi*180,arclen)
xlabel('\theta degrees')
ylabel('Arc length')
```

which gives

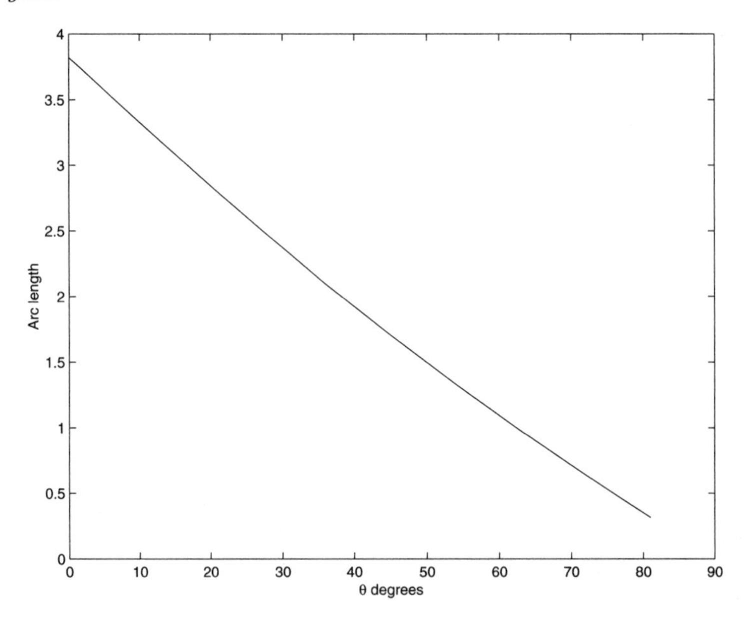

Solution 7.9 *Firstly we give details of the analytical solution:*

$$\int_0^{10} \frac{\cos x}{x^{1/2}}\,\mathrm{d}x = \int_0^{\epsilon} \frac{\cos x}{x^{1/2}}\,\mathrm{d}x + \int_{\epsilon}^{10} \frac{\cos x}{x^{1/2}}\,\mathrm{d}x$$

For the first of these integrals we approximate $\cos x$ *by* $1 - x^2/2$ *(that is the first two terms in its Taylor series).*

$$\int_0^{\epsilon} \frac{1}{x^{1/2}} - \frac{x^{3/2}}{2}\,\mathrm{d}x = \left[2x^{1/2} - \frac{x^{5/2}}{5}\right] = \left[2\epsilon^{1/2} - \frac{\epsilon^{5/2}}{5}\right].$$

We can now use the code

```
clear all
epsil = input('Epsilon :');
int1 = 2*epsil^(0.5)-epsil^(2.5)/5;
N = 100;
x = linspace(epsil,10,N);
h = x(2)-x(1);
f = cos(x)./sqrt(x);
int2 = (sum(f)-f(1)/2-f(N)/2)*h;
int = int1+int2;
```

The first integral gives a significant contribution.

Solution 7.10 *The quadratic through the three points is*

$$y(x) = a_0 + (x - x_0)\Delta a_1 + (x - x_0)(x - x_1)a_2$$

where the constants are

$$\begin{aligned} a_0 &= f_0 \\ a_1 &= \frac{f_1 - f_0}{x_1 - x_0} \\ a_2 &= \frac{(f_2 - f_0)(x_1 - x_0) - (f_1 - f_0)(x_2 - x_0)}{(x_1 - x_0)(x_2 - x_0)(x_2 - x_1)} \end{aligned}$$

Now integrating

$$\int_{x=x_0}^{x_2} y(x)\,\mathrm{d}x \int_{x=x_0}^{x_2} a_0 + (x - x_0)a_1 + (x - x_0)(x - x_1)a_2\,\mathrm{d}x$$

Solution 7.11 *Here we need to use the code*

```
function [f] = fxlnx(x)
f = x.*log(x);
```

and then use the code `quad('fxlnx',1,2)`. *This gives* 0.63629536463993 *(using* `format long`*). The exact value can be calculated using integration by parts*

$$\int_{x=1}^{2} q\ln q\,\mathrm{d}q = \left[\frac{1}{2}q^2\ln q\right]_{x=1}^{2} - \int_{x=1}^{2}\frac{q}{2}\,\mathrm{d}q$$
$$= 2\ln 2 - \left[\frac{q^2}{4}\right]_{x=1}^{2}$$
$$= 2\ln 2 - \frac{3}{4}.$$

The value of this expression agrees very well with that above.

C.8 Solutions for Tasks from Chapter 8

Solution 8.1 *Let us firstly find the exact solution. Start by dividing the equation through by y and then integrate with respect to t which gives*

$$\int \frac{1}{y}\frac{\mathrm{d}y}{\mathrm{d}t}\,\mathrm{d}t = -\int \sqrt{t}\,\mathrm{d}t,$$

hence we have

$$\ln y = -\frac{2}{3}t^{3/2} + C.$$

This can be rearranged to give

$$y = A\mathrm{e}^{-\frac{2}{3}t^{\frac{3}{2}}}$$

and the particular solution can be found by setting $y(0) = 1$, which gives $A = 1$.

In order to obtain the numerical solution the code should be modified to

```
dt = 0.05;
t = 0.0:dt:1.0;
y = zeros(size(t));
y(1) = 1;
for ii=1:(length(t)-1)
      y(ii+1) = y(ii) + dt * (-y(ii)*sqrt(t(ii)));
end
exact = exp(-2/3*(t).^(3/2));
plot(t,y,t,exact,'--')
```

This produces

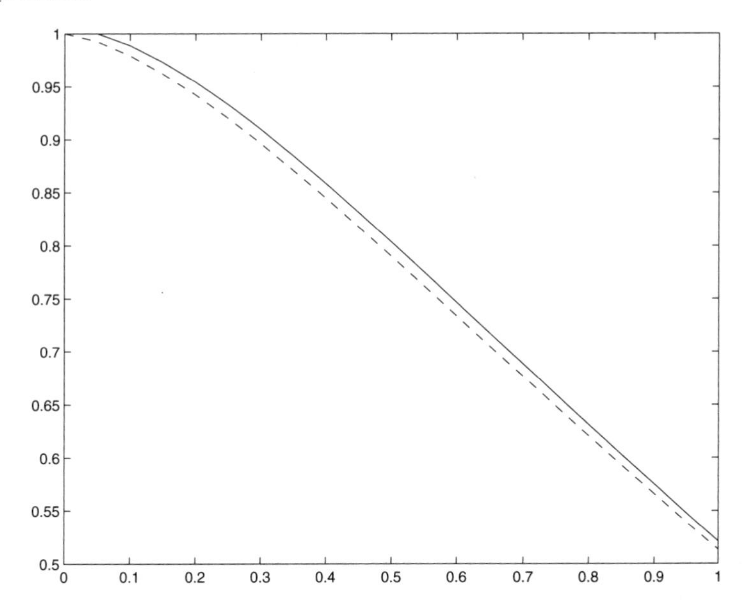

Notice that although this solution is "reasonable" it can be improved by reducing the value of `dt`.

Solution 8.2 *Let us start by considering the exact solutions to both differential equations. Again start by dividing through by y and integrating with respect to t, which gives*

$$\int \frac{1}{y}\frac{\mathrm{d}y}{\mathrm{d}t}\,\mathrm{d}t = \int \pm t\,\mathrm{d}t,$$

hence

$$\ln y = \pm\frac{t^2}{2} + C \qquad \Longrightarrow \qquad y = A\mathrm{e}^{\pm t^2/2}.$$

In each calculation the value of the constant is unity, hence we have the solutions

$$y = \mathrm{e}^{t^2/2} \qquad \text{and} \qquad y = \mathrm{e}^{-t^2/2}.$$

The scheme for the solution of the equations is written as

$$\frac{y_{n+1} - y_n}{\Delta t} = \pm t_n y_n,$$

which can be rearranged to give

$$y_{n+1} = y_n + \pm\Delta t\, t_n y_n = y_n \left(1 + \pm n\Delta t^2\right),$$

where we have used the fact that $t_n = n\Delta t$.

Let us consider the first case. Start with $n = 0$

$$y_1 = 1$$

and now $n = 1$, *etc.*

$$\begin{aligned} y_2 &= 1\left(1 + \frac{1}{16}\right) = \frac{17}{16}, \\ y_3 &= \frac{17}{16}\left(1 + \frac{2}{16}\right) = \frac{153}{128}, \\ y_4 &= \frac{153}{144}\left(1 + \frac{3}{16}\right) = \frac{2907}{2028} \approx 1.419\,. \end{aligned}$$

The exact answer is $\mathrm{e}^{1/2} \approx 1.648$, *hence the absolute error is* $|1.419 - \mathrm{e}^{1/2}| \approx 0.229$ *and the relative error is* $|1.419 - \mathrm{e}^{1/2}|/\mathrm{e}^{1/2} \approx 0.139$ *(or this can be written as* 13.9%*).*

Now we can repeat the calculation for the other case

$$y_1 = 1$$

and now $n = 1$, *etc.*

$$\begin{aligned} y_2 &= 1\left(1 - \frac{1}{16}\right) = \frac{15}{16} \\ y_3 &= \frac{15}{16}\left(1 - \frac{2}{16}\right) = \frac{105}{128} \\ y_4 &= \frac{105}{128}\left(1 - \frac{3}{16}\right) = \frac{1365}{2048}. \end{aligned}$$

Here the absolute error is ≈ 0.21378 *whereas the relative error is* 0.352 *(or around* 35%*). Despite the absolute errors being comparable the relative errors are different (due to the magnitude of the answers involved).*

It is up to the individual as to which error is best to use and this generally comes with experience.

Solution 8.3 *The code for this task is*

```
dt = pi/10;
t = 0.0:dt:10.0*pi;
y = zeros(size(t));
y(1) = 0;
for ii=1:(length(t)-1)
      y(ii+1) = y(ii) + dt * (sin(t(ii))+sin(y(ii)));
end
```

This produces the result

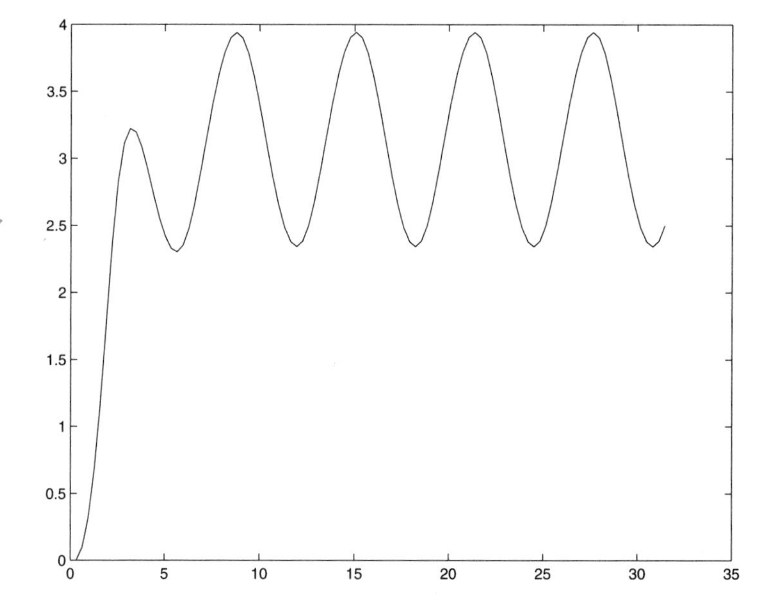

You can now change the step length, simply by changing the **dt=** *line (it might be a good idea to add a* **clear all** *statement at the top of the code as well).*

Solution 8.4 *First let us construct the exact solution to the equation. We need to multiply through by an integrating factor, namely* $\mathrm{e}^{t/3}$*, which gives*

$$\begin{aligned}\mathrm{e}^{t/3}\frac{\mathrm{d}y}{\mathrm{d}t} + \frac{1}{3}\mathrm{e}^{t/3}y &= -\frac{1}{2}\mathrm{t}\mathrm{e}^{t/3} \\ \frac{\mathrm{d}}{\mathrm{d}t}\left(y\mathrm{e}^{t/3}\right) &= -\frac{1}{2}\mathrm{t}\mathrm{e}^{t/3}\end{aligned}$$

and now integrating with respect to t *we find that*

$$y\mathrm{e}^{t/3} = A - \int \frac{1}{2}\mathrm{t}\mathrm{e}^{t/3}\,\mathrm{d}t.$$

Now integrating the right hand side by parts gives

$$ye^{t/3} = A - \left\{\left[\frac{3}{2}te^{t/3}\right] - \int \frac{3}{2}e^{t/3}\,dt\right\},$$

$$ye^{t/3} = A - \left\{\frac{3}{2}te^{t/3} - \frac{9}{2}e^{t/3}\right\}.$$

Hence we have the solution

$$y = Ae^{-t/3} - \left\{\frac{3}{2}t - \frac{9}{2}\right\}.$$

Now applying the boundary condition gives $A = -9/2$. *The solution is*

$$y = \frac{9}{2}\left(1 - e^{-t/3}\right) - \frac{3t}{2}.$$

Now consider the discretised form of the equation, which is

$$\frac{y_{n+1} - y_n}{\Delta t} = -\frac{t_{n+1}}{2} - \frac{y_{n+1}}{3},$$

which can be rearranged to give

$$y_{n+1} = \frac{1}{1 + \frac{\Delta t}{3}}\left(y_n - \Delta t\frac{t_{n+1}}{2}\right).$$

Now with $\Delta t = 1/3$ *and* $n = 0$ *this gives*

$$y_1 = \frac{1}{1 + \frac{1}{9}}\left(0 - \frac{1}{3}\frac{1}{3 \times 2}\right) = \frac{9}{10}\left(-\frac{1}{18}\right) = -\frac{1}{20},$$

and now with $n = 1$

$$y_2 = \frac{9}{10}\left(-\frac{1}{20} - \frac{1}{3}\frac{2}{3 \times 2}\right) = -\frac{29}{200},$$

and finally for $n = 2$ *which gives* $y_3 = y(1)$

$$y_3 = \frac{9}{10}\left(-\frac{29}{200} - \frac{1}{3}\frac{3}{3 \times 2}\right) = -\frac{561}{2000}.$$

The code to produce this and the other required solutions is

```
clear all
dt = 1/3;
t = 0.0:dt:1;
y = zeros(size(t));
y(1) = 0;
for ii = 1:(length(t)-1)
      y(ii+1) = 1/(1+dt/3)*(y(ii)-dt*t(ii+1)/2);
end
ts = t; ys = y;
dt = 1/1000;
t = 0.0:dt:1;
y = zeros(size(t));
y(1) = 0;
for ii = 1:(length(t)-1)
      y(ii+1) = 1/(1+dt/3)*(y(ii)-dt*t(ii+1)/2);
end
exact = 9/2*(1-exp(-t/3))-3/2*t;
plot(ts,ys,'*',t,exact,t,y)
```

This gives the picture

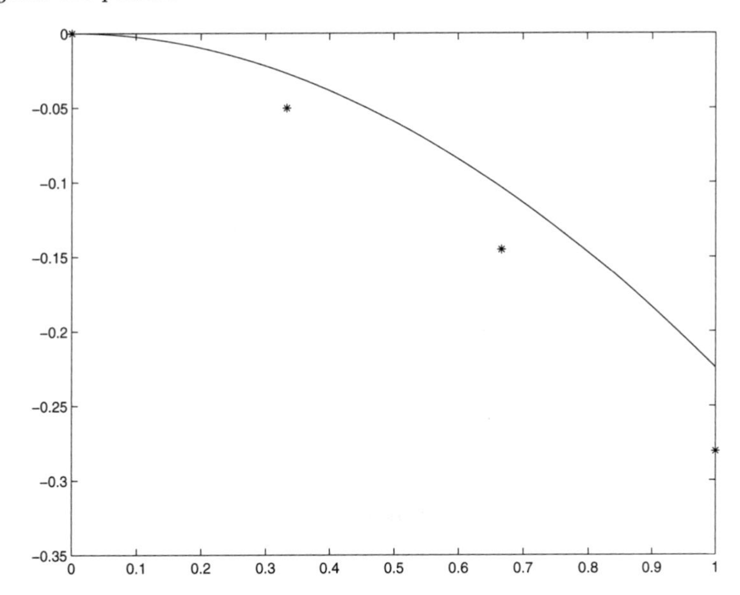

where the solution above is shown using stars.

Solution 8.5 *The codes now become*

```
function [value] = odes(t,y)
value = t^2-y^2;
```

and

```
y0 = 0;
tspan = [0 2];
[t,y] = ode45('odes',tspan,y0);
```

These give

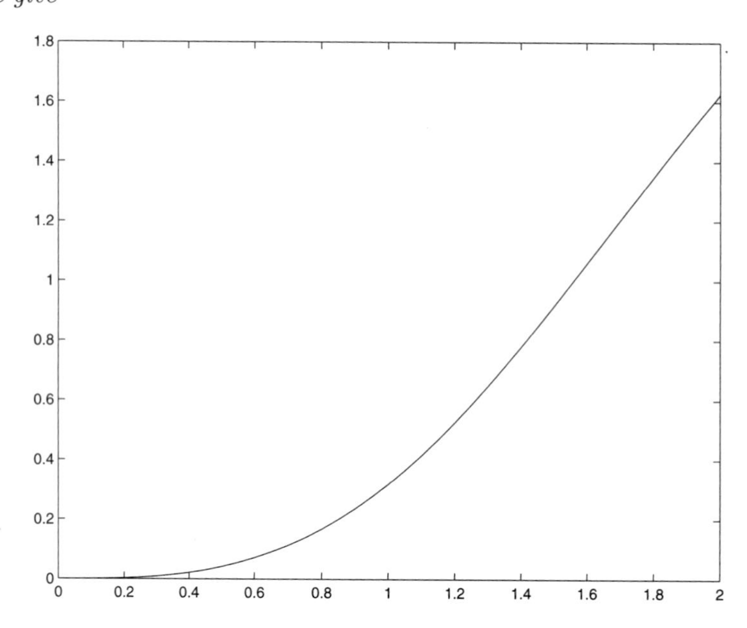

Solution 8.6 *The solution to this equation can be obtained by multiplying by the integrating factor* e^t *and then integrating by parts. After application of the initial condition we find that*

$$y(t) = t^2 - 2t + 2 - \mathrm{e}^{-t}.$$

The numerical solution can be determined using:

```
N = 20;
t = linspace(0,2,N);
dt = t(2)-t(1);
y(1) = 1;
for j = 1:(N-1)
    y(j+1) = y(j)+dt*(-y(j)+t(j)^2);
end
ex = t.^2-2*t+2-exp(-t);
```

This gives:

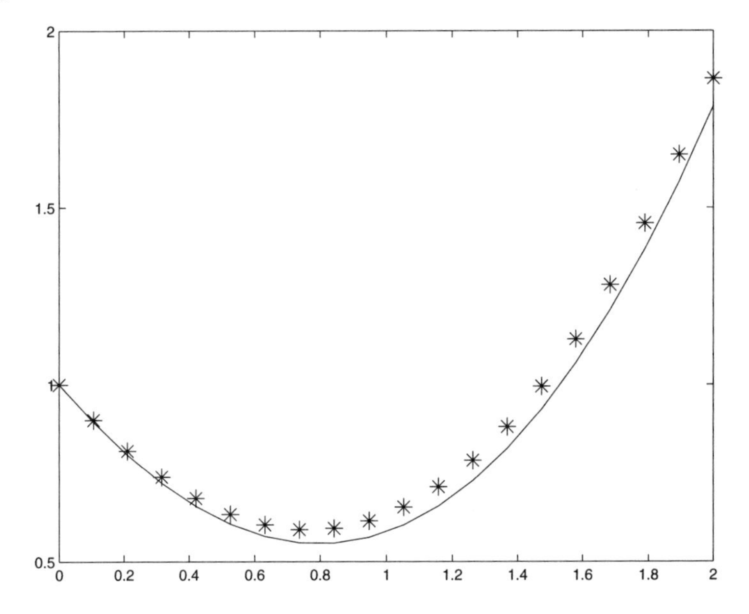

which is a reasonable match (the exact solution is shown with the asterisks).

Solution 8.7 *1. The solution of this problem is:*

$$y(x) = 2\sin x - x\cos x + x(-2\sin 1 + \cos 1 - 1) + 1.$$

```
% Set up system
x = 0.0:0.1:1.0;
N = length(x);
h = x(2)-x(1);
a = 1/h^2*ones(size(x));
b = -2/h^2*ones(size(x));
c = 1/h^2*ones(size(x));
r = x.*cos(x);
a(1) = 0; b(1) = 1; r(1) = 1;
c(N) = 0; b(N) = 1; r(N) = 0;
% Forward sweep
for j = 2:N
    b(j) = b(j)-c(j)*a(j-1)/b(j-1);
    r(j) = r(j)-c(j)*r(j-1)/b(j-1);
end
% Final equation
y(N) = r(N)/b(N);
for j = (N-1):-1:1
    y(j) = r(j)/b(j)-a(j)*y(j+1)/b(j);
end
```

which gives

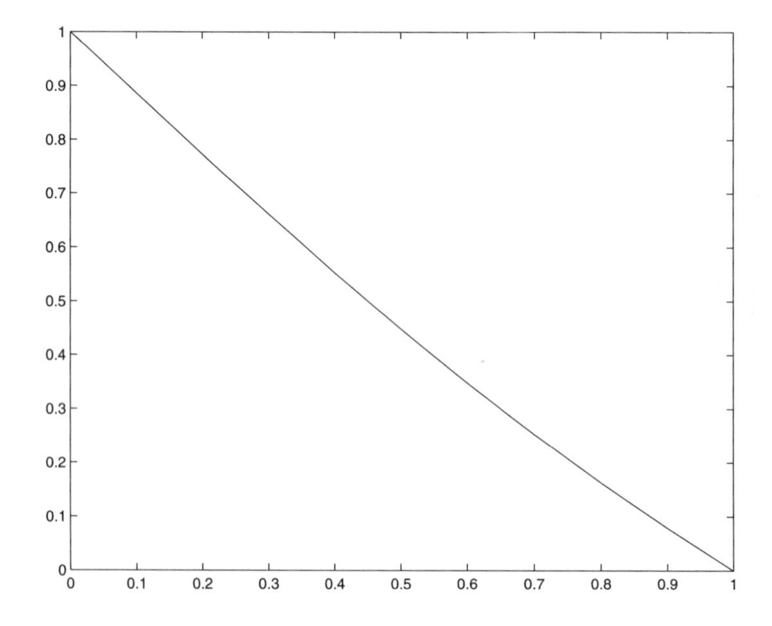

2. *The analytic solution here is:*

$$y(x) = 2\sin x - x\cos x - x - 2\sin 1 + \cos 1 + 1.$$

```
% Set up system
x = 0.0:0.1:1.0;
N = length(x);
h = x(2)-x(1);
a = 1/h^2*ones(size(x));
b = -2/h^2*ones(size(x));
c = 1/h^2*ones(size(x));
r = x.*cos(x);
a(1) = -1; b(1) = 1; r(1) = 0;
c(N) = 0; b(N) = 1; r(N) = 0;
% Forward sweep
for j = 2:N
    b(j) = b(j)-c(j)*a(j-1)/b(j-1);
    r(j) = r(j)-c(j)*r(j-1)/b(j-1);
end
% Final equation
y(N) = r(N)/b(N);
for j = (N-1):-1:1
    y(j) = r(j)/b(j)-a(j)*y(j+1)/b(j);
end
```

which gives

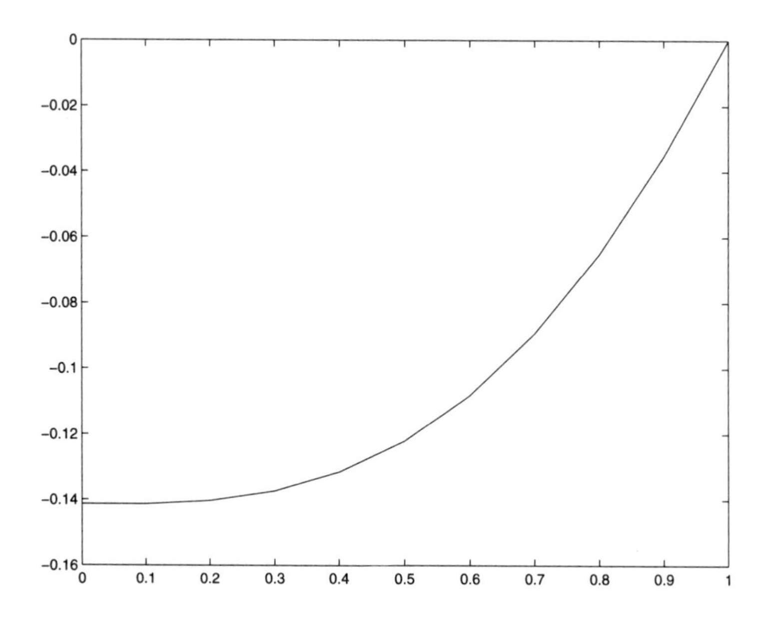

3. *The solution here is*

$$y(x) = \frac{e^{-2x} - 1}{e^{-2} - 1}.$$

```
% Set up system
x = 0.0:0.1:1.0;
N = length(x);
h = x(2)-x(1);
a = (1/h^2+2/(2*h))*ones(size(x));
b = -2/h^2*ones(size(x));
c = (1/h^2-2/(2*h))*ones(size(x));
r = 0;
a(1) = 0; b(1) = 1; r(1) = 0;
c(N) = 0; b(N) = 1; r(N) = 1;
% Forward sweep
for j = 2:N
    b(j) = b(j)-c(j)*a(j-1)/b(j-1);
    r(j) = r(j)-c(j)*r(j-1)/b(j-1);
end
% Final equation
y(N) = r(N)/b(N);
for j = (N-1):-1:1
    y(j) = r(j)/b(j)-a(j)*y(j+1)/b(j);
end
```

which yields

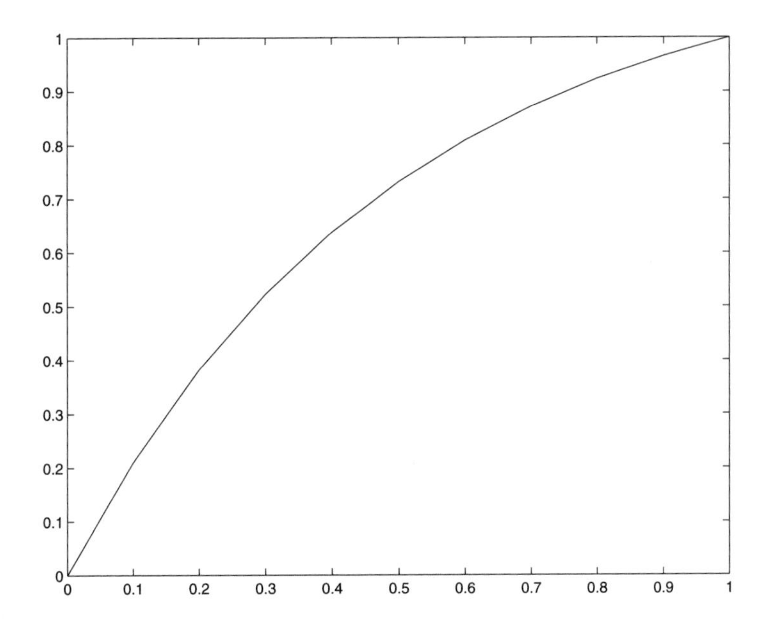

Solution 8.8 *The exact solution here is*

$$y(t) = \frac{1}{3}\left(t - \frac{1}{\sqrt{3}} \sin \sqrt{3}t\right).$$

By discretising the equation we have

$$y_{n+1} = 2y_n - y_{n-1} + \Delta t^2 \left(-3y_n + t_n\right).$$

The initial conditions can be realised by setting $y_1 = 0$ *and* $y_2 = 0$ *as a result of the fact that* $y'(0) \approx (y_2 - y_1)/\Delta t = 0$.

```
N = 20;
t = linspace(0,1,N);
dt = t(2)-t(1);
y(1) = 0;
y(2) = 0;
for j = 2:N-1
    y(j+1) = 2*y(j)-y(j-1) ...
     +dt^2*(3*y(j)+t(j));
end
ex = (t-sin(sqrt(3)*t)/sqrt(3))/3;
```

This gives

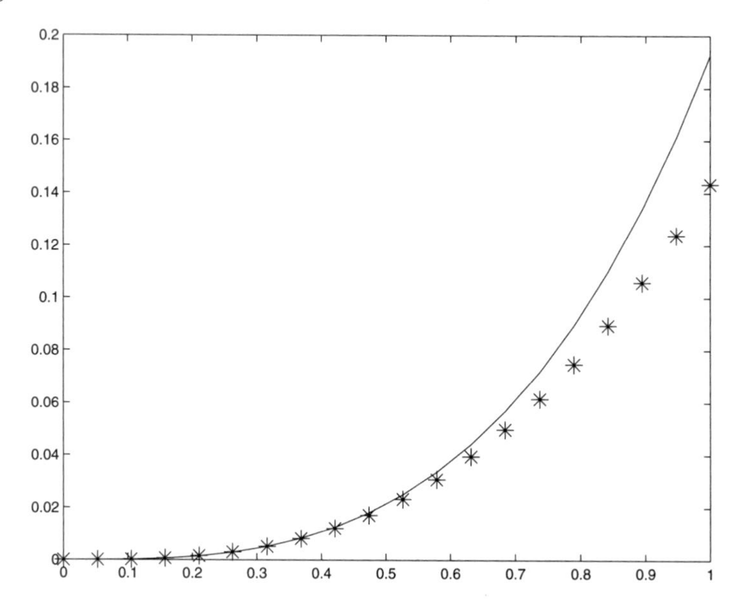

The numerical solution does reasonably initially: however as t increases the solution diverges.

Solution 8.9 *This equation can be solved in a similar manner to the previous task:*

```
N = 20;
t = linspace(0,2,N);
dt = t(2)-t(1);
y(1) = 1;
y(2) = 1;
for j = 2:N-1
    y(j+1) = 2*y(j)-y(j-1) ...
          +dt^2*(t(j)*y(j)+sin(t(j)));
end
```

This gives:

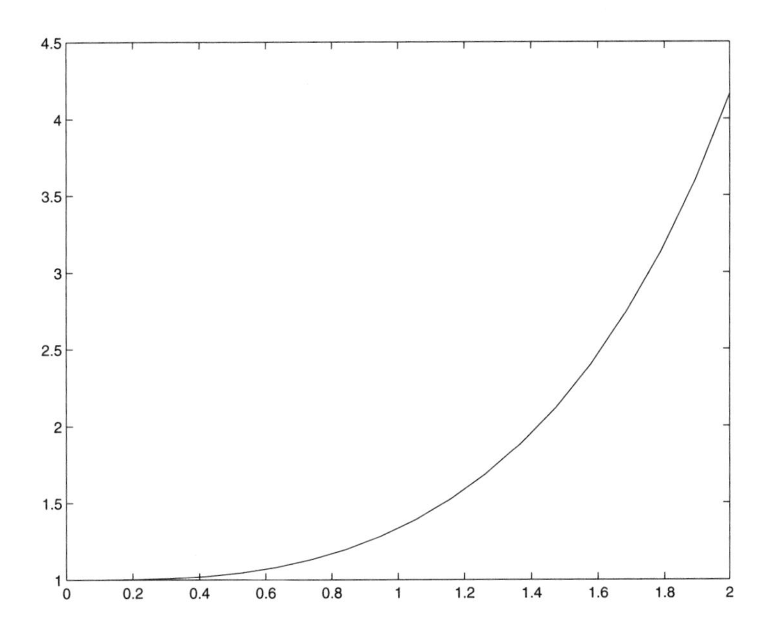

(the solution is expressible in terms of Airy functions, but this does not represent a great advantage to us).

This second problem can be solved by setting up the discretised system

$$\frac{y_{n+1} - 2y_n + y_{n-1}}{\Delta t^2} + t_n y_n = \sin t_n$$

with the conditions that $y_1 = 0$ *(*$y(0) = 0$*)* $y_N = 0$ *(*$y(2) = 0$*).*

```
N = 20;
t = linspace(0,2,N);
dt = t(2)-t(1);
A = zeros(N);
A = diag(-2/dt^2*ones(N,1)+t',0) ...
     +diag(1/dt^2*ones(N-1,1),-1) ...
     +diag(1/dt^2*ones(N-1,1),1);
r = transpose(sin(t));
A(1,:) = 0;
A(1,1) = 1; r(1) = 0;
A(N,:) = 0;
A(N,N) = 1; r(N) = 0;
sol = A\r;
```

This gives:

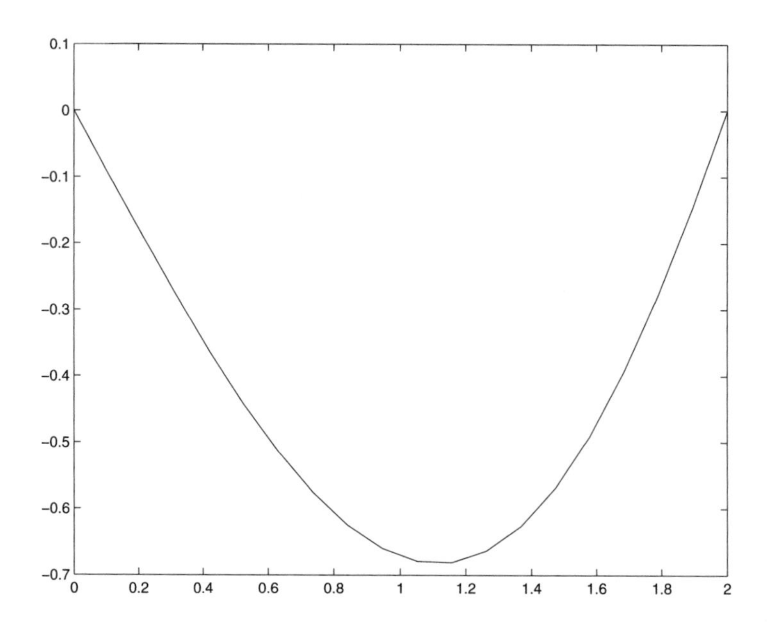

Solution 8.10 *This equation can be integrated directly by dividing through by $y(t^2+1)$. This gives*

$$\frac{y'}{y} = -\frac{2t}{1+t^2}$$

so that

$$y(t) = \frac{1}{1+t^2}.$$

The equation can be solved numerically using

```
N = 50;
t = linspace(0,5,N);
dt = t(2)-t(1);
y(1) = 1;
for j = 1:(N-1)
  y(j+1) = y(j)-dt*2*t(j)*y(j)/(1+t(j)^2);
end
ex = 1./(1+t.^2);
```

This yields

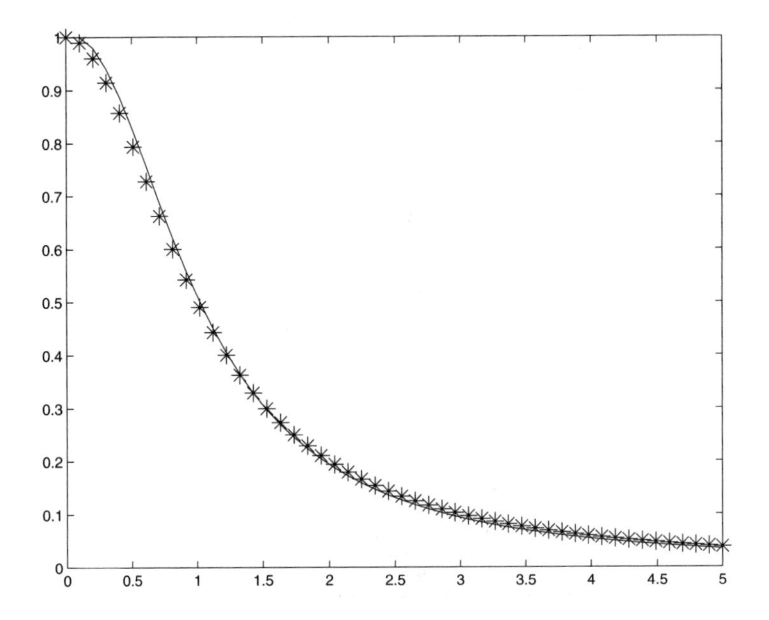

Solution 8.11 *This system can be written as*

$$\mathbf{y}' = \begin{pmatrix} 0 & 1 & 0 \\ 0 & 0 & 1 \\ 2 & 1 & -2 \end{pmatrix} \mathbf{y},$$

where $\mathbf{y} = (y, y', y'')$. *Using MATLAB we find that*

```
>> A = [0 1 0; 0 0 1; 2 1 -2];
>> [V,D] = eig(A)

V =


   -0.5774   -0.2182    0.5774
    0.5774    0.4364    0.5774
   -0.5774   -0.8729    0.5774



D =


   -1.0000         0         0
         0   -2.0000         0
         0         0    1.0000
```

And hence we know that

$$\mathrm{e}^{\mathbf{A}t} = \mathbf{V}\mathrm{e}^{\mathbf{D}t}\mathbf{V}^{-1}.$$

For convenience we shall use the non-normalised form of $\mathbf{V}$ *so that*

$$\tilde{\mathbf{V}} = \begin{pmatrix} -1 & 1 & 1 \\ 1 & -2 & 1 \\ -1 & 4 & 1 \end{pmatrix} \text{ and } \tilde{\mathbf{V}}^{-1} = \begin{pmatrix} -1 & \frac{1}{2} & \frac{1}{2} \\ -\frac{1}{3} & 0 & \frac{1}{3} \\ \frac{1}{3} & \frac{1}{2} & \frac{1}{6} \end{pmatrix}$$

Hence

$$\begin{aligned}
\mathrm{e}^{\mathbf{A}t} &= \begin{pmatrix} -1 & 1 & 1 \\ 1 & -2 & 1 \\ -1 & 4 & 1 \end{pmatrix} \begin{pmatrix} \mathrm{e}^{-t} & 0 & 0 \\ 0 & \mathrm{e}^{-2t} & 0 \\ 0 & 0 & \mathrm{e}^{t} \end{pmatrix} \begin{pmatrix} -1 & \frac{1}{2} & \frac{1}{2} \\ -\frac{1}{3} & 0 & \frac{1}{3} \\ \frac{1}{3} & \frac{1}{2} & \frac{1}{6} \end{pmatrix} \\
&= \begin{pmatrix} -1 & 1 & 1 \\ 1 & -2 & 1 \\ -1 & 4 & 1 \end{pmatrix} \begin{pmatrix} -\mathrm{e}^{-t} & \frac{1}{2}\mathrm{e}^{-t} & \frac{1}{2}\mathrm{e}^{-t} \\ -\frac{1}{3}\mathrm{e}^{-2t} & 0 & \frac{1}{3}\mathrm{e}^{-2t} \\ \frac{1}{3}\mathrm{e}^{t} & \frac{1}{2}et & \frac{1}{6}\mathrm{e}^{t} \end{pmatrix} \\
&= \begin{pmatrix} \mathrm{e}^{-t} - \frac{1}{3}\mathrm{e}^{-2t} + \frac{1}{3}\mathrm{e}^{t} & -\frac{1}{2}\mathrm{e}^{-t} + \frac{1}{2}\mathrm{e}^{t} & -\frac{1}{2}\mathrm{e}^{-t} + \frac{1}{3}\mathrm{e}^{-2t} + \frac{1}{6}\mathrm{e}^{t} \\ -\mathrm{e}^{-t} + \frac{2}{3}\mathrm{e}^{-2t} + \frac{1}{3}\mathrm{e}^{t} & \frac{1}{2}\mathrm{e}^{-t} + \frac{1}{2}\mathrm{e}^{t} & \frac{1}{2}\mathrm{e}^{-t} - \frac{2}{3}\mathrm{e}^{-2t} + \frac{1}{6}\mathrm{e}^{t} \\ \mathrm{e}^{-t} - \frac{4}{3}\mathrm{e}^{-2t} + \frac{1}{3}\mathrm{e}^{t} & -\frac{1}{2}\mathrm{e}^{-t} + \frac{1}{2}\mathrm{e}^{t} & -\frac{1}{2}\mathrm{e}^{-t} + \frac{4}{3}\mathrm{e}^{-2t} + \frac{1}{6}\mathrm{e}^{t} \end{pmatrix}.
\end{aligned}$$

The solution is obtained by multiplying $(0, 0, 1)^T$ *(the initial conditions) by this matrix*

$$\mathrm{e}^{\mathbf{A}t} \begin{pmatrix} 0 \\ 0 \\ 1 \end{pmatrix} = \begin{pmatrix} -\frac{1}{2}\mathrm{e}^{-t} + \frac{1}{3}\mathrm{e}^{-2t} + \frac{1}{6}\mathrm{e}^{t} \\ \frac{1}{2}\mathrm{e}^{-t} - \frac{2}{3}\mathrm{e}^{-2t} + \frac{1}{6}\mathrm{e}^{t} \\ -\frac{1}{2}\mathrm{e}^{-t} + \frac{4}{3}\mathrm{e}^{-2t} + \frac{1}{6}\mathrm{e}^{t} \end{pmatrix}.$$

Finally we have the answer

$$y(t) = -\frac{1}{2}e^{-t} + \frac{1}{3}e^{-2t} + \frac{1}{6}e^{t}.$$

Solution 8.12 *This is solved using the finite difference code:*

```
N = 20;
x = linspace(0,pi,N);
h = x(2)-x(1);
% Only the internal points
A = diag(-2/h^2*ones(N-2,1) ...
   +sin(x(2:(N-1)))',0) ...
   +diag(1/h^2*ones(N-3,1),1) ...
   +diag(1/h^2*ones(N-3,1),-1);
[V,D] = eigs(A,3,'SM');
for j = 1:3
    ti = ['\lambda = ' num2str(D(j,j))];
    subplot(1,3,j)
    plot(x,[0; V(:,j); 0])
    title(ti,'FontSize',14)
end
```

This creates

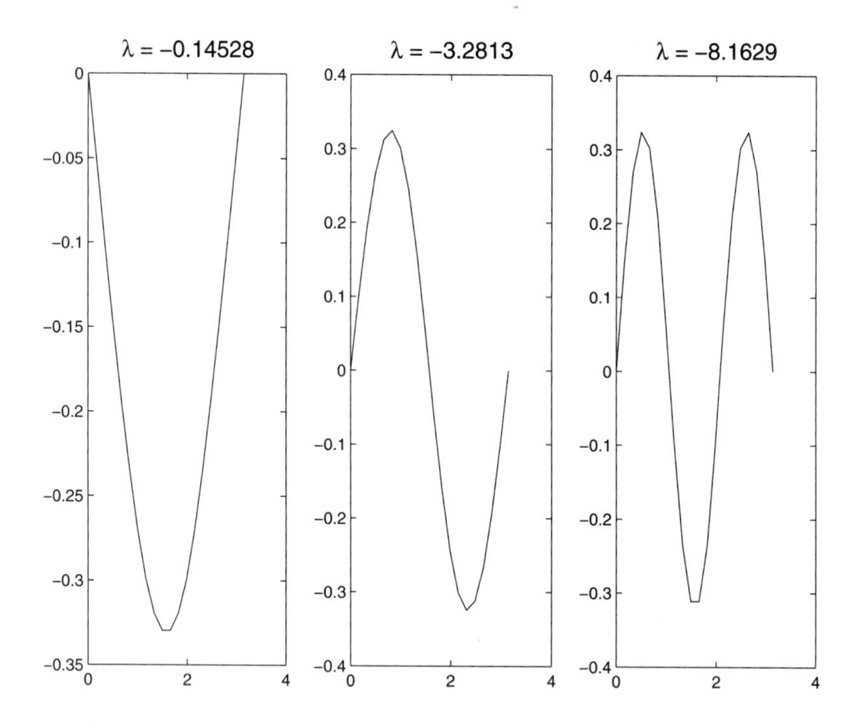

C.9 Solutions for Tasks from Chapter 9

Solution 9.1 *This can be achieved using*

```
x = [3 2 4 5 6 -1 5 6 7 8 2];
y = [2 -6 3 2 0 1 4 5 6 7 8];
mean(x)
mean(y)
median(x)
median(y)
var(x)
var(y)
A = cov(x,y);
A(1,2)
A = corrcoef(x,y);
A(1,2)
```

This gives

```
ans =

    4.2727


ans =

    2.9091


ans =

     5


ans =

     3


ans =

    6.8182


ans =

   15.0909


ans =

    4.2273


ans =

    0.4167
```

respectively.

Solution 9.2 *We note that if $Y = aX + b$ then $\bar{y} = a\bar{x} + b$ which can be seen by substitution into the formula for the mean. Similarly substituting this into the expression for the correlation shows that*

$$\sigma_{XY} = \frac{a}{\sqrt{a^2}} = \frac{a}{|a|} = sign(a).$$

Solution 9.3 *These two can be calculated using*

```
xb=mean(x);
vx=sum((x-xb).^2)/length(x);
std_x=sqrt(vx);
skew=sum(((x-xb)/std_x).^3)/length(x);
kurt=sum(((x-xb)/std_x).^4)/length(x)-3;
```

Solution 9.4 *In order to solve this problem we shall use a basic loop structure*

```
global mu
mu = 0.4;
x(1) = 0.25;
for i = 1:9
     x(i+1) = map(x(i));
end
```

The value this code returns is 4.3420e-05 (notice if you get MATLAB to write out `x` *it will appear that this entry is zero, depending on the current format).*

Solution 9.5 *One value which works is $\mu = -2$. Then use*

```
mu = -2;
co = [-mu^3 2*mu^3 -mu^2*(1+mu) (mu^2-1) 0];
[r] = roots(co);
x1 = r(3);
x2 = map(x1);
x3 = map(x2);
disp([x1 x2 x3])
```

Note that `x2` *is also a period 2 point and its image is* `x1`*. These values are actually given by*

$$\frac{1}{2}\frac{\mu^2 + \mu \pm \sqrt{\mu^4 - 2\mu^3 - 3\mu^2}}{\mu^2}$$

The real values of these roots occur for $\mu < -1$ *and* $\mu > 3$.

Solution 9.6 *The code for this map is*

```
function [xn,yn] = map3(xo,yo)
xn = mod(xo+2*yo,1);
yn = mod(3*xo-2*yo,1);
```

Solution 9.7 *The solution of the Hénon map for* $\cos\alpha = 0.24$ *gives*

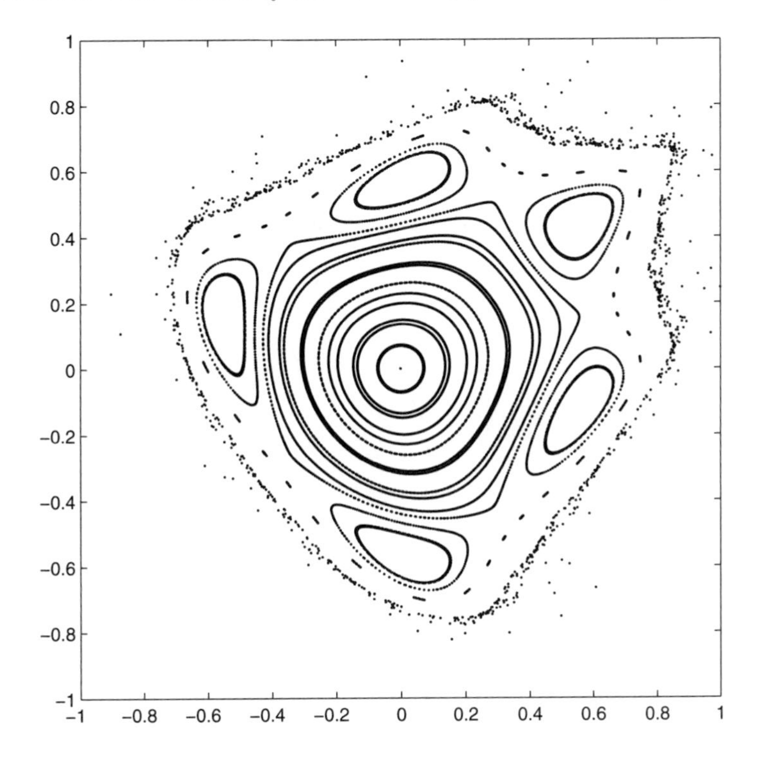

Index